Energy and Environmental Engineering

Energy and Environmental Engineering

Edited by **Chuck Lancaster**

New York

Published by Callisto Reference,
106 Park Avenue, Suite 200,
New York, NY 10016, USA
www.callistoreference.com

Energy and Environmental Engineering
Edited by Chuck Lancaster

© 2015 Callisto Reference

International Standard Book Number: 978-1-63239-309-8 (Hardback)

This book contains information obtained from authentic and highly regarded sources. Copyright for all individual chapters remain with the respective authors as indicated. A wide variety of references are listed. Permission and sources are indicated; for detailed attributions, please refer to the permissions page. Reasonable efforts have been made to publish reliable data and information, but the authors, editors and publisher cannot assume any responsibility for the validity of all materials or the consequences of their use.

The publisher's policy is to use permanent paper from mills that operate a sustainable forestry policy. Furthermore, the publisher ensures that the text paper and cover boards used have met acceptable environmental accreditation standards.

Trademark Notice: Registered trademark of products or corporate names are used only for explanation and identification without intent to infringe.

Printed in the United States of America.

Contents

Preface IX

Chapter 1 **Investigation on Inorganic Pollution Level in Surface Sediments of Naples and Salerno Bay** 1
Menghan Wang, Benedetto De Vivo, Stefano Albanese, Annamaria Lima, Wanjun Lu, Flavia Molisso, Marco Sacchi

Chapter 2 **An Experimental Investigation of Temperature Distribution in Different Urban Locations in Aswan, Egypt of Hot and Dry Climate** 6
Soubhi A. Hassanein, Osama K. Osman, Waleed A. Abd-Fadeel

Chapter 3 **Fundamentals of Direct Inverse CFD Modeling to Detect Air Pollution Sources in Urban Areas** 13
Mahmoud Bady

Chapter 4 **On the Short-Term Optimisation of a Hydro Basin with Social Constraints** 25
Gloria Hermida, Edgardo D. Castronuovo

Chapter 5 **Considerations on Research of Environmental Risk Prevention and Control in "12th Five-Year Plan" Period** 37
Pengli Xue, Xiaofeng Sun, Yun Song, Yanjun Cheng, Dezhi Sun

Chapter 6 **Modeling and OLAP Cubes for Database of Ground and Municipal Water Supply** 41
Taskeen Zaidi, Annapurna Singh, Vipin Saxena

Chapter 7 **Adsorption Characteristics of Zinc (Zn^{2+}) from Aqueous Solution by Natural Bentonite and Kaolin Clay Minerals: A Comparative Study** 47
Tushar Kanti Sen, Chi Khoo

Chapter 8 **Digital Interactive Kanban Advertisement System Using Face Recognition Methodology** 53
Feng-Yi Cheng, Chu-Ja Chang, Gwo-Jia Jong

Chapter 9 **Adsorption of Methylene Blue by NaOH-modified Dead Leaves of Plane Trees** 58
Lei Gong, Wei Sun, Lingying Kong

Chapter 10 **Production of Hydrogen by Electrolysis of Water: Effects of the Electrolyte Type on the Electrolysis Performances** 65
Romdhane Ben Slama

Chapter 11 **Refrigerator Coupling to a Water-Heater and Heating Floor to Save Energy and to Reduce Carbon Emissions** 70
Romdhane Ben Slama

Chapter 12 **Time-Effect Relationship of Toxicity Induced by Roundup® and Its Main Constituents in Liver of Carassius Auratus** 79
Jinyu Fan, Jinju Geng, Hongqiang Ren, Xiaorong Wang

Chapter 13 **Protecting Water Quality and Public Health Using a Smart Grid** 85
Ken Thompson, Raja Kadiyala

Chapter 14 **Explosion of Sun** 93
Alexander Bolonkin, Joseph Friedlander

Chapter 15 **New Approach to the Diagnosis of Control Hydraulic Systems** 107
Noura Rezika Bellahsene Hatem, Mohamed Mostefai, Oum El-Kheir Aktouf

Chapter 16 **Silver Nanoparticle Adsorption to Soil and Water Treatment Residuals and Impact on Zebrafish in a Lab-scale Constructed Wetland** 113
Angela Ebeling, Victoria Hartmann, Aubrey Rockman, Andrew Armstrong, Robert Balza, Jarrod Erbe, Daniel Ebeling

Chapter 17 **Effect of Ambient Temperature on PUF Passive Samplers and PAHs Distribution in Puerto Rico** 123
Nedim Vardar, Ziad Chemseddine, Juan Santos

Chapter 18 **Atmospheric Dispersion and Deposition of Radionuclides (^{137}Cs and ^{131}I) Released from the Fukushima Dai-ichi Nuclear Power Plant** 128
Soon-Ung Park, Anna Choe, Moon-Soo Park

Chapter 19 **Effect of pH and Dissolved Silicate on the Formation of Surface Passivation Layers for Reducing Pyrite Oxidation** 136
Shengjia Zeng, Jun Li, Russell Schumann, Roger Smart

Chapter 20 **Growing Public Health Concerns from Poor Urban Air Quality: Strategies for Sustainable Urban Living** 142
Bhaskar Kura, Suruchi Verma, Elena Ajdari, Amrita Iyer

Chapter 21 **Feasibility Study for Power Generation during Peak Hours with a Hybrid System in a Recycled Paper Mill** 151
Adriano Beluco, Clodomiro P. Colvara, Luis E. Teixeira, Alexandre Beluco

Chapter 22 **Are there Monthly Variations in Water Quality in the Amman, Zarqa and Balqa Regions, Jordan?** 162
Khaled A. Alqadi, Lalit Kumar

Chapter 23 **Variations in the Water Quality of an Urban River in Nigeria** 172
F. A. Oginni

Chapter 24 **Evaluation of the Impact of Government Policy on the Overuse of Groundwater in the Minqin Basin in China** 183
Lihua Yang

Chapter 25 **Voltage Stability Constrained Optimal Power Flow Using NSGA-II** 193
Sandeep Panuganti, Preetha Roselyn John, Durairaj Devraj, Subhransu Sekhar Dash

Chapter 26 **Introspections and Suggestions on the Amount Fixing of Administrative Penalty for Environmental Pollution** 201
Xiaohong Zheng

Permissions

List of Contributors

Preface

Computational water engineering is a field of study that embraces various stages of high performance modelling, simulation, and knowledge based tools to plan and manage a specific nation's water environment and engineering structure. Fashioning of rivers, river basins, contaminant transport, climate change, flood risks, water conservation and groundwater flow are few of the research arenas that come under this field. There is another field of engineering, namely energy engineering that is a recent and upcoming field of study. Matters such as energy efficiency, plant engineering, energy services and alternative energy technologies come under its umbrella. This field usually combines the areas of physics, math and chemistry with environmental and economic practices and the main goal is to find the most energy efficient and sustainable method to operate manufacturing processes. This often results in new innovations like better insulation, better and efficient heating and advanced lighting in building. On the other hand, environmental engineering is one of the fastest growing fields of engineering. Environmental engineering integrates engineering and science for a chance to improve the natural environment and to provide healthy environments for human habitation as well as clean up polluted sites. It also searches for plausible solutions to problems in the field of public health. It involves control of air pollution, waste water management, waste disposal and recycling as well as industrial hygiene and radiation protection.

This book attempts to bring these divergent, yet still related fields of engineering together in the form of their research. I am thankful for the hard work of those who were behind these studies.

Editor

1

Investigation on Inorganic Pollution Level in Surface Sediments of Naples and Salerno Bay

Menghan Wang[1], Benedetto De Vivo[1], Stefano Albanese[1], Annamaria Lima[1], Wanjun Lu[2], Flavia Molisso[3], Marco Sacchi[3]

1Dipartimento di Scienze della Terra Università di Napoli Federico II, Napoli, Italy

[2]Department of Marine Science, Faculty of Earth Resource, China University of Geosciences, Wuhan, China

[3]C.N.R. Istituto Geomare Sud, Napoli, Italy

ABSTRACT

In this study, superficial marine sediments collected from 96 sampling sites were analyzed for 53 inorganic elements. Each sample was digested in aqua regia and analyzed by ICP-MS. A developed multifractal inverse distance weighted (IDW) interpolation method was applied for the compilation of interpolated maps for both single element and factor scores distributions. R-mode factor analysis has been performed on 23 of 53 analyzed elements. The 3 factor model, accounting 84.9% of data variability, were chosen. The three elemental associations obtained have been very helpful to distinguish anthropogenic from geogenic contribution. The aim of this study is to distinguish distribution patterns of pollutants on the sea floor of Naples and Salerno bays. In general, local lithologies, water dynamic and anthropogenic activities determine the distribution of the analyzed elements. To estimate pollution level in the area, Italian guidance, Canadian sediment quality guidance and Long's criteria are chosen to set the comparability. As the result shows, arsenic and lead may present highly adverse effect to living creatures.

Keywords: Pollution Level; Compositional Data Analysis; Factor Analysis; Napoli and Salerno Gulf

1. Introduction

Naples bay is a 10-mile wide gulf located in the south western coast of Italy, while Salerno bay is a gulf of Tyrrhenian Sea and separated from Naples bay by Sorrento Peninsula. Industrial complexes, intense commercial and transport activities insist on this area, which makes it potentially a heavily polluted coastal district and in need of remediation activities. Sediments are considered as a suitable medium to distinguish contamination and geochemical background of marine environment, since they are the pool of different deposition source and are a more stable medium than sea water. Moreover, toxic contaminants prefer to adsorb on sediments surface especially hydrophobic organics such as PAHs and PCBs. The aim of this study is to accomplish a comprehensive investigation of inorganic elements concentration on sediment surface and illustrate their distribution patterns.

2. Materials and Method

2.1. Sampling

Surface sediment samples (following the directives of the national program for assessment of marine pollution of highly contaminated Italian coastal areas) were collected from 96 locations (**Figure 1**) of Naples and Salerno bays in May 2000. A differential global positioning system (DGPS) was used to identify each location precisely. 23 samples were collected using a box-corer with an inner diameter of 25 cm, of which we have used the superficial sediments to be analyzed. 63 samples were collected by grab. Each sample was divided into three and stored in 4°C freezer.

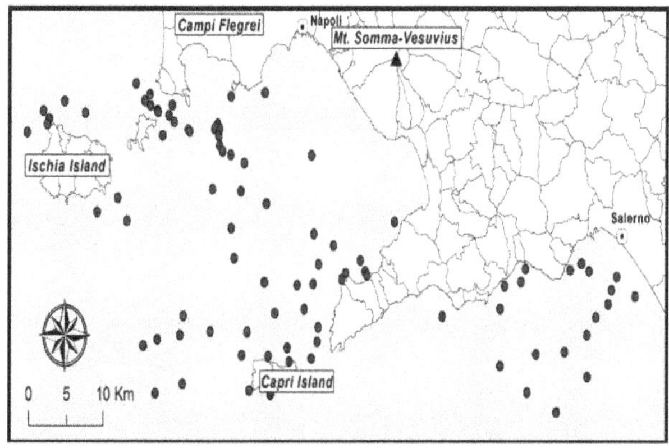

Figure 1. Study area and samples locations.

2.2. Chemical Analyses and Quality Control

The all air-dried sediment samples were sieved and 30 g of the < 150 μm fraction was retained for analysis of 53 elements (**Table 1**). Analyses were carried out by Acme Analytical Laboratories Ltd. (Vancouver, Canada), through its Italian affiliate (Norwest Italia Srl, Napoli). Each sample was digested in a modified aqua regia solution and analyzed by inductively coupled plasma–mass spectrometry (ICP-MS) and atomic emission spectrometry (ICP-AES). Specifically, a 15 g split of the pulp was digested in 45 ml of the aqua regia mixture (1 part concentrated hydrochloric acid to 1 part nitric acid to 1 part deionised water) at 90˚C for 1 h. The solution was taken to a final volume of 300 ml with 5% HCl. Aliquots of sample solution were aspirated into a Jarrel Ash Atomcomp 975 ICP-Emission Spectrometer and a Perkin Elmer Elan 6000 ICP-Mass Spectrometer.

Table 1. Rotating component matrix.

Element	Component		
	1	2	3
Cu	.424	.786	.345
Pb	.392	.299	.788
Zn	.251	.655	.676
Ag	-.005	-.141	.838
Ni	-.006	.981	.035
Co	.167	.920	.063
U	.884	-.161	.122
Au	.105	-.048	.944
Th	.853	.307	.147
Bi	.126	.408	.781
V	.609	.541	.163
La	.894	-.056	.122
Cr	.024	.671	.665
Ti	.722	-.309	.015
Al	.846	.483	.116
Na	.808	.196	.147
K	.916	.133	.134
Sc	-.033	.946	.005
Tl	.824	.353	.257
Hg	.163	.104	.927
Sn	.293	.086	.878
Be	.891	.279	.207
Li	.186	.900	.090

2.3. Statistical and Spatial Analysis

For single element interpreting, all the data should be transformed with proper log-ratio method to avoid negative effect of compositional property [8]. All the information was managed in a GIS georeferenced environment, using ArcGis 9.3 software package. Geodetic reference system is the Universal Transverse Mercator (fuse 33) projection on the ellipsoid World Geodetic System (WGS, 1984). All the geochemical maps were generated using the Multifractal Inverse Distance Weighted (IDW) algorithm as an interpolation method. Factor analysis performed with the IBM SPSS Statistics 19 software package was applied to reduce the number of dimensions and extracting synthetic information about the distribution of elements in the studied environment [1].

3. Results and Discussion

Heavy metals concentrations of study area were compared to 3 different marine sediment quality guidelines [5] with the purpose of illustrating contamination level of Naples and Salerno bays (**Table 2**). Long [7] initializes the estimation of adverse biological effect by collect and summary publications. In his work, ERL represents effect-range low which means below it rarely adverse effects, and ERM represents effect-range median which means above it frequently associated with adverse effects. Canadian sediment quality guidelines [5], shares same idea with Long, and include results recently. In CCME's guideline, IQSG represents interim marine sediment quality guideline and PEL represent probably effect level. Italian 367 [2] set the environment tolerated value based on Italian law.

Pollution levels are compared in **Figure 2** by presenting percentage of different category samples. Concentrations of Zn, Ag, Cd and Cr in sediment seldom exceed adverse effect thresholds in the area, while Ni, Hg, Pb and Cu have median polluted value which may cause adverse effect to living creatures as well as human beings. Only Arsenic shows values highly dangerous, reflecting probably mostly influence of volcanic sediments from Neapolitan volcanoes (Vesuvius, Campi Flegri, Ischia Island) rather than anthropogenic source.

Figure 3 shows the distribution pattern of selected analyzed elements. Most of heavy toxic metals have a similar distribution. Ag, Hg and Pb aggregates close to Napoli metropolitan area, indicating that intense industrial, agricultural and commercial activities affect these elements distribution pattern. Arsenic is mostly concentrated around Pozzuoli bay, where hydrothermal activity related to Campi Flegrei is documented as being very rich in As [2]. The distributions of Ni, Zn, and Cu indicate the water energy decrease from coastal to deep sea and cause finer sediment deposit off gulf. Distribution

Table 2. Heavy metals concentration comparison with different environmental guidelines.

	Cu(ppm)	Pb(ppm)	Zn(ppm)	Ag(ppb)	Ni(ppm)	Cr(ppm)	Hg(ppb)	As(ppm)	Cd(ppm)
Average	26.4	39.24	68.56	103.28	20.29	26.28	97.65	16.85	0.1
10.00%	7.61	20.13	35.64	31	6.98	8.92	21.7	9.68	0.05
25.00%	14.45	26.18	63.23	46.5	11.8	16	42.25	12	0.07
50.00%	26.62	35.5	71	71	21.6	27.8	79	14.35	0.09
75.00%	37.81	47.47	84.43	109.5	27.7	35.75	116.25	17.45	0.11
Median	29	69.7	106	519	7.5	38.6	435	12.3	0.11
MAD	12	34.4	35.8	450	14.1	12.6	356	3.05	0.03
Minimium	3	9	16	17	2	3	3	5	0.02
Maxmium	72	128	178	1012	40	74	864	74	0.19
ER-L	34	46.7	150	1000	20.9	81	150	8.2	1.2
ER-M	270	218	410	3700	51.6	370	710	70	9.6
ISQG	18.7	30.2	124			52.3	130	7.24	0.7
PEL	108	112	271			160	700	41.6	4.2
Italian 367		30			30	50	300	12	0.3

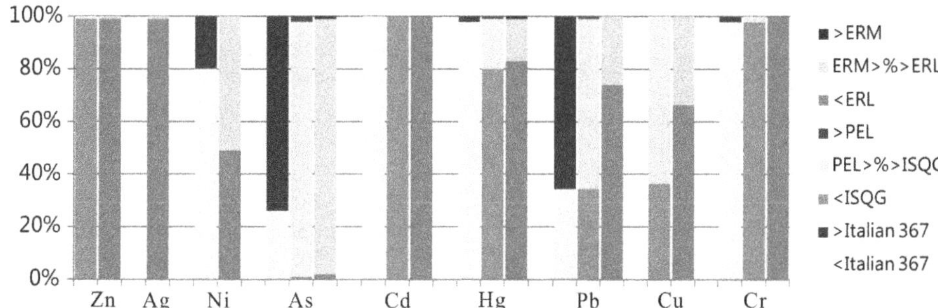

Figure 2. Percentage of different sediment category divided by environmental guidelines. Brown bar shows the percentage of samples that exceed ERM, pink bar shows that between ERL and ERM while green bar shows that below ERL. Red bar represents that exceed PEL, yellow one shows that between PEL and ISQG and blue one represents that below ISQG. Dark grey bars shows percentage exceed Italian 367, light grey shows that below it.

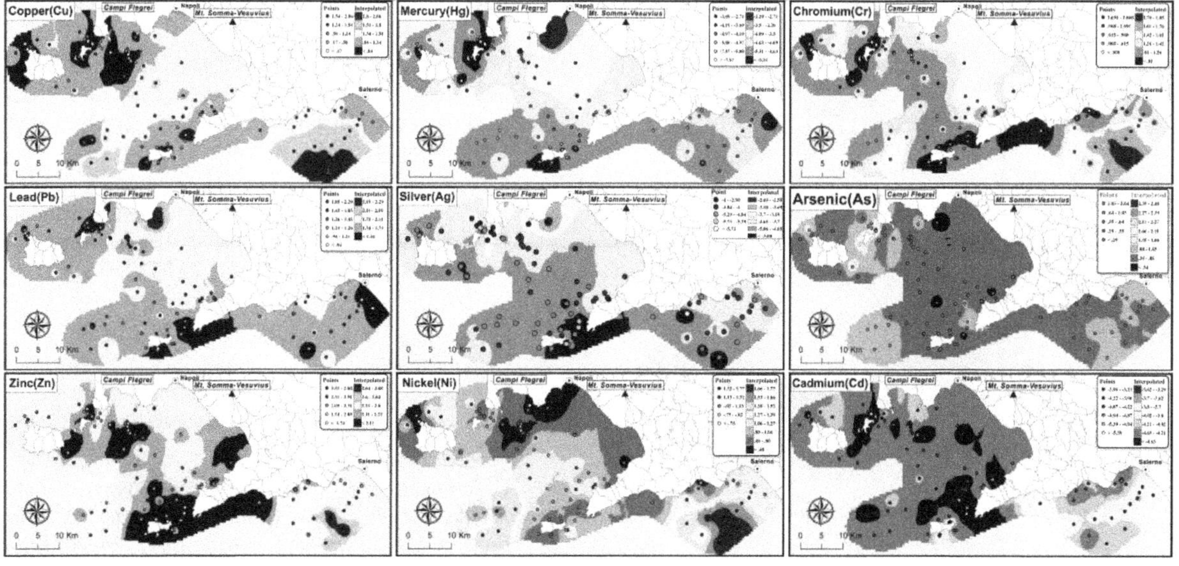

Figure 3. Single elements distribution in Napoli and Salerno Gulf.

pattern of Chromium indicate combination of anthropogenic effect and water energy effect.

To understand the distribution modes of the different heavy metals and discriminate the different sources, R-mode factor analysis (FA) on 23 of 53 analyzed elements was carried out. The factor model three, accounting 84.9% of data variability, have been chosed. The elements are considered to describe effectively the composition of factors if the loading is over 0.51. The associations of the three-factor model are F1 (K, La, Be, U, Th, Al, Tl, Na, Ti, V) accounted for 32.5% of data variability, F2 (Ni, Sc, Co, Li, Cu, Cr, Zn, V) accounted for 27.4% of data variability and F3 (Au, Hg, Sr, Ag, Pb, Bi, Zn, Cr) accounted for 24.9% of data variability. **Figure 4** shows the distribution pattern of the three association factor scores.

F1 association represents the elements whose distribution is mostly of geogenic source, meaning that human activities have little, control on their behavior in Naples and Salerno bays. F2 represents elements mainly influenced by water energy. F3 is the most anthropogenic factor, showing intense human activities of the Naples metropolitan area as the main source for this 9 elements distribution patterns. The latter is in agreement with the results obtained by Cicchella et al., 2005 for volcanic soils of the metropolitan and provincial areas of Napoli.

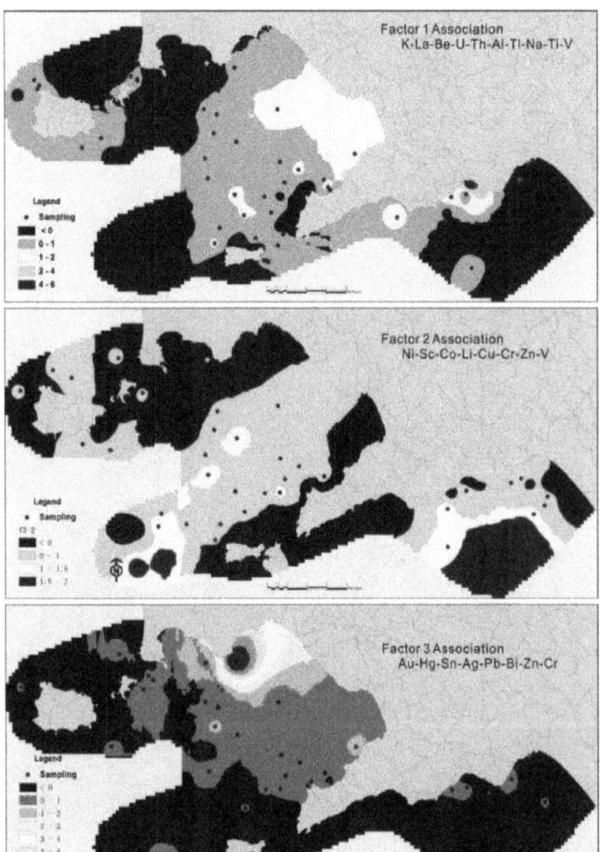

Figure 4. Factor score association maps.

4. Acknowledgements

Authors thank the C.N.R. Istituto Geomare Sud a Napoli to provide samples and corresponding data, and Dr. Monica Capodanno for sediment storage and disposal. This preliminary study is part of a more comprehensive PhD program of M. Wang, aimed at the study as well of Persistent Organic Pollutants (POP), such as PAH and PCBs, by SERS laboratory and in situ experimental researchs to understand the behaviour of POP with time.

REFERENCES

[1] S. Albanese, B. De Vivo, A. Lim and D. Cicchella, "Geochemical Background and Baseline Values of Toxic Elements in Stream Sediments of Campania Region (Italy)," *Journal of Geochemical Exploration*, Vol. 93, No. 1, 2007, pp. 21-34.

[2] S. Albanese, et al., "Geochemical Baselines and Risk Assessment of the Bagnoli Brownfield Site Coastal Sea Sediments (Naples, Italy)," *Journal of Geochemical Expltion,* Vol. 105, No. 1-2, 2010, pp. 19-33.

[3] D. Cicchella, "Palladium and Platinum Concentration in Soils from the Napoli Metropolitan Area, Italy: Possible Effects of Catalytic Exhausts," *The Science of The Total Environment*, Vol. 308, No. 1-3, 2003, pp. 121-131.

[4] D. Cicchella, B. De Vivo and A. Lima, "Background and Baseline Concentration Values of Elements Harmfull to Human Healt in the Volcanic Soils of the Metropolitan Provincial Area of Napoli (Italy)," *Geochemistry: Exploration-Environment-Analyses,* Vol. 5, 2005, pp. 29-40.

[5] CCME and C.C.C.o.M.o.t., "Canadian Sediment Quality Guidelines for the Protection of Aquatic Life. Canadian Environmental Quality Guidelines," In: Environment, C.C.o.M.o.t., Ed., MB, Winnipeg, 2002.

[6] E. R. Long, L. J. Field and D. D. MacDonald, "Predicting Toxicity in Marine Sediments with Numerical Sediment Quality Guidelines," *Environmental Toxicology andChemistry*, Vol. 17, No. 4, 1998, pp. 714-727.

[7] E. R. Long, D. D. MacDonald, S. L. Smith and F. D. Calder, "Incidence of Adverse Biological Effects within Ranges of Chemical Concentrations in Marine and Estuarine Sements," *Environmental Management*, Vol. 19, No. 1, 1995, pp. 81-97.

[8] C. Reimann, *et al.*, "The Concept of Compositional Data Analysis in Practice - Total Major Element Concentrations in Agricultural and Grazing Land Soils of Europe," *Science of the Total Environment*, Vol. 426, 2012, pp.

196-210.

[9] M. Sprovieri, *et al.*, "Heavy metals, Polycyclic Aromatic Hydrocarbons and Polychlorinated Biphenyls in Surface Sediments of the Naples Harbour (Southern Italy). Chemosphere," Vol. 67, No. 5, 2007, pp. 998-1009.

2

An Experimental Investigation of Temperature Distribution in Different Urban Locations in Aswan, Egypt of Hot and Dry Climate

Soubhi A. Hassanein[1*], Osama K. Osman[2], Waleed A. Abd-Fadeel[1]

[1]Department of Mechanical Engineering, Aswan Faculty of Energy Engineering, Aswan University, Aswan, Egypt

[2]Department of High Voltage Networks Engineering, Aswan Faculty of Energy Engineering, Aswan University, Aswan, Egypt

ABSTRACT

This paper describes the measurements and analysis of an experimental campaign performed in different urban street in Aswan, Egypt. The present study is focused on the experimental investigation of thermal characteristics during summer 2012 of five different regions location (building on Nile river shore, building in front of a mountain, building under high power transmission line, two building opposite other in resident region, finally ship in Nile river) aiming at the investigation of the impact of urban location on the potential of natural and hybrid ventilation under hot weather conditions. The temporal and spatial distribution of air and surface temperatures is examined. Emphasis was given on the vertical distribution of air and surface temperatures. The results showed that the measured surface temperature across the street was the highest value than the air and wall temperature where temperature difference between street and air temperature could reach 35°C and this favored the overheating of lower air levels. Buoyancy generated mainly from asphalt-street heating resulted in the development of the predominant recirculation inside the street canyon. The results also show that air temperature for two building opposite other has a lower value followed by building on Nile river shore followed by building at front of mountain.

Keywords: Experimental; Temperature Distribution; Urban; Hot and Dry Climate

1. Introduction

With rapid urbanization, there has been a tremendous growth in population and buildings in cities. The high concentration of hard surfaces actually triggered many environmental issues. The urban heat island (UHI) effect, one of these environmental issues, is a phenomenon where air temperatures in densely built cities are higher than the suburban rural areas. The primary root of heat island in cities is due to the absorption of solar radiation by mass building structures, roads, and other hard surfaces during daytime. The absorbed heat is subsequently re-radiated to the surroundings and increases ambient temperatures at night.

High temperatures in urban street canyons do create health impacts to people. It is, therefore, important to understand the temperature characteristics in street canyons under real urban meteorological conditions, to formulate effective strategies for control and urban planning.

The main factors which govern the temperature distribution in streets are: emissions from vehicles in the street, street configuration, and meteorological conditions. Pollution from traffic is often the most substantial. However, emissions from power plants, industries and domestic use are important as well. The street geometry, *i.e.* height of the surrounding buildings and street width, influences the temperature distribution inside the street. Also important is the orientation of the street compared to the prevailing wind direction. The city of Aswan has been chosen due to the large difference in meteorological conditions.

Air circulation and temperature distribution within urban canyons is of high significance for pedestrian comfort, pollutant dispersion, radiated and energy studies and for the potential of natural and hybrid ventilation in urban buildings. as well as, for the evaluation of street canyon models that focus on this field. Several theoretical but only few experimental urban street canyon studies are reported in the literature.

In the past few years, along with the rapid urbanization,

*Corresponding author.

the energy conservation in the civilian sector, which reflects the improvement of living standard, plays an important role in the sustainable development of the city. as well as enough attention to the environmental protection, more and more attention is paid to the reformation of urban energy system. Measurement and analysis of an experimental campaign were performed by K. Niachou *et al.* [1] in an urban street canyon in Athens, Greece. A number of field and indoor experimental procedures were organized during summer 2002 aiming at the investigation of the impact of urban environment on the potential of natural and hybrid ventilation. The influence of trafic-induced pollutants (e.g. CO, NO, NO_2 and O_3) on the air quality of urban areas was investigated by W. Kuttler and A. Strassburger [2] in the city of Essen, North Rhine-Westphalia (NRW), Germany. Twelve air hygiene profile measuring trips were made to analyze the trace gas distribution in the urban area. A comparison was by Elisabetta Vignati *et al.* [3] between pollution levels in a street in Copenhagen characterized by frequent high wind conditions and a street in Milan with generally low wind speed conditions. The analysis has shown that the differences in pollution levels in these two cities are governed mainly by the meteorological conditions and especially wind speed. A mobile survey was conducted by Nyuk Hien Wong and Chen Yu [4] to explore both the severity of UHI effect and cooling impacts of green areas at macro-level in Singapore. Island wide temperature distribution was mapped relying on data derived from the mobile survey. Air flow and pollutant dispersion characteristics in an urban street canyon are studied by Yun-Wei Zhang *et al.* [5] under the real time boundary conditions. A new scheme for realizing real-time boundary conditions in simulations is proposed. Berlian I. Idris *et al.* [6] have investigated differences in smoking prevalence between urban and non-urban area of residence in six Western European countries (Sweden, Finland, Denmark, Germany, Italy and Spain), and smoking prevalence trends over the period 1985-2000. An analytical model has been developed by Hongbo Ren *et al.* [7] for estimating an economically efficient installation and operation pattern for the distributed energy system for the urban area in china. The data analyzed by Jing Yuan and Akula Venkatram [8] are relevant to dispersion from near surface urban sources at distances where the vertical plume spread is less than the average height of the buildings. An attempt is made by Vali Kalantar [9] to study the cooling performance of a wind tower numerically in hot and arid region. A three-dimensional computational fluid dynamics (CFD) simulation coupled with radiation and conduction analysis was carried out by Hong Huang *et al.* [10] to analyze the pollutant dispersion under non-isothermal conditions within an objective area in Kawasaki city, Japan, in winter.

In the present study, a number of field experimental procedure were performed in an urban street, in Aswan, aiming at the investigation of the thermal characteristics during hot weather conditions for different urban location. The aim is to present the detailed experimental data within a real urban street during hot summer weather conditions. The results of this experiment can be very useful for the understanding of the impact of the urban street location on the potential of natural and hybrid ventilation in urban buildings, as well as, for the evaluation of street canyon models that focus on this field.

2. Experimental Investigation

A number of experimental procedures were organized in different urban street canyon location as shown in **Figure 1**, located at a high density residential area, near the center of Aswan. The experiments were performed inside and outside the canyon during the period 9:00 am to 2:00 pm for 19-24 June 2012. Less attention has been paid to analyzing the air temperature of urban areas with various types of land use and their interactions with high spatial resolution. Meteorological data consisting of air and surface temperature measurements were recorded on a continuous basis. A detailed description of the field measurements is given in **Table 1**. The temperature measure-

Building at front of mountain

Building at Nile river shore

Two building opposite other

Ship in Nile river

Street under and far from power transmission line

Figure 1. Photo of different urban location studied at Aswan.

Table 1. Experimental runs.

Run number	Location	Run time	Air temperature	Surface temperature	
				Street temperature	Wall temperature
1	Building at front of mountain	Day	Yes	Yes	Yes
2	Building at front of mountain	Night	Yes	Yes	Yes
3	Building at Nile river shore	Day	Yes	Yes	Yes
4	Two building opposite each others	Day	Yes	Yes	Yes
5	Ship in Nile river	Day	Yes	Yes	Yes
6	Under power line transmission	Day	No	Yes	No

ments consisted of: the performed air and surface temperature measurements in five locations.

3. Measuring Instrumentation

3.1. Air Temperature

The air temperature at different positions in the urban location has been measured by digital air velocity and temperature instrument as shown in **Figure 2**. It has the capability to be connected to a computer and/or a printer for accessing data and printing it. The specifications of the air velocity and temperature are listed in **Table 2**. All measuring devices have been monitored throughout the day and their outputs have been recorded every 60 min.

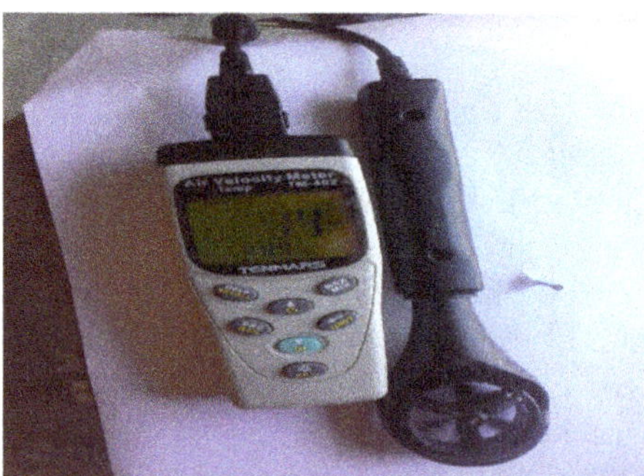

Figure 2. Digital air velocity and temperature instrument.

3.2. Surface Temperature

Surface temperatures (as wall and street temperatures) ere also measured with Radiation Pyrometer ARDOCELL PZ Profibus 7MC3060-····-Z-A40 as shown in **Figure 3**. Measurements were performed from the ground level up to the fourth floor, every 1 h.

Figure 3. Radiation Pyrometer for surface temperature.

4. Description of Field Measurements

The studied canyons (**Figure 1**), having a range of height to width ratio (H/W) of (1.5 - 2). The canyons are oriented with its long-axis in an N-W direction. The building walls are made of concrete. The vertical distribution of the surface temperatures on the buildings opposite mountain was studied during day and night, since it is important to understand the transfer phenomena between them and the adjacent air.

Table 2. Specification of the motive air thermo-anemoter air meter unit.

Measured quantity	Velocity	Temperature
Range	0.2 - 25 m/sec	−20˚C to 60˚C
Accuracy	(±2% + 0.2)	±0.8˚C
Resolution	0.1 m/sec	0.1˚C

5. Results and Discussion

In this part, the obtained experimental results will be comprehensively discussed. The results include air temperature and surface temperature (street and wall temperatures) distribution along five urban locations. Finally a comparison between temperature distribution for different locations has been done. Inside the urban street canyon during June 2012 in Aswan, Egypt.

5.1. Air Temperature near Canyon Facades

In order to understand the mechanisms that determine the distribution of air temperature inside a canyon, air temperature measurements have been performed in close to opposite building facades. The air temperature distribution has been analyses in order to investigate the impact of the street layout and orientation, as well, of the surface temperatures due to convection heat transfer phenomena.

The air temperature distribution across the canyon is of great interest. As temperature in the middle of canyon and near the ground level air temperature was more dependent upon the flux divergence in air volume including that of the horizontal transport. The measured air temperature differences between the two facades vary, as a function of the canyon layout and the surface characteristics.

Air temperature near building walls for different location urban was measured at a distance of 0.5 m from the exterior walls as shown in **Figure 4**. Measurements were performed from the ground level up to the fourth or sixth floor, every 1 h depend on geometry of building. **Figure 5** showed air temperature of different urban at Aswan. It could be seen from **Figure 5** that the air temperature at 0.5 m from the exterior walls for most urban location for the same floor number during day time is increase with time to maximum value near a noon then decrease with time as does the general trend of solar radiation. Only the air temperature at 0.5 m from the exterior walls for building facing a mountain for the same floor number during night time is decreasing with time as could be explained by losing heat by convection during night. Also It could be seen that the air temperature at 0.5 m from the exterior walls for most urban location (except ship in Nile river) at the same time decrease with floor number *i.e.* decrease as far up from land as this go in line with general trend of ambient temperature with vertical distance from land. Also the air temperature at 0.5 m from the exterior walls for ship at Nile river was found to be increased with floor number *i.e.* increase as far vertically from water surface. Finally for two building opposite each other the measured air temperature differences between the two facades vary, as a function of the canyon layout and the surface characteristics.

As it was expected, the air temperature close to west facade was higher than near east facade during time range (9:00 am-12:00 am). And vice verse the air temperature close to west facade was lower than near east facade during time range (1:00 pm-2:00 pm).

Comparison between air temperature of building at front of mountain during day and night it could be seen that temperature values for day is more than for night due

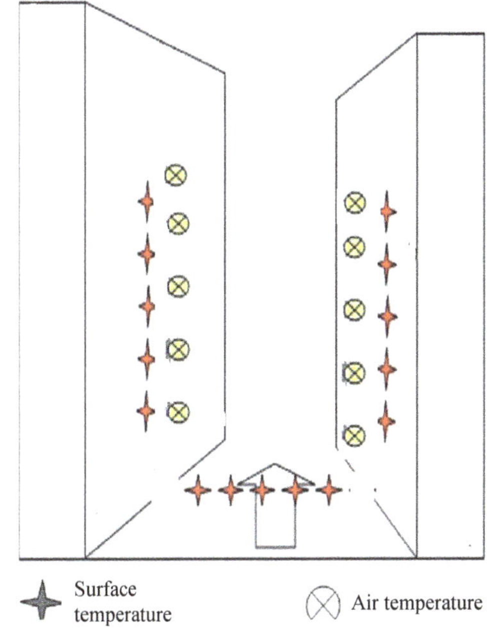

Figure 4. A schematic representation of temperature measurements performed.

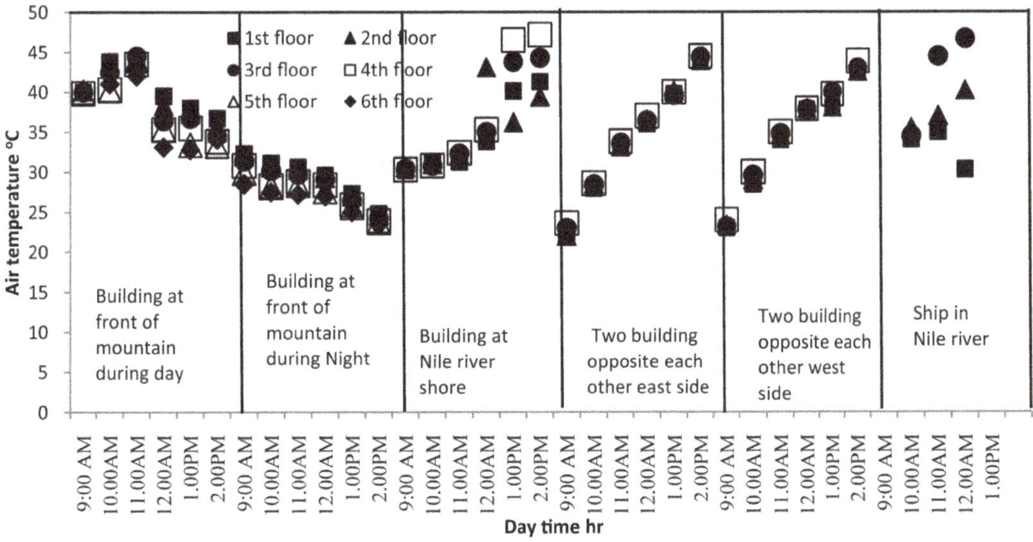

Figure 5. Hourly distributions of air temperatures for different urban locations.

to solar radiation in day. Finally comparing of air temperature for all urban location it could be seen that air temperature for two building opposite each other has a lower value followed by building on Nile river shore followed by building at front of mountain.

5.2. Surface Temperatures

Surface temperatures were conducted on the ground by an infrared thermometer equipped with a laser beam, every one hour. The surface temperatures were recorded across the canyon on the asphalt street and on the sidewalks on each side of the road. The emissivity value used for the measured materials was considered.

Wall Temperature

The vertical distribution of the surface temperatures on the different buildings shown in **Figure 4** was studied during day. Since it is important to understand the transfer phenomena between them and the adjacent air. Surface temperatures were measured with the infrared thermometer. Measurements were performed from the ground level up to the fourth or sixth floor, every 1 h.

The surface temperatures on the different buildings wall were reported in **Figure 6**. It could be seen from **Figure 6** that the surface temperature of most building location for the same floor during day increase with time to maximum value then decrease as do solar radiation with time. But the surface temperature for building opposite mountain during night decrease with time due to heat losses during night. Also it could be seen from **Figure 6** that the surface temperature for most building location decrease with increase floor number (far from land) as this could be explained as a result of the higher ground temperatures and the reduced wind effect, also this could be explained by the fact that lower level surfaces have lower sky view factors and thus irradiative losses to the sky are smaller. Except surface temperature at ship increase with floor number *i.e.* as far from water surface. **Figure 6** also could show that surface temperatures during night has a lower values than during day for building opposite mountain as this could be due to heat losses by convection during night. Comparing surface temperature for different location it could be seen that building opposite mountain has higher value followed by ship in Nile river followed by two building opposite others followed by building on Nile river shore. It could be seen that although ship was in river it has a higher wall temperature comparing to other building as this could be explained that its wall was from metal which absorb heat thus higher its temperature.

As expected, the thermal behavior of the two opposite walls is more complex due to parameters affecting the thermal balance of building materials (physical properties and canyon geometry) and due to incident solar and emitted infrared radiation. In general, it has been observed that during day period, the west facade presented higher temperatures than the opposite wall as this could be explained due to facing sun for more time than other wall.

5.3. Surface Temperature (Street Temperature)

Surface temperatures were conducted on the ground by an infrared thermometer equipped with a laser beam, on an hourly basis, at a number of some points across the asphalt street. The surface temperatures were recorded across the canyon on the asphalt street for different urban

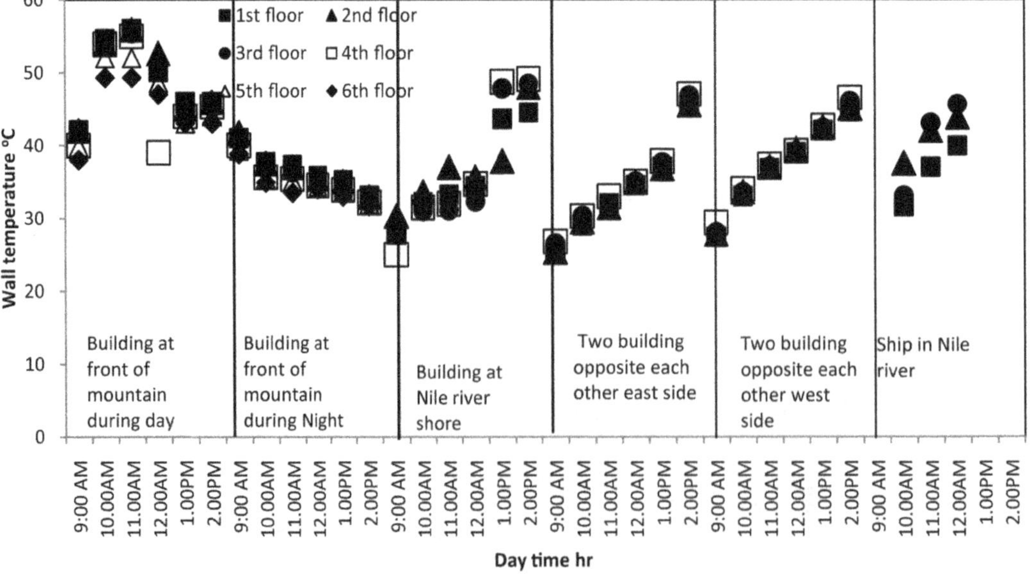

Figure 6. Hourly distributions of wall temperatures for different urban locations.

location in **Figure** 7. It could be seen from the figure that the street temperature for all building location during day increase with time to maximum value then decrease as do solar radiation with time. But street temperature at building opposite mountain decrease with time during night. Also it could be seen from **Figure** 7 that the street temperature increase as far from building as this could be due to shade. **Figure** 7 could be seen that street temperature increase toward western wall for two building opposite other as this could be explain to facing sun. There are a little bits difference in street temperature under and far from power transmission line as this could be explained for shade or difference in place. Finally the street temperature values are higher than both air and surface temperatures.

5.4. Comparison between (Air, Wall and Street) Temperatures

Comparison of (air, wall and street) temperatures have shown in **Figure 8** during daytime shows that in almost all cases the street temperature is highest followed by the wall temperature followed by the air temperature, in some cases, the air temperature is higher than the corresponding surface temperatures which was explained probably due to the warming of the air volume by the combined effects of turbulent sensible heat. The results showed that there were less air temperature variations compared with the surface temperatures due to street geometry and sky view factor.

6. Conclusions

The analysis of temperature distribution in different urban street canyon has been performed in Aswan Egypt under hot summer conditions. The analysis of the field measurements has led to the following observations:

1) The air temperature, wall temperature, and street temperature for most urban location during day time increase with time to a maximum value, then decrease with time. And continuous decreasing during night.

2) The air temperature and wall temperature for most urban location (except ship in Nile river) at the same time decrease with floor number *i.e.* decrease as far up from land.

3) The air and wall temperatures for ship at Nile river was found to be increased with floor number.

4) For the two building opposite each other the measured air and wall temperatures differences between the two facades vary, as a function of the canyon layout and the surface characteristics.

5) Comparison between air temperature during day and night it could be seen that temperature values for day is more than for night

6) Comparing of air temperature for all urban location it could be seen that air temperature for two building opposite each other has a lower value followed by building on Nile river shore followed by building at front of mountain.

7) Comparison of the hourly surface temperatures measured on the opposite building walls from the ground level up to the fourth floor has shown that the west facade presented greater temperatures than the opposite (east facade). The measured temperature differences between the opposite canyon walls during the day were higher at the fourth floor, because of the increased solar radiation and at the ground floor, as a result of the higher ground

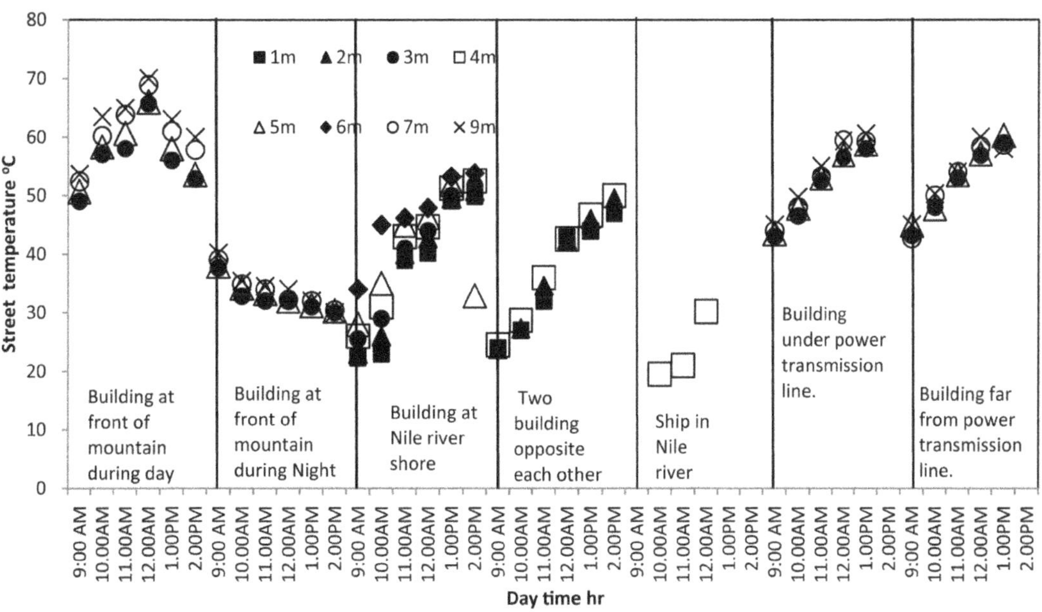

Figure 7. Hourly distributions of street temperatures for different urban locations.

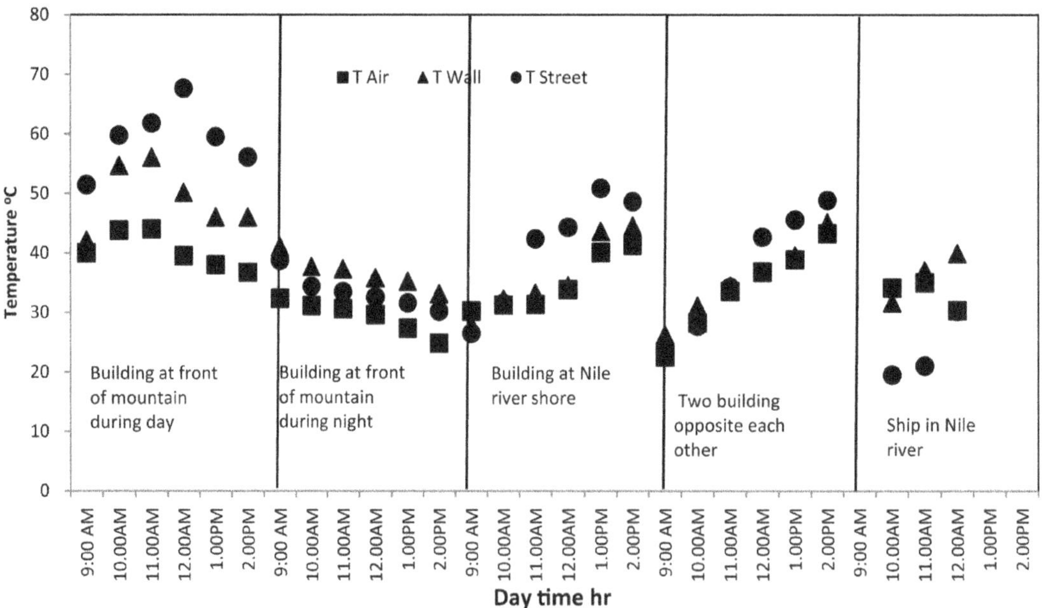

Figure 8. Hourly distributions of (air, wall temperatures at first floor and average street) temperatures for different urban locations.

temperatures and the reduced wind effect.

8) Analysis of the street temperatures across a section fromeasttowest facing mountain revealed the existence of maximum surface temperatures on the asphalt street up to 68.9°C, during midday, due to the vertical incidence of solar radiation.

9) The above results indicate that the air temperature distribution inside a street canyon is the function of canyon geometry and orientation, as well as, of the optical and thermal properties of building and street materials andambient weather conditions. Furthermore, the understanding of the specific thermal characteristics is essential for the understanding of airflow inside the canyons and for the studies of natural and hybrid ventilation in the urban environment.

REFERENCES

[1] K. Niachou, I. Livada and M. Santamouris "Experimental Study of Temperature and Airflow Distribution inside an Urban Street Canyon during Hot Summer Weather Conditions Part I: Air and Surface Temperatures," *Building and Environment*, Vol. 43, No. 8, 2008, pp. 1383-1392.

[2] W. Kuttler and A. Strassburger, "Air Quality Measurements in Urban Green Areas, a Case Study," *Atmospheric Environment*, Vol. 33, No. 24, 1999, pp. 4101-4108.

[3] E. Vignati, R. Berkowitz and O. Hertel, "Comparison of Air Quality in Streets of Copenhagen and Milan, in View of the Climatological Conditions," *Science of the Total Environment*, Vol. 189-190, 1996, pp. 467-473.

[4] N. H. Wong and C. Yu, "Study of Green Areas and Urban Heat Island in a Tropical City," *Habitat International*, Vol. 29, No. 3, 2005, pp. 547-558.

[5] Y.-W. Zhang, Z.-L. Gu, Y. Cheng and S.-C. Lee, "Effect of Real-Time Boundary Wind Conditions on the Air Flow and Pollutant Dispersion in an Urban Street Canyond Large Eddy Simulations," *Atmospheric Environment*, Vol. 45, 2011, pp. 3352-3359.

[6] B. I. Idris, K. Giskesa, C. Borrell, J. Benach, G. Costa, B. Federico, S. Helakorpi, U. Helmert, E. Lahelma, K. M. Moussa, P.-O. O¨stergren, R. Pra¨tta¨la¨ , N. Kr. Rasmussen, J. P. Mackenbach and A. E. Kunsa, "Higher Smoking Prevalence in Urban Compared to Non-Urbanareas: Time Trends in Six European Countries," *Health & Place*, Vol. 13, No. 3, 2007, pp. 702-712.

[7] H. B. Ren, W. S. Zhou, K. Nakagami, W. J. Gao and Q. Wu, "Feasibility Assessment of Introducing Distributed Energy Resources in Urban Areas of China," *Applied Thermal Engineering*, Vol. 30, 2010, pp. 2584-2593.

[8] J. Yuan and A. Venkatram, "Dispersion within a Model Urban Area," *Atmospheric Environment*, Vol. 39, No. 26, 2005, pp. 4729-4743.

[9] V. Kalantar, "Numerical Simulation of Cooling Performance of Wind Tower (Baud-Geer) in Hot and Arid Region," *Renewable Energy*, Vol. 34, No. 1, 2009, pp. 246-254.

[10] H. Huang, R. Ooka, H. Chen, S. Katoa, T. Takahashia and T. Watanabec, "CFD Analysis on Traffic-Induced Air Pollutant Dispersion under Non-Isothermal Condition in a Complex Urban Area in Winter," *Journal of Wind Engineering and Industrial Aerodynamics*, Vol. 96, 2008, pp. 1774-1788.

3

Fundamentals of Direct Inverse CFD Modeling to Detect Air Pollution Sources in Urban Areas

Mahmoud Bady
Department of Environmental Engineering, Egypt-Japan University of Science and Technology (E-JUST), New Borg El-Arab City, Alexandria, Egypt

ABSTRACT

This paper presents the fundamentals of direct inverse modeling using CFD simulations to detect air pollution sources in urban areas. Generally, there are four techniques used for detecting pollution sources: the analytical technique, the optimization technique, the probabilistic technique, and the direct technique. The study discusses the potentialities and limits of each technique, where the direct inverse technique is focused. Two examples of applying the direct inverse technique in detecting pollution source are introduced. The difficulties of applying the direct inverse technique are investigated. The study reveals that the direct technique is a promising tool for detecting air pollution source in urban environments. However, more efforts are still needed to overcome the difficulties explained in the study.

Keywords: Inverse Modeling; Outdoor Environments; Reverse Simulation; CFD

1. Introduction

Nowadays, inverse modeling seems to be a promising topic in terms of environmental research. The importance of reverse modeling arises from the increased numbers of pollution sources due to continuing industrial expansion coupled with population growth, especially in large cities. In addition, terrorist attacks are becoming more frequent and deadlier, such as the sarin gas attacks on the Tokyo subway in 1995, which resulted in 12 deaths and over 6000 injured [1,2], and the terrorist attacks on New York City and Washington DC on September 11, 2001 and the following anthrax dispersion by mail. Moreover, in the US, fire accidents occur in structures at the rate of one every 61 s, and in particular residential fires occur every 79 s. Nationwide, there was a civilian fire-related death every 156 minutes and a civilian fire-related injury every 28 minutes [3]. All of these events have confirmed that the terroristic attacks are no longer hypothesis but a reality. From this perspective, the ability to predict pollution sources characteristics: location, strength, and release time has become a necessity in order to create a complete picture of the air quality conditions within the release domain and to ensure the public's safety. Inverse modeling using Computational Fluid Dynamics (CFD) simulations is one efficient and promising tool in this regard.

Traditionally, CFD is used to explore the cause-effect relationship that is called forward-time modeling. Inverse modeling is to find the causal characteristics such as the pollutant source location and strength from the finite effectual information like the distributions of airflow and pollutant concentrations. Generally, in modeling process, there are three different types of problems [4]:

- The forward-time problem: given input and system parameters, find out output of the model.
- The reconstruction problem: given system parameters and output, find out which input has led to this output.
- The identification problem: given input and output, determine the system parameters that agree with the relationship between input and output.

Type 1 problem is an oriented cause-effect sequence. In this sense, type 2 and 3 problems are inverse problems because they are the problems of finding unknown causes of known consequences.

In order to determine the pollution source within a domain through the use of inverse modeling, some measurements such as concentration distributions are needed to infer the values of the inputs or parameters that characterize the system [5]. In reality, the concentration field is obtained from data measured by concentration sensors. The conventional CFD codes predict airflow based on given inputs (boundary conditions) and system parameters (building and system characteristics), which is a forward-time problem. Then, the reverse simulation is carried out through the solution of the reversed transport equation using the measured concentration field and the solved air flow field.

By carrying out inverse CFD simulations, one can detect air pollution source (or release) characteristics. Identification of pollution sources immediately after release is a matter of urgency in order to ensure the public's safety. Since, inverse modeling is one of the efficient and promising tools in this regard; this paper sheds light on the existing inverse techniques and explores the feasibility of using these techniques for urban environment study. Generally, there are four techniques used for detecting pollution sources the analytical technique, the optimizition technique, the probabilistic technique, and the direct inverse technique. First, the present study discusses the potentialities and limits of each technique. Then, the direct inverse modeling technique, as a promising tool to detect air pollution sources, is focused.

2. Different Techniques of Inverse Modeling

In the present study, the existing inverse modeling techniques that have been employed to detect pollutant sources in urban areas are reviewed. **Figure 1** shows classification of inverse CFD modeling techniques. Each technique has its own advantage and disadvantage. In the following sections, the four techniques will be discussed in details.

2.1. The Analytical Approach

The analytical approach requires analytical solution of the distributions of airflow and pollutant concentrations. The causal characteristics are then inversely solved. The analytical approach has been successfully applied to multi-dimensional heat conduction problems [6]. In groundwater transport, the analytical approach has been used to solve contaminant transport in one-dimensional flow [7] or in 2-D uniform flow [8]. The analytical approach has also been used to solve an inverse atmospheric transport problem in three-dimensional uniform flow [9]. Using analytical and graphical techniques, Islam [10,11] worked out a method for determining an unknown emission source in urban areas based on the well known Gaussian Plume Model (GPM). He assumed that the GPM describes the dispersion process fully.

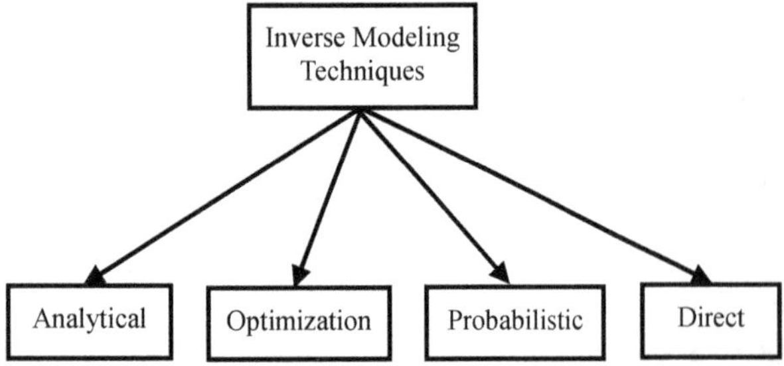

Figure 1. Classification of inverse CFD modeling techniques.

The above discussion shows that the analytical approach can be accurate and efficient. However, it is only used for simple problems. Accordingly, the application of the analytical approach in problems of complex geometry is very limited.

2.2. The Optimization Approach

In the optimization approach, forward-time modeling is used to obtain the effectual data (such as distributions of pollutant concentrations) based on all possible causal characteristics. Then the approach optimizes a solution that is best-fitted with the corresponding measured data. This approach has been widely applied in identifying groundwater pollution source as linear optimization method [12], maximum likelihood method [13], and nonlinear optimization method [14]. As an example, Arvelo *et al.* (2002) [6] have used a modified multi-zone model to study the optimal placement of chemical/biological warfare agent sensors in a building with nine offices and a hallway. The optimal sensor locations should be in the spaces where most possible sources of chemical/biological warfare agent could be located so the sensors can detect the agent in least amount of time. Different from the previous researchers, they used the genetic algorithm to interpret the computed data to locate the sources. Because plausible combinations of possible causal characteristics are huge, this approach involves a large amount of modeling.

2.3. The Probabilistic Approach

The probabilistic approach also does forward-time modeling. The approach uses probability to express a possible causal aspect. In groundwater transport, Bagtzoglou *et al.* (1992) [15] calculated the possibility of a contaminant source in groundwater by reversing only the convective contaminant transport. Other researchers [16] used Bayesian theory to interpret the possibility of each contaminant source. In enclosed environment, Sohn *et al.* (2002) [17] also used Bayesian probability model to identify the contaminant source in a five-room building. They used a multi-zonal model to calculate the airflow and contaminant transport. Since the multi-zone model can only provide some macroscopic information about the contaminant transport, it is necessary to run CFD simulations if more accurate and detailed information is needed. Accordingly, both the probability and optimization approaches need a huge amount of forward-time modeling.

2.4. The Direct Inverse Approach

The direct inverse approach solves inverse problems by reversing directly the governing equations that describe cause-effect relations. Kato *et al.* (1992) applied the reverse tracking of flow field over time, to obtain the residual lifetime of air at a point. In addition, Kato *et al.* [18-20] used the same technique to assess local pollution from upwind regions with backward trajectory analysis of the flows in an atmospheric environment. They reversed only the pollutant transport by convection and neglected the effect of diffusion. The method has also been used in groundwater contaminant transport [15,21], where convection transport of pollutants is solved with reversed velocity field and the diffusion was left unchanged.

In the following sections, the direct inverse modeling technique is investigated in details. The fundamentals of the technique are introduced. Then, the difficulties in applying the technique are discussed. Finally, two examples of applying the direct inverse technique in detecting pollution source in urban area are introduced.

3. Fundamentals of Direct Inverse Modeling

Pollutant concentrations are predicted based on the convection-diffusion equation including meteorological data, transport diffusion, and the relevant emissions (S). It can be written as follows [22]:

$$\frac{\partial(c)}{\partial t}+\frac{\left[\partial(u_i)\,c\right]}{\partial x_i}=\frac{\partial}{\partial x_i}\left(\frac{K}{\rho}\frac{\partial c}{\partial x_i}\right)+\frac{S}{\rho}\quad t\to\infty \tag{1}$$

where

c is the pollutant concentration (kg/kg),
t is the time (s),
u_i is the Cartesian components of the velocity (m/s),
x_i is the Cartesian coordinates (m),
K is the concentration diffusivity coefficient (kg/m/s),
ρ is the air density (kg/m^3), and
S is the pollutant source strength (kg/m^3/s).

In order to explain the principles of inverse modeling, Equation (1) can be simply integrated based on one dimensional flow as shown in **Figure 2** with using upwind scheme for space:

$$\begin{aligned}&\int_w^e\int_t^{t+\Delta t}\frac{\partial c(t)}{\partial t}\,\mathrm{d}t\cdot\mathrm{d}x\\&=-\int_t^{t+\Delta t}\int_w^e\frac{\partial\left(uc(t)\right)}{\partial x}\mathrm{d}x\cdot\mathrm{d}t\\&+\int_t^{t+\Delta t}\int_w^e\frac{\partial}{\partial x}\left(\frac{K}{\rho}\frac{\partial c(t)}{\partial x}\right)\mathrm{d}x\cdot\mathrm{d}t\\&+\int_t^{t+\Delta t}\int_w^e\frac{S}{\rho}\,\mathrm{d}x\cdot\mathrm{d}t\end{aligned} \tag{2}$$

Assuming K and ρ to be constant, the integration result becomes:

$$\begin{aligned}&\left\{c_P(t+\Delta t)-c_P(t)\right\}\cdot\Delta x\\&=-\int_t^{t+\Delta t}\left[u_e c_P-u_w c_W\right]\cdot\mathrm{d}t\\&+\frac{K}{\rho}\int_t^{t+\Delta t}\left[\left(\frac{\partial c}{\partial x}\right)_e-\left(\frac{\partial c}{\partial x}\right)_w\right]\cdot\mathrm{d}t\\&+\int_t^{t+\Delta t}\int_w^e\frac{S}{\rho}\,\mathrm{d}x\cdot\mathrm{d}t\end{aligned} \tag{3}$$

Rearranging Equation (3), then:

$$\begin{aligned}c_P(t+\Delta t)=c_P(t)&-\frac{1}{\Delta x}\int_t^{t+\Delta t}\left[u_e c_P-u_w c_W\right]\cdot\mathrm{d}t\\&+\frac{K}{\rho(\Delta x)^2}\int_t^{t+\Delta t}\left[c_E-2c_P+c_W\right]\cdot\mathrm{d}t\\&+\frac{1}{\Delta x}\int_t^{t+\Delta t}\int_w^e\frac{S}{\rho}\,\mathrm{d}x\cdot\mathrm{d}t\end{aligned} \tag{4}$$

To perform the integration with respect to time, the concentration-time relation is needed. Usually, the following equation is assumed [23]:

$$\int_t^{t+\Delta t}c\cdot\mathrm{d}t=\left[f\cdot c(t+\Delta t)+(1-f)\cdot c(t)\right]\cdot\Delta t \tag{5}$$

where f is a weighting factor used to combine the concentration value at the new time step with both of new and old concentration values.

By applying fully implicit scheme (*i.e.* $f=1$), Equation (5) can be rearranged to the form:

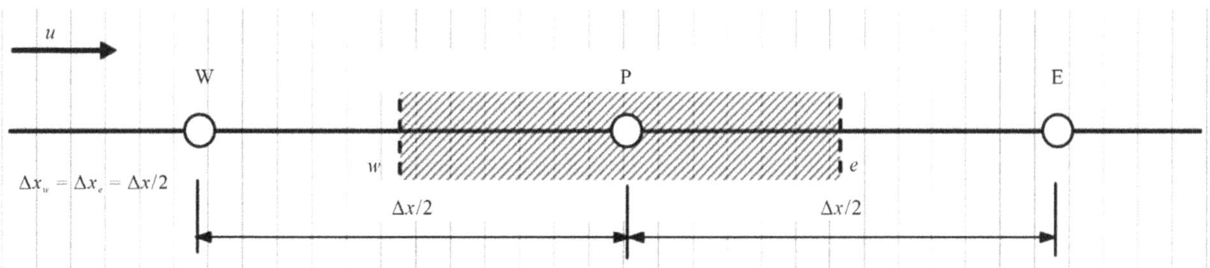

Figure 2. Typical control volume for the one-dimensional flow.

$$c_P(t+\Delta t)=\frac{1}{1+\frac{u_e\Delta t}{\Delta x}+\frac{2K\Delta t}{\rho(\Delta x)^2}}\cdot c_P(t)+\frac{\frac{K\Delta t}{\rho(\Delta x)^2}+\frac{u_w\Delta t}{\Delta x}}{1+\frac{u_e\Delta t}{\Delta x}+\frac{2K\Delta t}{\rho(\Delta x)^2}}\cdot c_W(t+\Delta t)+\frac{\frac{K\Delta t}{\rho(\Delta x)^2}}{1+\frac{u_e\Delta t}{\Delta x}+\frac{2K\Delta t}{\rho(\Delta x)^2}}\cdot c_E(t+\Delta t)+\frac{\frac{1}{\Delta x}\int_t^{t+\Delta t}\int_w^e\frac{S}{\rho}\mathrm{d}x\cdot\mathrm{d}t}{1+\frac{u_e\Delta t}{\Delta x}+\frac{2K\Delta t}{\rho(\Delta x)^2}} \tag{6}$$

Clearly, in order to calculate the concentration at previous time, one can simply apply a negative time step in Equation (1). However, the negative time step makes the coefficient of $c_P(t)$ greater than one, which amplifies the calculation errors as time elapsed in negative direction. Thus, Equation (1) is unstable in inverse modeling. Accordingly, in order to solve the reverse transport equation with numerical stability, another approach has to be used. One of these approaches is to use a slightly different equation instead of the original equation [24-26].

The inverse direct technique requires less-efforts compared with the probabilistic technique and the optimization technique. However, the instability problem caused by the negative diffusion term represents the main difficulty when applying such technique. This will be investigated in the following section.

4. Solution Instability in Direct Approach

Since the reversed governing equations are unstable due to the negative diffusion term, many researchers have worked out to improve the solution stability through imposing a bound on the solution. The regularization technique (or the stabilization technique) has been used to improve the solution stability. The solution is obtained by minimizing the objective function with a regularized term. The stabilization technique introduces some stabilization terms into the reversed governing equations or solves some auxiliary equations to improve the solution stability. For example, Zhang and Chen (2007) [26] and Liu *et al.* (2011) [27] changed the diffusion term in the reversed transport equation as (*i.e.* Equation (1)):

$$\frac{\partial(c)}{\partial t}+\frac{\partial[(u_i)\,c]}{\partial x_i}=\varepsilon\frac{\partial^2}{\partial x_i^2}\left(\frac{\partial^2 c}{\partial x_i^2}\right) \tag{7}$$

where ε is a stabilization coefficient.

The equation is numerically stable compared with Equation (6), since the coefficient of c_P can be smaller than one, where:

$$c_P(t+\Delta t)=\frac{1}{1+\frac{u_e\Delta t}{\Delta x}-\frac{6\varepsilon\Delta t}{(\Delta x)^4}}\cdot c_P(t)+\frac{-\frac{4\varepsilon\Delta t}{(\Delta x)^4}+\frac{u_w\Delta t}{\Delta x}}{1+\frac{u_e\Delta t}{\Delta x}-\frac{6\varepsilon\Delta t}{(\Delta x)^4}}\cdot c_W(t+\Delta t)+\frac{\frac{-4\varepsilon\Delta t}{(\Delta x)^4}}{1+\frac{u_e\Delta t}{\Delta x}-\frac{6\varepsilon\Delta t}{(\Delta x)^4}}\cdot c_E(t+\Delta t) \tag{8}$$

The required condition is just to make ε larger than $u_e\Delta x^3/6$, which is easy to be satisfied. Then the numerical scheme becomes stable and the reversed transport equation becomes dispersive.

By comparing the stabilization and the regularization methods, Skaggs and Kabala (1995) [25] concluded that the stabilization method used significantly less computational effort than the regularization method, although the stabilization method might provide slightly inferior results.

In the present study, in order to make the equation solvable with numerical stability, this study proposes the use of a filter to deal with negative concentration gradients in order to avoid unrealistic solutions. Then, Equation (1) can be rewritten as:

$$\frac{\partial(c)}{\partial t}+\frac{\partial[(-u_i)c]}{\partial x_i}=\left[-\frac{\partial}{\partial x_i}\left(\frac{\hat{K}}{\rho}\frac{\partial c}{\partial x_i}\right)\right]\quad t\to\infty \tag{9}$$

where the symbol "^" means that this term is filtered.

The filter for the diffusion term is set in such a way that the negative concentration values are changed to the minimum positive value of the concentration within the solution domain. This can be expressed as:

$$\text{if }(c<0)\,c=c_{\min} \tag{10}$$

Indeed, the filtration process of the diffusion term in the reversed transport equation affects the solution accuracy and decreases the method's effectiveness in identifying the sources of pollution. This will be investigated in the following examples.

5. Examples of Applying Direct Inverse Modeling in Identifying Source Locations

5.1. Simple Laminar Flow

A 3-D numerical model for modeling a simple laminar

flow is designed as shown in **Figure 3**. The dimensions of the computational domain are 90 m × 30 m × 30 m (L × W × H). The minimal mesh size above the ground is set as 0.18 m, and the vertical growing factor for mesh points is 1.05. The total number of cells is 594,000. A pollution source of strength 0.01 kg/m^3/s and dimensions 0.5 m × 0.5 m × 0.5 m is considered at a location 12 m from the inlet face at a height of 2 m.

5.2. Turbulent Flow around a Single Building

The second example for identifying the source of pollution in urban areas using direct inverse modeling is shown in **Figure 4**, in which wind flows around a single building. The building dimensions are 10(x) × 10(y) × 30(z) m and the computational domain dimensions are 300(x) × 110(y) × 70(z) m with a mesh size of 444,800 cells.

Two different locations for the source are considered in order to examine the effect of pollutant source location on the prediction accuracy of the reverse technique: location (I), which is 2.5 m upwind the building, and location (II), which is 2.5 m downwind the building. In both cases, the source strength is set at 0.01 kg/m^3/s with dimensions of 0.5 m × 0.5 m × 0.6 m.

5.3. Numerical Simulation

Numerical simulations are carried out using the CFD code Star-CD, based on the finite-volume discretization method. During forward-time simulation, steady-state analysis is adopted for the flow field and the Monotone Advection and Reconstruction Scheme (MARS) [28] is applied to the spatial difference. The standard k-ε model is used to simulate the turbulence effects and the pressure/velocity linkage is solved via the SIMPLE algorithm [23].

At the inflow boundary, a constant flux layer is assumed

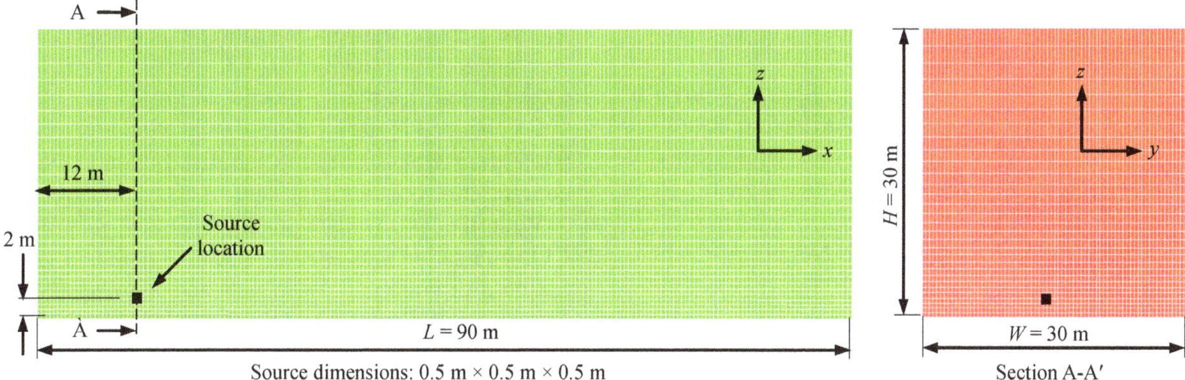

Figure 3. Computational domain and mesh arrangement in the case of laminar flow.

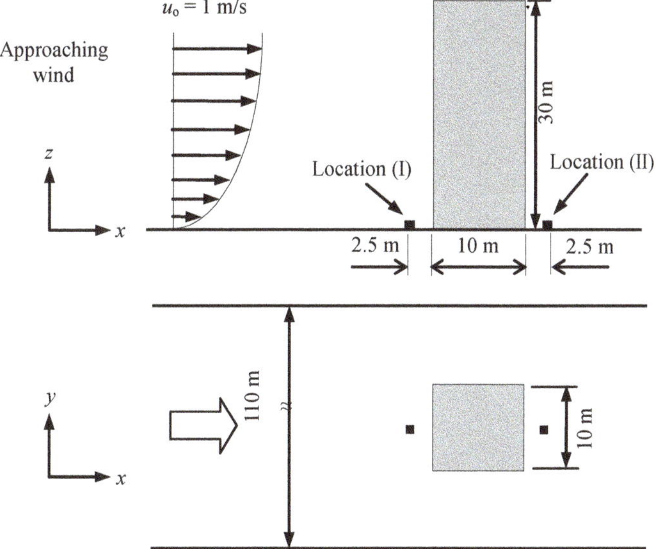

Figure 4. Geometry of wind flow around a single block and the locations of pollution sources (I) and (II), (a) front view; (b) top view.

for the turbulent energy, and turbulent intensity is assumed to be 10% of the inflow wind velocity at a representative height (z_o) of 74.6 m. Free slip condition is applied to the top and side boundaries. The generalized logarithmic law with parameter $E = 9$ (m) is applied to building walls and the ground surface as smooth walls.

Tables 1 and **2** summarize the parameters used in the numerical simulations for both examples, together with the applied boundary conditions.

Once the flow field is solved, it is considered steady. In fact, airflow characteristics within outdoor environments are unsteady due to fluctuations in both speed and direction. However, in the present stage of this study, fluctuations in the applied wind are not considered in the analysis. Accordingly, the wind flow is treated as steady. After solving the flow field, the concentration fields in both forward-time and reverse simulations are solved against time. The wind flow distribution together with the pollutant concentrations of the forward-time simulation are used as initial conditions for the reverse simulation. A time step of $\Delta t = 0.1$ s is used with the implicit scheme.

In forward-time simulation, a pollutant source of strength 0.01 kg/m^3/s is released in the period from $t = 0$ to 30 s. Then, the solution of the transport equation is continued with the absence of the source until $t = 100$ s

Table 1. Simulation parameters in the case of laminar flow.

Differential schemes	Convection term: MARS [28] Diffusion term: CD scheme Concentration: First order UD scheme Temporal term: Implicit scheme
Inflow	Constant velocity: $u = 0.01$ and 0.5 m/s
Outflow	Zero normal derivatives
Sides and sky	Free slip
Wall and ground	Generalized logarithmic law ($E = 9$)

Table 2. Simulation parameters in the case of laminar flow.

Turbulence model	The standard k-ε model
Differential schemes	Convection term: MARS [28] Diffusion term: CD scheme Concentration: First order UD scheme. Temporal term: Implicit scheme
Inflow	$u = u_o(z/z_o)^{0.25}$, $u_o = 1$ m/s & $z_o = 74.6$ m $k = 1.5(u_o \times I)^2$, $I = 0.1$ $\varepsilon = C_\mu^{1/2} \times k \times \frac{\partial u}{\partial z}$, $C_\mu = 0.09$
Outflow	Zero normal derivatives
Sides and sky	Free slip
Wall and ground	Generalized logarithmic law ($E = 9$)

for the case of laminar flow and until $t = 200$ s for the case of wind flow around a single building. **Figure 5** shows the characteristics of the pollutant release as a function against time.

In reverse simulation, the flow field calculated by forward-time simulation is reversed at first and the source term is set to zero. Then, the transport equation is solved from the moment of $t = 100$ s in the reverse time direction. The time at which the reversed simulation is stopped is not known in reality. In other words, starting from the moment of solving the transport equation in the reverse time direction, the time $t = 0$ at which the calculations have to be stopped is not known in advance. Since the present examples are introduced just to explain the technique of direct inverse modeling, the end time is assumed.

6. Results and Discussions

6.1. Case of Laminar Flow

Figure 6 shows the concentration fields for two different constant inflow velocities of 0.01 and 0.5 m/s. In subplots (a) and (b), the concentration fields at $t = 1$ s and 100 s obtained by forward-time simulation are presented, while the subplot (c) shows the distribution of pollutant concentrations obtained through reverse simulation.

With the steady-state airflow pattern, forward-time CFD simulation was used to calculate pollutant concentration distribution at $t = 100$ s, where a pollution source was released from $t = 0$ to 30 s. The distributions of steady-state airflow and transient pollutant concentration at $t = 100$ s were used as initial data for the inverse CFD modeling. The inverse CFD modeling calculated backwards pollutant transport from $t = 100$ s to 0 s is shown in **Figure 6(c)**.

In order to determine the pollutant release location, the distribution of pollutant concentration should be in a small region around the release source as shown in **Figure 6(c)**, and by using the maximum pollutant concentration over all locations at this instance (the peak pollutant

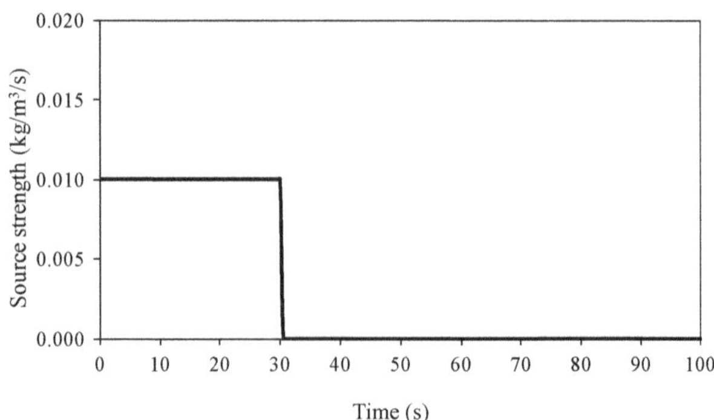

Figure 5. Pollutant source strength as a function against time.

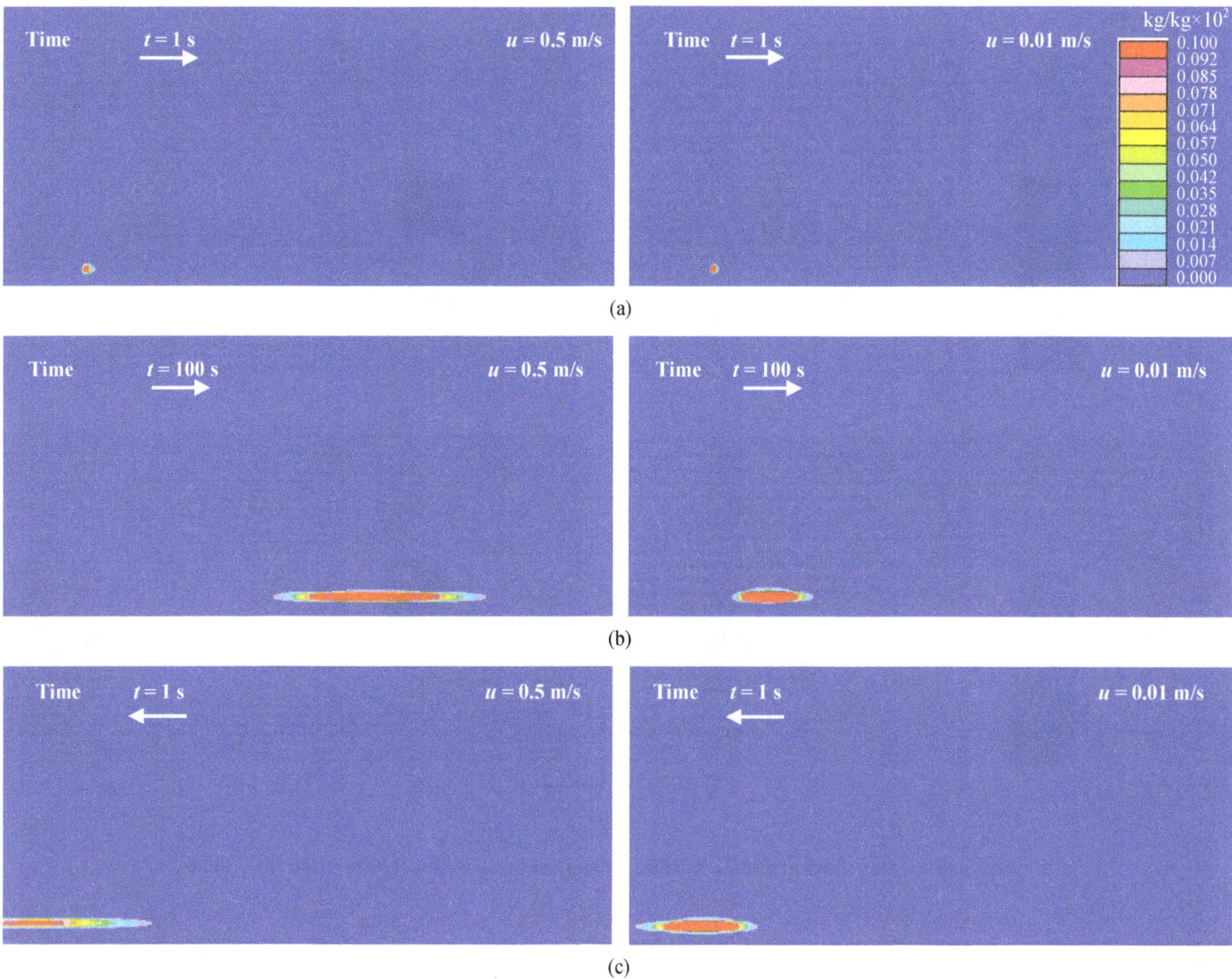

Figure 6. Concentration fields for the case of laminar flow, (a) at $t = 1$ s obtained by direct simulation; (b) at $t = 100$ s obtained by direct simulation; (c) at $t = 1$ s obtained by reverse simulation.

concentration), one could identify the pollution source. The peak concentration computed by inverse CFD modeling in **Figure 6(c)** clearly shows the position of the pollutant source.

By comparing the two concentration fields of $u = 0.01$ m/s and 0.5 m/s given in **Figure 6(c)**, the case where $u = 0.5$ m/s shows a more dispersive concentration field compared with the case of $u = 0.01$ m/s. This can be attributed to the increased plume size with the increase in wind velocity. However, the direct reverse simulation appears to effectively identify the release sources in both cases.

6.2. Case of Turbulent Flow around a Single Building

Figure 7 shows the computed airflow pattern around a single building under steady state conditions. **Figure 7(a)** shows the flow field calculated by the forward-time simulation and **Figure 7(b)** shows the reversed flow field which was used to carry out the reverse simulation. In **Figure 7(a)**, a symmetrical flow field is shown around the building, where two identical circulation regions are formed around the building.

It is expected that the effectiveness of inverse direct modeling in identifying the pollution source is affected by its location. So, as mentioned before, two locations for the pollutant source were examined here, namely locations (I) and (II). The first location is upwind of the building and the second location is in the wake region downwind of the building.

Source location (I):

The concentration fields around the building obtained for the source at location (I) are shown in **Figure 8**. At $t = 1$, shown in **Figure 8(a)**, the plume starts to diffuse where the concentration field makes contact with the upwind face of the building. The plume then travels with the wind downwind of the building. The pollutant continues to be released until the moment $t = 30$ s. At that time, the emission source was stopped and the scalar transport equation was solved against time using inverse

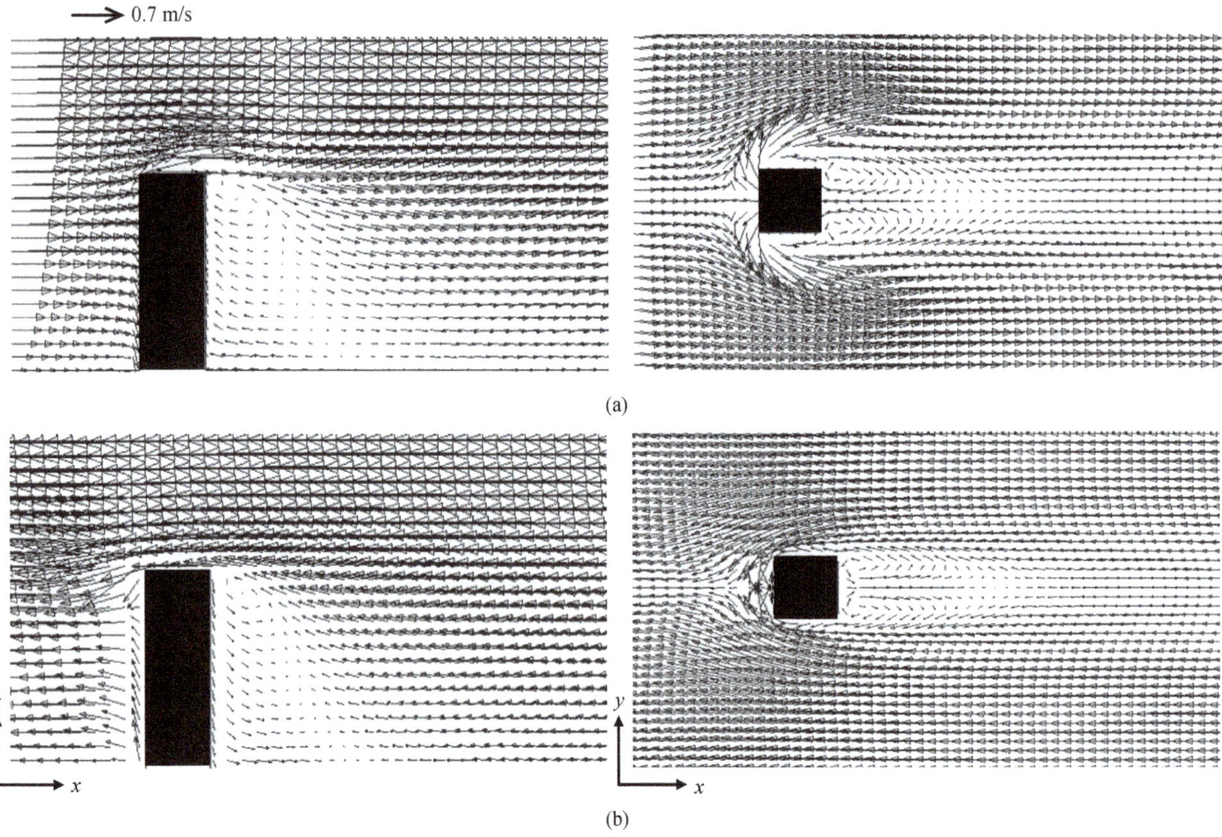

(a)

(b)

Figure 7. Wind flow field around a building, (a) direct flow field; (b) reversed flow field.

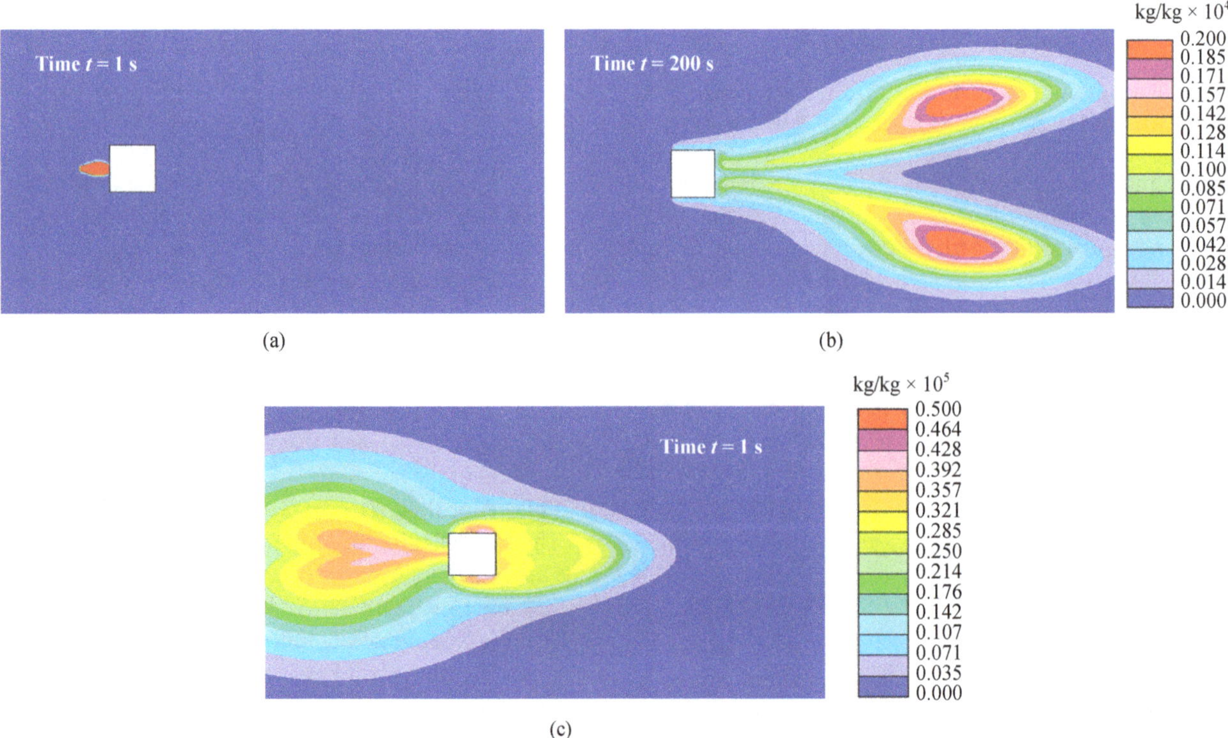

(a) (b)

(c)

Figure 8. Concentration fields for the source is in location (I), (a) at $t = 1$ s obtained by forward-time simulation; (b) at $t = 200$ s obtained by forward-time simulation; (c) at $t = 1$ s obtained by reverse simulation.

direct simulation without the source. At t = 200 s, two identical regions were formed far from the building. The conditions at such time were used as the initial conditions for the inverse modeling.

Figure 8(c) shows the concentration field obtained by the inverse modeling. In such figure, the concentration field area is wider than that of the original forward-time simulation. By detecting the location of the maximum concentration overall in the domain volume, the source location is clearly determined. It is located downwind of the building along the domain centerline. However, the figure implies two problems. The first one is that the maximum concentration occupies a wide area in front of the block, which means that estimations for the location of the source will not be 100% accurate. It is thought that the wide spread of the concentration field is attributed to what is called "false diffusion", introduced by the first order approximation of the advection term in the upwind difference scheme [29]. The second problem is that, compared to the peak concentration obtained with the forward-time simulation (see **Figure 8(b)**), the source strength identified by the reverse simulation (see **Figure 8(c)**) is more dispersive. The reason is that the reversed transport equation is not exactly the same as the governing transport equation due to the presence of a filter. Indeed, these two problems affect the prediction accuracy of the inverse modeling and this appears clearly in the wide area of the maximum concentration, which gives a wide range of possibilities for the pollution source location. This is not considered to be the ideal distribution when a gas release position is required. At the same time, the dispersive property of the reversed transport equation renders the estimated source strength inaccurate.

Source location (II):

Figure 9 shows the concentration fields around the building when the pollutant source is at location (II). **Figure 9(a)** shows the source location, which is 2.5 m downwind of the building's rear wall. As with the case for location (I), the pollutant is emitted in the period from t = 0 to 30 s, and then the transport equation was solved in the absence of the source term until the moment t = 200 s. In subplot (b), two high concentration regions are formed behind the building. However, diffusion of the pollutant in this case is limited to a narrow region compared to the case for location (I).

Figure 9(c) shows the concentration field obtained using inverse modeling. The figure demonstrates that the concentration increases in the direction of the reversed wind from right to left. Also, the figure shows that the peak concentration region occupies a narrow region compared with the case for location (I). This can be attributed to the presence of the source in the wake region where a lower wind velocity exists and convection is weak. In such case, the location of the pollution source as estimated by the reverse simulation is not clear. In **Figure 9(c)**, there are two peak concentration regions near the edges of the block. So the location of the source is not accurately identified. This indicates that the prediction

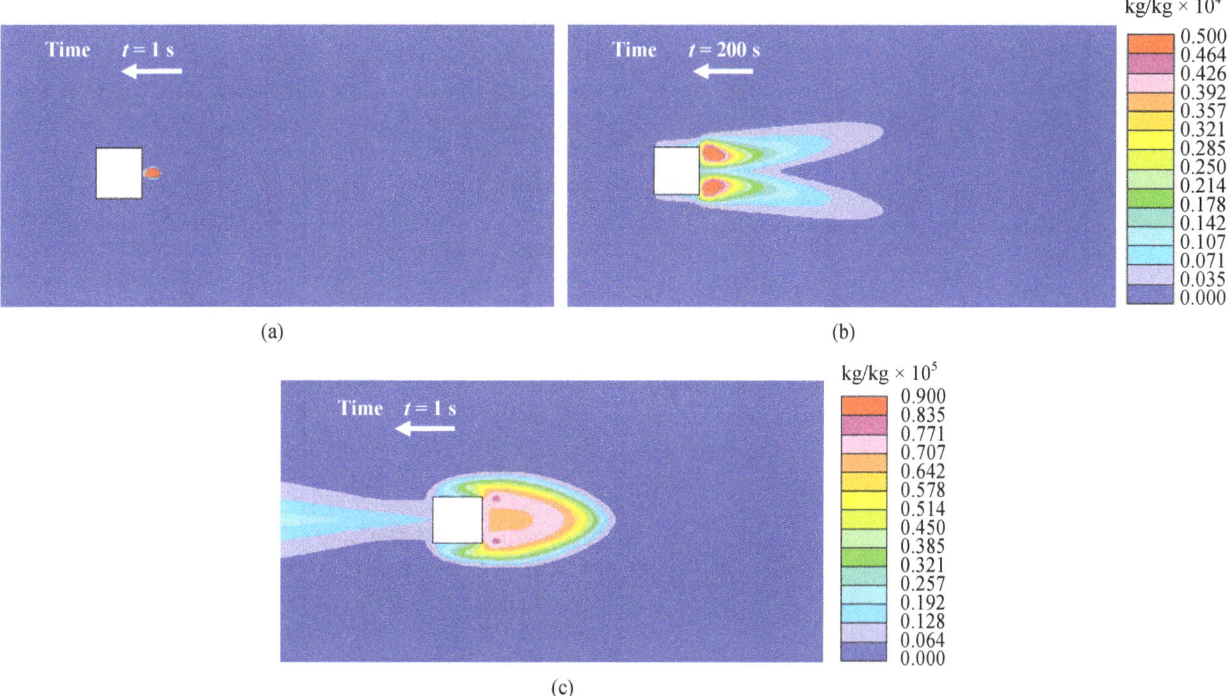

Figure 9. Concentration fields for the source is in location (II), (a) at t = 1 s obtained by forward-time simulation; (b) at t = 200 s obtained by forward-time simulation; (c) at t = 1 s obtained by reverse simulation.

accuracy of the inverse modeling is diminished in case of weak convection.

6.3. Prediction Accuracy Improvement

From the results of the above examples, it is clear that the accuracy of the method is limited due to the wide diffusion fields around the source location caused by the filter. Accordingly, a technique for improving the prediction accuracy of the method is needed. The proposed technique used what is called a "sink" in the reversed transport equation in order to decrease the widespread of the concentration field around the source and render identification of the source more easily. The reversed transport equation is then rewritten as:

$$\frac{\partial(c)}{\partial t}+\frac{\partial\left[(-u_i)c\right]}{\partial x_i}=\left[-\frac{\partial}{\partial x_i}\left(\frac{K}{\rho}\frac{\partial c}{\partial x_i}\right)\right]-\frac{S}{\rho} \quad (11)$$

where the last term on the right-hand side is the sink term.

In order to apply the sink term technique to identify pollution source characteristics, some criteria are needed. Such criteria have to satisfy the following requirements:

- Sufficiently general to be applied to any case.
- It should show where to put the sink within the study domain.
- The criteria should give a reasonable value for the sink strength to carry out the calculations.
- Then, the following procedure is applied which couples CFD with optimization approach.
- Setting the sink at any location within the domain with any strength (S).
- Carrying out CFD simulation to solve the scalar transport equation with any release time (T)

$$C=\int_{vol} c\mathrm{d}V \quad (12)$$

- At time (T), we calculate the whole concentration in the domain, which is:
- Then, the optimization approach is used to find the minimum value of the concentration within the domain.
- If the initial guess doesn't give the minimum domain concentration, another trial with new sink location and new values for S and T selected by the optimization technique is used, until the minimum concentration within the domain is obtained.

A flow chart for the above mentioned procedure is shown in **Figure 10**.

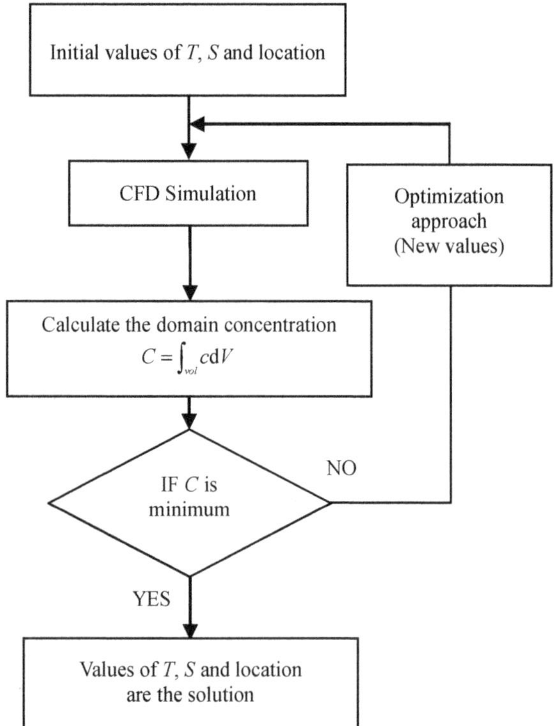

Figure 10. Flow chart of the proposed technique.

7. Conclusions

This paper discusses the advantages and feasibility of using inverse CFD modeling to detect air pollution sources in urban environments. The study reviews various inverse modeling approaches and categorizes them into four techniques: the analytical technique, the optimization technique, the probabilistic technique, and the direct inverse technique. Each technique has its own advantage and limitations. The analytical approach requires analytical solution of the distributions of airflow and pollutant concentrations. However, it is only used for simple problems, which means that the applications of such technique are very limited. The optimization approach uses forward-time modeling to obtain the effectual data based on all possible causal characteristics. The limitation of this approach is that it needs a large amount of forward-time modeling. The probabilistic approach uses probability to express a possible causal aspect. In such technique all possible causal characteristics should be known before doing the modeling. The direct inverse approach solves inverse problems by reversing directly the governing equations that describe cause-effect relations. The instability problem caused by the negative diffusion term represents the main difficulty when applying such technique. The study proposed the use of a filter to account overcome the instability caused by the negative diffusion term in the scalar transport equation. In addition, in order to decrease the wide diffusion field around the source location caused by the filter, the study proposed a "sink-term" technique in the reversed transport equation to render identification of the source more easily. As a conclusion, the study reveals that the direct inverse modeling is a promising tool for detecting air pollution source in urban environments. However, more efforts are still needed to overcome the difficulties explained in the study.

REFERENCES

[1] K. Yokoyama, "Our Recent Experiences with Sarin Poisoning Cases in Japan and Pesticide Users with References to Some Selected Chemicals," *Neurotoxicology*, Vol. 28, No. , 2007, pp. 364-373.

[2] T. Okumura, N. Takasu, S. Ishimatsu, S. Miyanoki, A. Mitsuhashi, K. Kumada, *et al.*, "Report on 640 Victims of the Tokyo Subway Sarin Attack," *Annals of Emergency Medicine*, Vol. 28, No. 2, 1996, pp. 129-135.

[3] M. Karter, "Fire Loss in the United States during 2002," Fire Analysis & Research Division, National Fire Protection Association, Quincy, 2003. http://www.nfpa.org/assets/files/pdf/OSfireloss02

[4] J. Friedr, "Stable Solution—Inverse Problems," Vieweg & Sohn, Verlagsgesellschaft mbH, Braunschweig, 1978.

[5] A. Tarantola, "Inverse Problem Theory and Methods for Model Parameter Estimation," Society for Industrial and Applied Mathematics, Philadelphia, 2004.

[6] O. Alifanov, "Inverse Heat Transfer Problems," Springer-Verlag, New York, 1994.

[7] S. Alapati and Z. J. Kabala, "Recovering the Release History of a Groundwater Contaminant Using a Non-Linear Least-Squares Method," *Hydrological Processes*, Vol. 14, No. 6, 2000, pp. 1003-1016.

[8] N. K. Ala and P. A. Domenico, "Inverse Analytical Techniques Applied to Coincident Contaminant Distributions at Otis Air Force Base, Massachusetts," *Groundwater*, Vol. 30, No. 2, 1992, pp. 212-218.

[9] P. Kathirgamanathan, R. Mckibbin and R. I. Mclachlan, "Source Term Estimation of Pollution from an Instantaneous Point Source," *Research Letters in Information Mathematic Science*, Vol. 3, 2002, pp. 59-67.

[10] M. Islam and G. Roy, "A Mathematical Model in Locating an Unknown Emission Source," *Water, Air, and Soil Pollution*, Vol. 136, No. 1-4, 2002, pp. 331-345.

[11] M. Islam, "Application of a Gaussian Plume Model to Determine the Location of an Unknown Emission Source," *Water, Air, and Soil Pollution*, Vol. 112, No. 3-4, 1999, pp. 241-245.

[12] S. M. Gorelick, B. E. Evans and I. Remson, "Identifying Sources of Groundwater Pollution: An Optimization Approach," *Water Resources Research*, Vol. 19, No. 3, 1983, pp. 779-790.

[13] B. J. Wagner, "Simultaneously Parameter Estimation and Contaminant Source Characterization for Coupled Groundwater Flow and Contaminant Transport Modeling," *Journal of Hydrology*, Vol. 135, No. 1-4, 1992, pp. 275-303.

[14] P. S. Mahar and B. Datta, "Identification of Pollution Sources in Transient Groundwater Systems," *Water Resources Management*, Vol. 14, No. 3, 2000, pp. 209-227.

[15] A. Bagtzoglou, D. Dougherty and A. Tompson, "Application of Particle Method to Reliable Identification of Groundwater Pollution Sources," *Water Resources Management*, Vol. 6, No. 1, 1992, pp. 45-23.

[16] M. F. Snodgrass and P. K. Kitanidis, "A Geo-Statistical Approach to Contaminant Source Identification," *Water Resources Research*, Vol. 33, No. 4, 1997, pp. 537-546.

[17] M. D. Sohn, P. Reynolds, N. Singh and A. J. Gadgil, "Rapidly Locating and Characterizing Pollutant Releases in Buildings," *Journal of the Air & Waste Management Association*, Vol. 52, No. 12, 2002, pp. 1422-1432.

[18] S. Kato, P. Pochanart and Y. Kajii, "Measurements of Ozone and Non-Methane Hydrocarbons at Chichi-Jima Island, a Remote Island in the Western Pacific: Long-Range Transport of Polluted Air from the Pacific Rim Region," *Atmospheric Environment*, Vol. 36, No. 34, 2002, pp. 385-390.

[19] S. Kato, P. Pochanart, J. Hirokawa, Y. Kajii, H. Akimoto, Y. Ozaki, K. Obi, T. Katsuno, D. G. Streets and N. P. Minko, "The Influence of Siberian Forest Fires on Carbon Monoxide Concentrations at Happo, Japan," *Atmospheric Environment*, Vol. 36, No. 2, 2002, pp. 385-390.

[20] S. Kato, Y. Kajii, R. Itokazu, J. Hirokawa, S. Koda and Y. Kinjo, "Transport of Atmospheric Carbon Monoxide, Ozone, and Hydrocarbons from Chinese Coast to Okinawa Island in the Western Pacific during Winter," *Atmospheric Environment*, Vol. 38, No. 19, 2004, pp. 2975-2981.

[21] J. L. Wilson and J. Liu, "Backward Tracking to Find the Source of Pollution," *Water Management Risk Remediation*, Vol. 1, 1994, pp. 181-199.

[22] J. Ferziger and M. Peric, "Computational Methods for Fluid Dynamics," 3rd Edition, Springer Verlag Publisher, New York, 2002.

[23] S. Patankar, "Numerical Heat Transfer and Fluid Flow," Hemisphere Publishing Corporation, New York, 1980.

[24] A. C. Bagtzoglou and J. Atmadja, "Marching-Jury Backward Beam Equation and Quasi-Reversibility Methods for Hydrologic Inversion: Application to Contaminant Plume Spatial Distribution Recovery," *Water Resources Research*, Vol. 39, No. 2, 2003, pp. SBH101-SBH1014.

[25] T. H. Skaggs and Z. J. Kabala, "Recovering the History of a Groundwater Contaminant Plume: Method of Quasireversibility," *Water Resources Research*, Vol. 31, No. 11, 1995, pp. 2669-2673.

[26] T. Zhang and Q. Chen, "Identification of Contaminant Sources in Enclosed Environments by Inverse CFD Modeling," *Indoor Air*, Vol. 17, No. 3, 2007, pp. 167-177.

[27] D. Liu, F. Zhao and H. Wanga, "History Recovery and Source Identification of Multiple Gaseous Contaminants Releasing with Thermal Effects in an Indoor Environment," *International Journal of Heat and Mass Transfer*, Vol. 55, No. 1-3, 2012, pp. 422-435.

[28] L. Bram, "Towards the Ultimate Conservative Difference Scheme, V. A Second Order Sequel to Godunov's Method," *Journal of Computational Physics*, Vol. 32, No. 1, 1979, pp. 101-136.

[29] T. Tsai and M. Nabil, "A Comparative Study of Central and Upwind Difference Schemes Using the Primitive Variables," *International Journal for Numerical Methods in Fluids*, Vol. 3, No. 3, 1983, pp. 295-305.

4

On the Short-Term Optimisation of a Hydro Basin with Social Constraints

Gloria Hermida, Edgardo D. Castronuovo
University Carlos III de Madrid, Madrid, Spain

ABSTRACT

In this paper, an optimisation problem for calculating the best energy bids of a set of hydro power plants in a basin is proposed. The model is applied to a real Spanish basin for the short-term (24-hour) planning of the operation. The algorithm considers the ecological flows and social consumptions required for the actual operation. One of the hydro plants is fluent, without direct-control abilities. The results show that the fluent plant can be adequately controlled by using the storage capacities of the other plants. In the simulations, the costs related to the social consumptions are more significant than those due to the ecological requirements. An estimate of the cost of providing water for social uses is performed in the study.

Keywords: Hydro Power Plants; Hydro Generation; Optimisation; Short-Term Planning; Social Resources

1. Introduction

Nowadays, the utilisation of water for electricity production is conditioned by many constraints. In Spain, primarily the Kyoto Agreements and the proposals of the European Commission to 2020 must be considered. The European Commission have specified a goal of 20% of the final energy consumption delivered from renewable sources by 2020 [1]. In Spain, 38.6% of the electricity generation comes from renewable resources, mainly from hydro (17.4%) and wind (16.6%) generation [2]. Because electricity generation has to compensate for other non-renewable energy consumptions, electricity production must increase its share of renewable generation. Hydro production is a mature renewable technology that can help reach the ambitious objectives proposed by the European Commission by 2020.

In addition, the exceptionally variable weather conditions of the past few years, most likely due to climate change, complicate the management of water for electricity production. The scarcity and the high variability of water resources have recently reduced the profits in several zones [3-6].

Many studies have been performed to calculate the optimal operation of a hydro basin. In long-term planning, Soares and Carneiro [7] consider the operation planning of a hydrothermal power system in Brazil. The paper highlights the importance on the control of the head hydro power plants (HPPs) in the basin. Granville *et al.* [8] consider the stochastic characteristics of the problem, including a representation of the market. The solution algorithm is based on stochastic dual dynamic programming. Cheng [9] applies particle swarm optimisation and dynamic programming for a large scale hydro system in China. Oliveira, Binato and Pereira [10] present two techniques: the extension of a binary disjunctive technique and screening strategies for planning studies in Brazil and Bolivia. Fosso *et al.* [11] give an overview of the planning tool used in Norway for long, medium and short horizons. Kanudia and Loulou [12] propose a stochastic version of the extended market allocation model for a hydro system in Québec, Canada.

In medium- and short-term planning, Habibollahzadeh and Bubenko [13] compare different mathematical methods: Heuristic, Benders and Lagrange methods for hydroelectric generation scheduling in the Swiss system. Castronuovo and Peças Lopez [14] describe economic profits of the coordination of wind and hydro energies. Zhao and Davison [15] analyse the inclusion of storage facilities in a hydro system, demonstrating the sensitive dependences between some of the parameters of the hydroelectric facility, the expected prices and water inflows. Pousinho, Mendes and Catalão [16] propose a mixed-integer quadratic programming approach for the short-term hydro scheduling problem, considering discontinuous operating regions and discharge ramping constraints. Simopoulos, Kavatza and Vournas [17] propose a decoupling method, dividing the hydrothermal problem into hydro and thermal sub-problems, which are solved independently. A Greek system is analysed in the study. Diniz and Piñeiro Maceira [18] use a four-dimensional piecewise linear model for the generation of a hydro plant as a function of storage, turbined and spilled outflows. Shawwash, Thomas and Denis Russell [19] discuss the optimisation model used in the British Columbia hydro system for hydrothermal coordination.

Most of the available reports about the optimal programming of hydro generation have been published in countries with abundant water (Norway [11], Brazil [10], Canada [15], USA [19]). In the algorithms reported by these studies, the restrictions on the social use of water and the ecological minimum flows are either minimally considered or not considered at all, aiming at improving the utilisation of the abundant resource in a strictly economical environment. In Spain, the focus of the present study, ecological flows and social uses of water must be considered for the optimal utilisation of the resource. Pérez-Díaz and Wilhelmi [20] want to assess the economic impact of environmental constraints in the operation of a short-term hydropower plant. For that purpose, a revenue-driven daily optimisation model based on mixedinteger linear programming is applied to calculate the optimal operation of a HPP in the northwest area of Spain. In a more recent paper, Pérez-Díaz *et al.* [21] propose adding a pumping capability to improve the economic feasibility of an HPP project, always fulfilling the environmental constraints imposed on the operation of the hydropower plant.

This paper presents an optimisation algorithm for calculating the optimal energy bids of a set of HPPs, including the economic objectives for energy generation and the regulations concerning the use of water in the region. The algorithm is applied to the upper Guadalquivir Basin, an area with scarce resources and variable flows, over a 24-hour horizon. Four HPPs are considered in the analysis. Three of them have storage capacity and the other one is run-of-the-river, without directly controllable alternatives. All of the plants are operated jointly with a unique owner or dispatcher (as in current practical operation). Actual data from real power plants and markets are considered in this study, including the travel times of the water (TTW) between the HPPs. The results show that the fluent plant can be controlled to achieve optimal operation by using the upstream HPPs. Moreover, an estimate of the costs of providing water for social uses (as a function of reductions in profits from selling the electricity produced in the market) is made in this study.

2. Rules Applicable to the Hydro Generation

2.1. Regulations Concerning the Use of Water for Electricity Generation

The Water Framework Directive [22] establishes a European Community framework for water protection and management. The objectives of this regulation are the prevention and reduction of pollution, promotion of sustainable water use, environmental protection, improvement in aquatic ecosystems and floods and drought mitigation. This norm was adapted to Spanish regulations by [23]. In this directive, the priorities regarding the use of water are fixed. Electricity generation is third in the order of precedence, after the use of water by the population and irrigation requirements. Additionally, this norm specifies the requirement of a Hydrological Plan for each basin or hydrological zone. In [24], the hydro regulations for the Andalucia region (the area considered in this study) are specified. The Guadalquivir Hydrographic Confederation (http://www.chguadalquivir.es) is the organisation designed to control the Guadalquivir basin. This organisation's website features historical data regarding affluences and other hydro information. The minimum levels of flows (ecological flows) are also specified for several points of the river.

2.2. The Daily Energy Market

In Spain, the electricity market has been deregulated since 1997 (Electricity Industry Act [25]). Some renewable productions have special incentives for their production (Royal Decree 661/2007 [26]). However, large or pre-existing hydro plants must auction their production in the conventional market without renewable bonuses and, practically, without special market regulation. This is the situation faced by the plants addressed in the present study.

The Spanish energy market is organised into the following sub-markets: futures market, daily market and several intra-daily markets. More than 95% of energy transactions and more than 80% of the economic volume are traded in the daily market [27]. There are also other markets that can affect hydroelectric production, such as the reserve and restriction management. For clarity, in this work, only daily market participation will be considered.

In the daily market, producers and consumers make their offers, in terms of energy quantity and prices for each hour of the D + 1 day. The Market Operator oversees the buying and selling of bids using a simple cassation model [28,29]. The present paper presents a method to calculate the optimal bids for energy over a 24-hour horizon of the hydro plants in the basin, assuming that the expected prices in these hours are known.

3. Mathematical Formulation

3.1. Flow Chart

In **Figure 1**, the flow chart of the algorithm is presented.

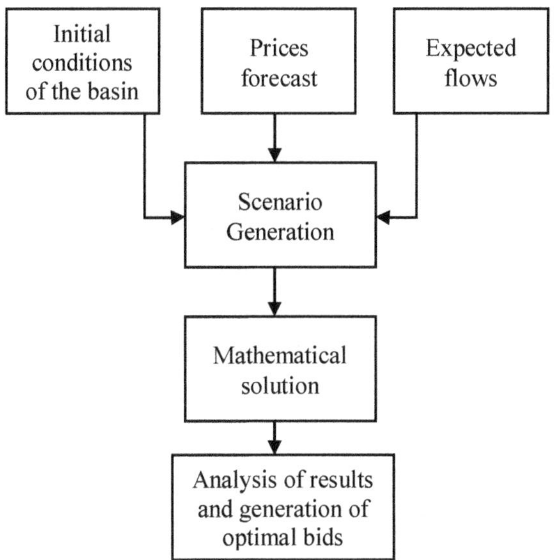

Figure 1. Flow chart of the proposed algorithm.

The initial conditions of the basin (level of stored water in the reservoirs, current flows, etc.) are known at the beginning of the study. Moreover, the expected flows in the analysed period can be considered known or estimated. The expected flows are depending also of the medium term planning for the operation of the basin. In the present study, an estimation of the prices in the market, for all the hours of the next day operation, is required. This prediction can be obtained from forecasting tools, outside the scope of the present study. With the knowledge of the initial condition, the price forecast and the expected flows, a scenario can be developed. In the present analysis, a determinist approach is used. However, the present method can be easily extended for considering uncertainties in the prices and/or in the expected flows, by solving many probable scenarios.

When the probable scenario is determined, the optimal solution for the operation in the hydro plants in the basin must be calculated. In the present case, ecological and social constraints are also included in the analysis. In the next section, a fully representation of the optimization problem is provided. After the calculation, the optimal flows of waters and the power and energy optimal bids are obtained. For achieving the profits presented in the analysis, it is considered that all the presented bids are accepted in the market, by offering the hydro production at low prices.

3.2. Mathematical Representation

The best operation of hydro plants in a basin can be calculated from the solution of an optimisation problem. In this problem, the restrictions to the operation are represented as mathematical constraints. The formulation of the problem is described by Equations (1)-(15).

Max.

$$\sum_{i=1}^{nr+nwr}\sum_{t=1}^{T}\left(C_t \cdot P_{i,t}\right) \quad (1)$$

s.t.

$$V_{i,t} = V_{i,t-1} + V_{i,t}^{AF} + V_{i-1,t} - V_{i,t}^{T} - V_{i,t}^{C} - V_{i,t}^{D} \quad i = 1,\cdots,nr \quad (2)$$

$$V_{i,t}^{AF} + V_{i-1,t} - V_{i,t}^{T} - V_{i,t}^{C} - V_{i,t}^{D} = 0 \quad i = 1,\cdots,nwr \quad (3)$$

$$V_{i-1,t} = \sum_{\alpha i}\left(V_{i-1,t-t_v}^{T} + V_{i-1,t-t_v}^{D}\right) \quad i = 1,\cdots,(nr+nwr) \quad (4)$$

$$V_{i,1} = V_{i,1}^{SP} \quad i = 1,\cdots,nr \quad (5)$$

$$V_{i,T} = V_{i,T}^{SP} \quad i = 1,\cdots,nr \quad (6)$$

$$P_{i,t} - \eta \cdot V_{i,t}^{T} \cdot g \cdot h_{i,t} = 0 \quad i = 1,\cdots,(nr+nwr) \quad (7)$$

$$h_{i,t} = k_{0,i} + k_{1,i}\cdot\left(V_i^U + V_i^0\right) + k_{2,i}\cdot\left(V_i^U + V_i^0\right)^2 + k_{3,i}\cdot\left(V_i^U + V_i^0\right)^3 \quad i = 1,\cdots,(nr+nwr) \quad (8)$$

$$\sum_{t=1}^{T} V_{i,t}^{C} \geq V_i^{CT\min} \quad i = 1,\cdots,(nr+nwr) \quad (9)$$

$$V_i^{C\min} \leq V_{i,t}^{C} \leq V_i^{C\max} \quad i = 1,\cdots,(nr+nwr) \quad (10)$$

$$V_{i,t}^{T} + V_{i,t}^{D} \geq V_i^{EC\min} \quad i = 1,\cdots,(nr+nwr) \quad (11)$$

$$0 \leq V_{i,t} \leq V_i^{\max} \quad i = 1,\cdots,nr \quad (12)$$

$$0 \leq V_{i,t}^{T} \leq V_i^{T\max} \quad i = 1,\cdots,nr \quad (13)$$

$$0 \leq V_{i,t}^{D} \leq 99 \quad i = 1,\cdots,nr \quad (14)$$

$$0 \leq h_{i,t} \leq h_i^{\max} \quad i = 1,\cdots,nr \quad (15)$$

$$t = 1,\cdots,T$$

where the variables indicate the following: $P_{i,t}$, the active power injection to the grid of hydro plant i at hour t; $V_{i,t}$, the useful volume stored in the reservoir of the hydro plant i in the period t; $V_{i-1,t}$, the affluence into reservoir i at period t, coming through the river from upstream plant (or plants); $V_{i,t}^{T}$, the turbined volume at hour t by plant i; $V_{i,t}^{D}$, the deviated (spilled) volume at hour t by plant i; $V_{i,t}^{C}$, the output water consumption for social uses delivered by plant i at hour t; and $h_{i,t}$, the height of reservoir i at hour t. The following are the parameters in the optimisation formulation: c_t, the expected market price of hour t; $V_{i,t}^{AF}$, the individual affluence into reservoir i at period t, not considering the flows coming through the river from the previous plant; t_V, the TTW between the considered HPPs; $V_{i,1}^{SP}$ and $V_{i,T}^{SP}$, the specified volumes at the beginning and at the end of the horizon (respectively) by plant i; η_i, the average efficiency of the hydro plant i; g, the acceleration of gravity; $k_{0,i}$, $k_{1,i}$, $k_{2,i}$ and $k_{3,i}$, the coefficients relating volume and height at reservoir i; V_i^U, the unused volume for electricity generation of reservoir i; $V_i^{CT\min}$, the minimum daily requirements of water for social uses in hydro plant i; $V_i^{C\min}$ and $V_i^{C\max}$, the minimum and maximum (respectively) hourly requirements of water for social uses, in plant i; $V_i^{EC\max}$, the minimum (ecological) volume to be maintained in the river downstream of reservoir i; $V_i^{\max}$ and $V_i^{T\max}$, the maximum useful reserve and capacity of production (respectively) of hydro plant i; and $h_i^{\max}$, the maximum height at plant i. In the equations, nr is the number of hydro plants with reservoirs, nwr is the number of fluent hydro plants (without reservoir), αi is the set of hydro plants upstream from the reservoir i and T is the number of discretisation steps.

The goal of the optimisation problem (1)-(15) is to calculate the optimal production of coordinated hydro plants in a basin in T periods and considering the expected prices in the market (1). Equality constraints (2) and (3) express the energy balances in the hydro plants with and without a reservoir, respectively. When the hydro plant has storage capacity (2), the useful volume in the reservoir can be increased by the individual affluence (rain, tributaries, etc.) and the flows coming from the immediately upstream hydro plants. Additionally, the energy stored in these plants can be reduced by electricity generation and social consumption. When large inflows enter the reservoir, a portion of the water can be deviated by using the spill way to preserve the security of the plant's operation. The amounts of useful energy at the reservoirs at the beginning and end of the programming horizon (5), (6) are pre-specified quantities. The hydro production efficiency for power production is expressed by using a third-order polynomial Equations (7), (8), as a function of the height. In hydro reservoirs with large nonlinear relationships between the height and the stored water (Equation (7)), partial approximations by using third order polynomial equations for each level of the reservoir can be adopted. In the present formulation, the social requirements for water are represented as minimum daily consumptions (9) and restrictions on hourly water flows (10). The operation of the hydrological system requires maintaining the minimum ecological levels of water flows into the basin (11). In Equations (12)-(15), the maximum capacities of the equipment of the hydro plants are expressed.

In the present analysis, the algorithm is solved by using Matlab [30]. Equations (1)-(15) constitute a large nonlinear optimisation problem requiring (T ($7nr$ + $6nwr$)) variables, ($4T$ (nr + nwr) + $2nr$) equality restrictions and (T ($16nr$ + $14nwr$)) inequality constraints.

4. The Test Case

The proposed optimisation problem (1)-(15) is applied to water management in the upper basin of the Guadalquivir River, Spain. **Figure 2** shows a map of the headwaters of the Guadalquivir River.

Figure 3 shows a schematic representation of four hydro power plants (HPPs). Three of them have a reservoir (HPP 1, Doña Aldonza; HPP 3, Guadalmena; and HPP 4, Marmolejo), and the other (HPP 2, Pedro Marín) is run-of-the-river. The TTW between the plants is shown in the diagram as T_V. Other important data related to the plants are presented in **Tables 3**, **4** of the **Appendix**.

In the present analysis, typical prices in the Daily Market in March 2011 (a month with medium hydro production) in Spain are used to simulate the optimal operation of the hydro system (**Figure 4**). The acceleration of gravity, g, is 9.81 m/s^2.

To analyse the effect of the constraints on electricity production, several cases are considered:

- Case A: base case, in which social consumptions and ecological flows are not represented. Therefore, the optimisation problem is solved without considering Equations (9)-(11).
- Case B: ecological flows are not considered. The optimisation problem is solved without Equation (11). In this case, the social consumptions are included in the formulation.
- Case C: social consumptions are not applied. The optimisation problem is solved without Equations (9) and (10). In this case, the ecological flows are included in the formulation.
- Case D: solution of the optimisation problem (1)-(15), considering both social consumptions and ecological flows.

In all of the cases, the same flow (7.944 Hm^3/day, the average flow of March 2011) is considered. The same flow (3.972 Hm^3/day in each HPP) is injected at the heads of the basin and uniformly distributed over 24 hours (0.1655 Hm^3/hour in each HPP). For simplicity in the analysis, no individual affluences ($V_{i,t}^{AF}$) in HPPs 2 and 4 are considered.

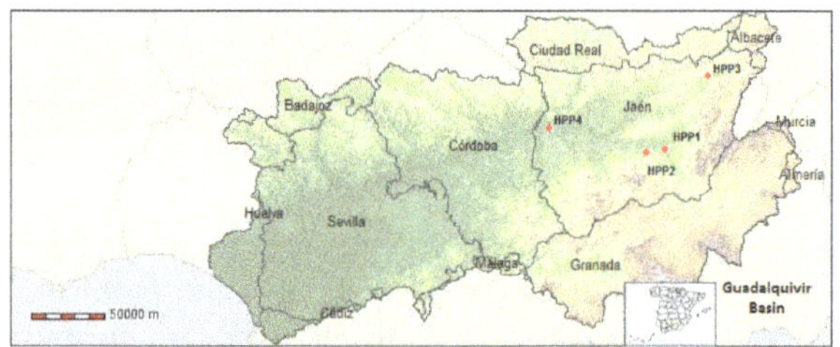

Figure 2. Geographical position of the Guadalquivir basin and relevant hydro power plants [31].

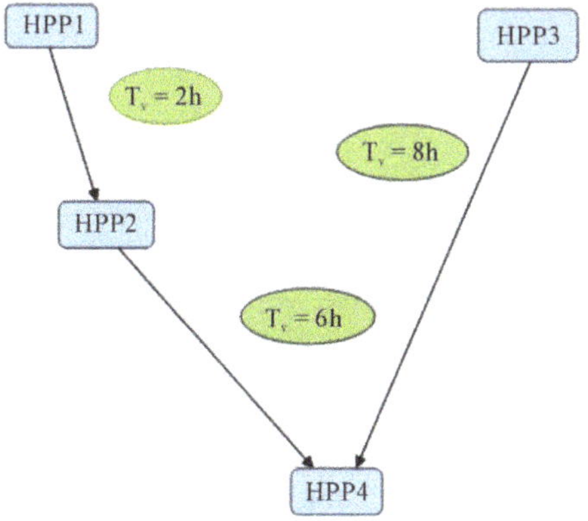

Figure 3. Spatial distribution of the reservoirs in the upper Guadalquivir basin.

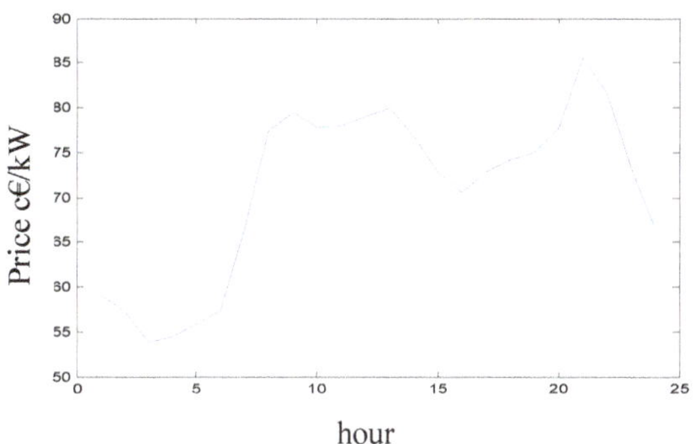

Figure 4. Typical spanish next-day market prices in march 2011.

For this sample basin, assuming 24 hours of operation and hourly discretisation, the formulation described by (1)-(15) implies 648 variables, 390 inequality constraints and 1488 inequality restrictions.

5. Results

5.1. Base Case, without Social Consumption and Ecological Flows

In **Figure 5**, the optimal production of the four hydro plants is shown. The hydro plants at the head of basin (HPPs 1 and 3) put the resources into circulation, if possible, during the high-price periods in the morning. However, the behaviour of these two plants is quite different due to the TTW between the plants in the basin and the type of plants downstream. The production of HPP 1 is limited by the capacity of the run-of-the-river HPP 2 located downstream. In this scheme, all of the water entering HPP 2 is turbined, obtaining the maximum possible profit in the combined operation. HPP3, with a controllable power plant downstream (HPP 4), generates electricity during the early hours of the day at the highest prices and full capacity. The resources coming from HPP 2 and HPP 3 reach HPP 4 in time to be turbined at full power during the hours of maximum daily price. A small quantity of water is turbined by HPP 3 at the hour of the maximum price of the day, hour 21, without reaching HPP 4 during the daily horizon.

As shown in **Figure 6**, hydro plants HPP 1 and HPP 3 (at the heads of the basin) use the water stored at the beginning of the day to increase production during the first hours. The inflows in the heads in the evening help recover the specified final values of stored energy at the end of the day. As expected, HPP 2 has no storage capacity. HPP 4 utilises its storage capabilities to wait for higher prices to sell its production in the market.

The reduced storage capacity of HPP 2 distributes the profits throughout the entire programming period (**Figure** 7). A higher generation capacity in the plants would centralise the revenue only at the peaks of the price curve. The profit of the joint operation is 165.6 M€.

5.2. Optimal Operation Considering only Social Consumption

In this case, the effect of social consumption is studied. Social-consumption values are required in all of the plants. The daily minimum consumption and the hourly limit at each plant are specified in **Table 3** of the **Appendix**, fifth and twelfth columns, respectively.

Figure 8 shows that at the beginning of the day HPP 1 turbines more than the maximum generation capacity of HPP 2, delivering water for social consumption to HPP 2 and HPP 4. This period has the lowest prices of the day. In the other head plant (HPP 3), social requests are supplied using water with less economic efficiency, eliminating HPP 3 generation at hour 21 (**Figure 5**). **Figure 9** shows the delivery of water for social uses for the four hydro plants. The upstream plants, HPPs 1, 2 and 3, transfer the volumes for social consumption at the beginning of the day, the period with lowest prices. HPP 4, without individual inflows, must yield to this restriction along the following minima of the price curve (hours 16 and 24). HPP 3, with the largest social consumption, also uses the minimum price at hour 24 to fulfil the social requirements. The profile of incremental profits is similar, considering (**Figure 10**) or without considering (**Figure** 7) social consumption. However, the final profits are

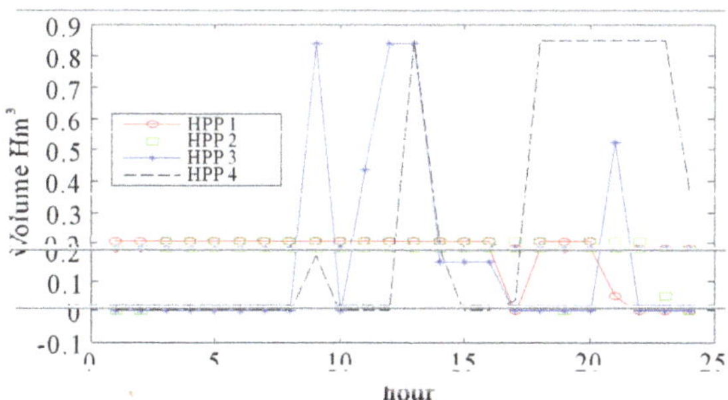

Figure 5. Production in the four hydro plants, Case A.

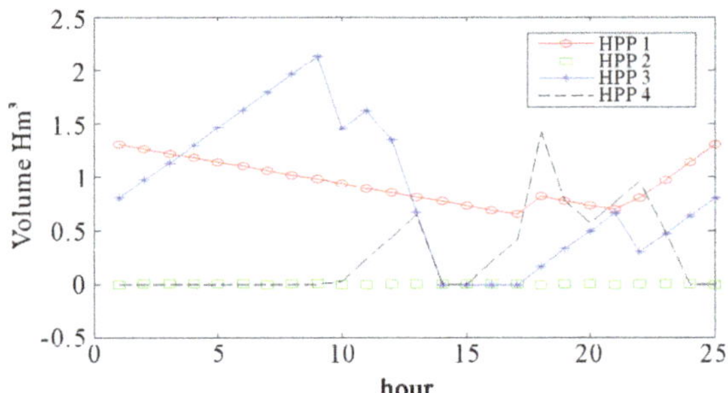

Figure 6. Energy storage in the hydro plants, Case A.

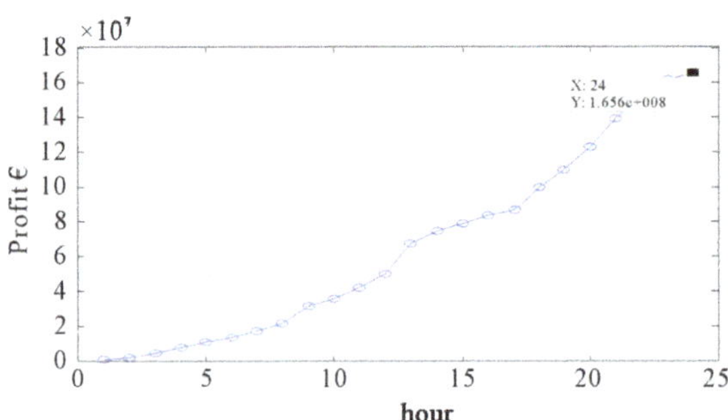

Figure 7. Incremental profits in the basin, Case A.

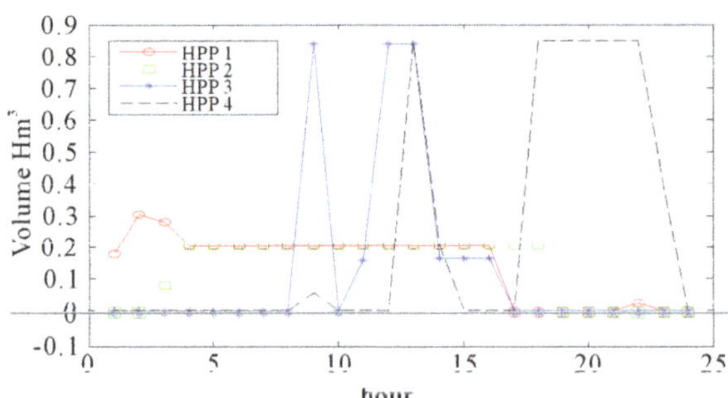

Figure 8. Production in the four hydro plants, Case B.

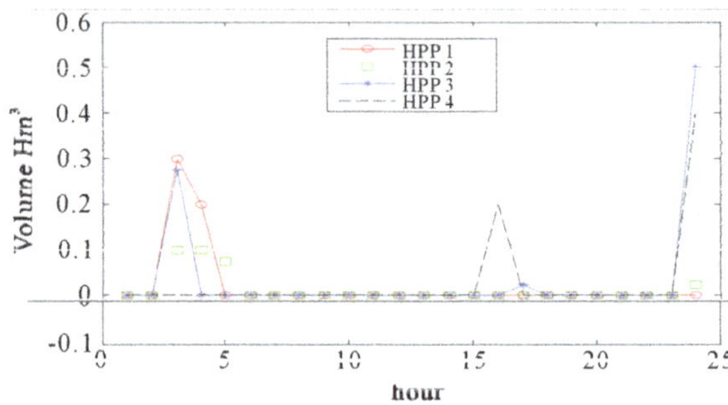

Figure 9. Energy storage in the hydro plants, Case B.

different. When considering social requirements, the total revenue is 137.09 M€, 17.20% lower than without human consumption in the basin.

5.3. Optimal Operation with only Ecological Constraints

In this case, the individual impacts of the environmental restrictions (minimum flows in the river) on the profits are analysed. In the present simulations, this restriction can only be imposed at the head plants (HPPs 1 and 3). A constant value of 16 m^3/s for each plant is considered. With this value, the minimum ecological flows in all of the basins can be maintained [32], considering TTW.

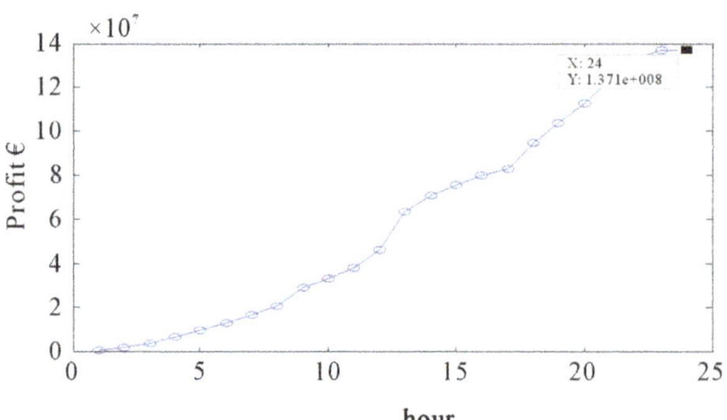

Figure 10. Incremental profits in the basin, Case B.

Figure 11 shows that the two head plants (HPPs 1 and 3) generate electricity at all hours of the day. As in Case A, the generation of HPP 1 is restricted by the limited capacity of HPP 2, and HPP 3 mainly generates electricity during the first high-price periods of the day. The ecological restrictions (minimum flow at all hours) make the slope of income almost constant (**Figure 12**). The profile of the volume turbined becomes flatter, and therefore, there are fewer resources for producing at the hours of maximum price. The optimal profit in this case reaches 163.14 M€ (1.5% less than that without ecological restrictions). In the present simulations, the restrictions on minimum flows in the river do not significantly reduce the profit of operation. It must be stressed that these restrictions are not consumptive; they only change the generation times of head HPPs 1 and 3. However, theincrease in the amount of ecological flow can reduce the total profits.

5.4. Optimal Operation with Social Consumption and Ecological Constraints

In this case, the effects of the two types of constraints (social consumption and minimum flows) are analysed. In this case (**Figure 13**), the optimal profiles of generation are similar to those observed in Case B (**Figure 8**). However, some differences must be highlighted. First, the ecological minimum flows require generation at HPPs 1 and 3 during all periods. The distribution of social consumption is also dissimilar (**Figure 14**). In Case B (with social consumption but without considering ecological restrictions, **Figure 9**), the volumes for social consumption are assigned to hours 2 to 5 in HPPs 1 and 2. The ecological flow requirement shifts the delivery of HPP 1 to hours 2 and 7 and the release of HPP 2 to the end of the day (hours 19 to 24). In HPP 3, delivery for

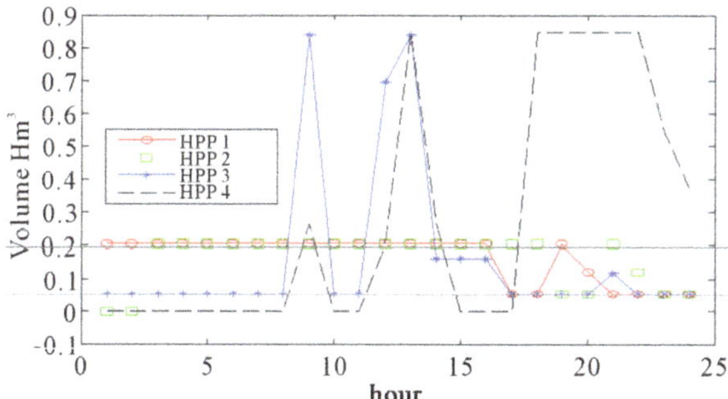

Figure 11. Production in the hydro plants, Case C.

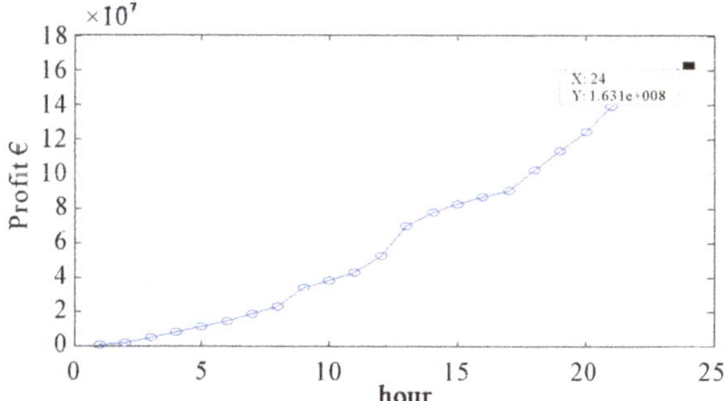

Figure 12. Incremental profits in the basin, Case C.

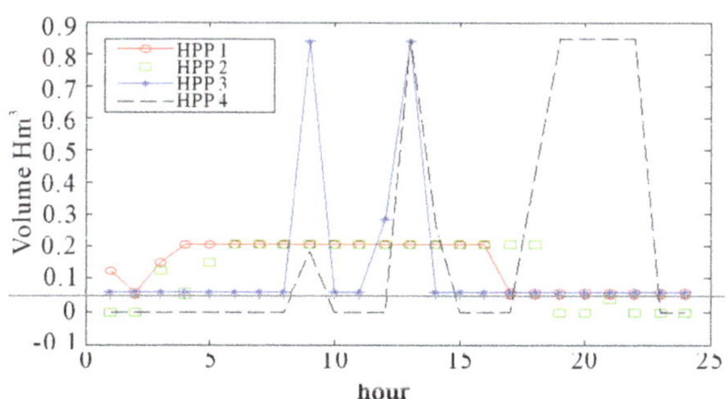

Figure 13. Production in the hydro plants, Case D.

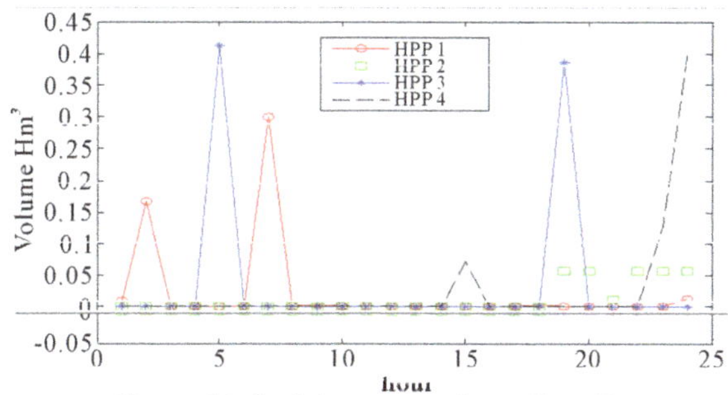

Figure 14. Social consumptions, Case D.

social consumption is increased at hour 19 and eliminated at hour 24. HPP 4 continues to provide for social consumption at the end of the day (hour 24) but shifts to small delivery from hour 16 to 15. These changes optimise the utilisation resources, increasing the combined profit of the operation. However, the optimal income in this case is 129.90 M€, 21.54% less than that of the base case (without social restrictions and ecological constraints).

5.5. Comparison of the Analysed Cases

As previously discussed, the economic results of the previous section depend on the type of restrictions added to the base case. Minimum flows in the river can be maintained without a loss of resources, only changing the time of generation. However, the social uses of water are consumptive constraints, extracting resources from the basin. Moreover, the economic results are a function of the amount of available resources. Therefore, three different scenarios are compared here: dry, medium and wet scenarios, for the two types of restrictions. The medium value coincides with the previous affluence (7.94 Hm^3/day). For comparison purposes, all of the results are obtained by maintaining the data previously used, in particular, the price profile shown in **Figure 3**.

5.1.1. Results Considering Only Ecological Constraints

In the present simulations, the ecological requirements of **Table 1** (1.6 m^3/s in HPPs 1 and 3) are maintained. However, the effect of the ecological constraints is evaluated in three different situations of affluence.

In **Table 1**, the first column shows the total inflow in the basin injected in head HPPs 1 and 3. The second and third columns show the optimal incomes obtained without considering or including the ecological constraints (Equations (9)-(11)), respectively. The economic difference between the two previous cases is represented in the fourth column. In the fifth column of the table, the relative cost of the ecological constraints, for each Hm^3 of inflow in the head HPPs, is calculated. Finally, the sixth column shows the relative cost of the ecological constraints, for each Hm^3 of minimum flow requested at the head HPPs of the basin. In this table, it can be seen that the cost of maintaining the ecological constraints depends on the amount of resources injected to the basin. In **Figure 15**, the curve of variation in the ecological cost (EC) as a function of the affluence is presented.

As shown in **Table 2** and **Figure 15**, the cost of maintaining the ecological requirements is far more important in dry scenarios. In fact, maintaining the same ecological flow of 3.42 Hm^3/day is relatively ten times more expensive than maintaining a flow of 12.47 Hm^3/day.

5.1.2. Results Considering only Social Consumptions

In the present section, the effect of social consumption (as specified in **Table 3**, **Appendix**) in the three previous scenarios of affluence is considered.

Table 2 has the same structure as **Table 1** but considers the costs of water delivered for social consumption.

Table 1. Costs of ecological requirements for different inflows.

Flow in HPPs 1 and 3 (Hm^3/da)	Income, Case A. (M€)	Income, Case C. (M€)	Income Gap. (M€)	Relative Ecological Costs, (€/Hm^3)	Relative Ecological Cost, (€/Hm^3)
12.47	228	227	1.4	112,549	507,646
7.94	166	163	3	305,253	877,073
3.42	80	67	13	3,801,169	4,947,837

Table 2. Social consumption costs for different inflows.

Flow in HPPs 1 and 3 (Hm^3/da)	Income, Case A. (M€)	Income, Case B. (M€)	Income Gap. (M€)	Relative Social Consumption Costs, (M€/Hm^3)	Relative Social Consumption Cost, (M€/Hm^3)
12.47	228	208	20	2	9
7.94	166	137	29	4	13
3.42	80	38	42	12	19

According to the two tables, the costs of water allocated for social uses are larger than those of maintaining the ecological constraints. In fact, for the medium scenario, the reduction in profit due to the social uses of water is 967% greater than the decrease in revenue due to the ecological constraints. Social uses extract resources from the basin; the ecological constraints only request a modi- fication in the profile of generation, but the resource re- mains in the river.

In **Figure 16**, the relative social consumption costs for the three scenarios of affluence are shown. The curve *SC, Social Consum.*, shows the cost of delivering 1 Hm^3 of water from the basin for social uses in the simulated scenarios. The values of this curve can be used to calculate the price of water allocated for human use in the basin as a function of the profits lost in electricity generation.

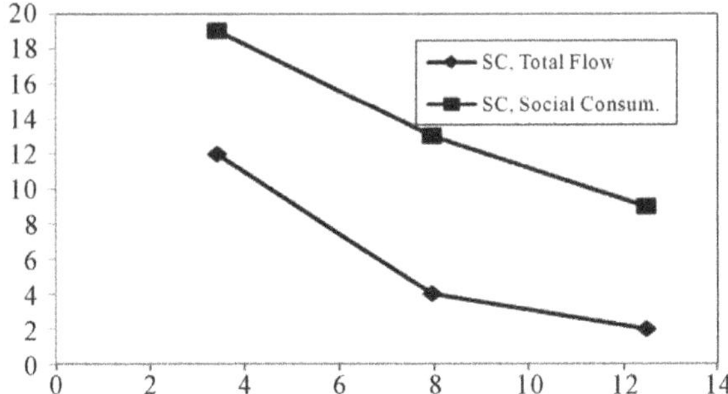

Figure 16. Social consumption (sc) costs for different inflows.

6. Conclusions

This paper presents an optimisation method to calculate the optimal operation of a basin with both controllable and non-controllable hydro power plants. This program considers both social and ecological restrictions, assessing the economic weight of each of them in the management of resources.

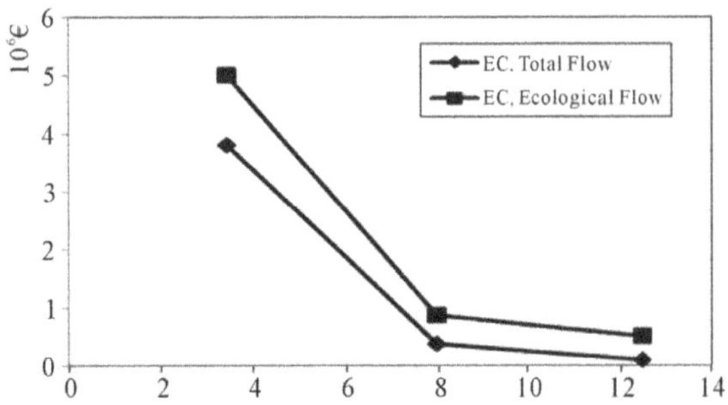

Figure 15. Cost of ecological requirements (EC) for different inflows.

The algorithm allows for control over the actions of fluent HPPs, modifying the operation of controllable HPPs. The method calculates the maximum profit electricity generation in the daily power market, considering ecological constraints and the social use of water.

The study of different inflow states shows that in this case the relative value of the social consumption of water is larger than that of maintaining ecological flows in the basin. Moreover, initial evaluations of the costs of providing water for social uses are performed. The proposed algorithm can be easily extended to consider other operational restrictions on the hydro systems.

7. Acknowledgements

The authors would like to acknowledge the Ministry of Science and Technology of Spain (Projects IT2009-0063, ENE2010-16074 and CENIT-CONSOLIDA) for supporting this work.

REFERENCES

[1] “Concerning Common Rules for the Internal Market in

Electricity and Repealing Directive 96/92/EC," European Parliament and of the Council, 1996. http://www.cne.es/cne/doc/legislacion/Directiva96_92.pdf

[2] "The Spanish Electric System 2010," Red Eléctrica de España (REE), Madrid, 2011.

[3] R. Sternberg, "Hydropower: Dimensions of Social and Environmental Coexistence," *Renewable Sustainable Energy Reviews*, Vol. 12, No. 6, 2008, pp. 1588-1621.

[4] M. Markoff and A. Cullen, "Impact of Climate Change on Pacific Northwest Hydropower," *Climate Change*, Vol. 87, No. 3/4, 2008, pp. 451-469.

[5] I. P. Holman, D. Tascone and T. M. Hess, "A Comparison of Stochastic and Deterministic Downscaling Methods for Modeling Potential Groundwater Recharge under Climate Change in East Anglia, UK: Implications for Groundwater Resource Management," *Hydrogeology Journal*, Vol. 17, No. 7, 2009, pp. 1629-1641.

[6] M. Marie, B. François, K. Stéphane and L. Robert, "Adaptation to Climate Change in the Management of a Canadian Water-Resources System Exploited for Hydropower," *Water Resources Management*, Vol. 23, No. 14, 2009, pp. 2965-2986.

[7] S. Soares and A. A. F. M. Carneiro, "Optimal Operation of Reservoirs for Electric Generation," *IEEE Transactions on Power Delivery*, Vol. 6, No. 3, 1991, pp. 1101-1107.

[8] S. Granville, G. C. Oliveira, L. M. Thome, N. Campodonico, M. L. Latorre, M. V. F. Pereira and L. A. Barroso, "Stochastic Optimization of Transmission Constrained and Large Scale Hydrothermal Systems in a Competitive Framework," *Proceedings of the Power Engineering Society General Meeting*, 2003.

[9] C. Cheng, S. Liao, Z. Tang and M. Zhao, "Comparison of Particle Swarm Optimization and Dynamic Programming for Large Scale Hydro Unit Load Dispatch," *Energy Conversion and Management*, Vol. 50, No. 12, 2009, pp. 3007-3014.

[10] G. C. Oliveira, S. Binato and M. V. F. Pereira, "Value-Based Transmission Expansion Planning of Hydrothermal Systems under Uncertainty," *IEEE Transactions on Power Systems*, Vol. 22, No. 4, 2007, pp. 1429-1435.

[11] O. B. Fosso, A. Gjelsvik, A. Haugstad, B. Mo and I. Wangensteen, "Generation Scheduling in a Deregulated System. The Norwegian Case," *IEEE Transactions on Power Systems*, Vol. 14, No. 1, 1999, pp. 75-81.

[12] A. Kanudia and R. Loulou, "Robust Responses to Climate Change via Stochastic MARKAL: The Case of Québec," *European Journal of Operational Research*, Vol. 106, No. 1, 1998. pp. 15-30.

[13] H. Habibollahzadeh and J. A. Bubenko, "Application of Decomposition Techniques to Short-Term Operation Planning of Hydrothermal Power System," *IEEE Transactions on Power Systems*, Vol. 1, No. 1, 1986, pp. 41-47.

[14] E. Castronuovo and J. A. P. Lopes, "Optimal Operation and Hydro Storage Sizing of a Wind-Hydro Power Plant," *International Journal of Electrical Power Energy Systems*, Vol. 26, No. 10, 2004, pp. 771-778.

[15] G. Zhao and M. Davison, "Optimal Control of Hydroelectric Facility Incorporating Pump Storage," *Renewable Energy*, Vol. 34, No. 4, 2009, pp. 1064-1077.

[16] H. M. I. Pousinho, V. M. F. Mendes and J. P. S. Catalão, "Scheduling of a Hydro Producer Considering Head-Dependency, Price Scenarios and Risk-Aversion," *Energy Conversion and Management*, Vol. 56, 2012, pp. 96-103.

[17] D. N. Simopoulos, S. D. Kavatza and C. D. Vournas, "An Enhanced Peak Shaving Method for Short Term Hydrothermal Scheduling," *Energy Conversion and Management*, Vol. 48, No. 11, 2007, pp. 3018-3024.

[18] A. L. Diniz and M. E. P. Maceira, "A Four-Dimensional Model of Hydro Generation for the Short-Term Hydrothermal Dispatch Problem Considering Head and Spillage Effects," *IEEE Transactions on Power Systems*, Vol. 23, No. 3, 2008, pp. 1298-1308.

[19] Z. Shawwash, K. Thomas, K. Siu and S. O. D. Russell, "The BC Hydro Short Term Hydro Scheduling Optimization Model," *IEEE Transactions on Power Systems*, Vol. 15, No. 3, 2000, pp. 1125-1131.

[20] J. I. Pérez-Díaz and J. R. Wilhelmi, "Assessment of the Economic Impact of Environmental Constraints on Short-Term Hydropower Plant Operation," *Energy Policy*, Vol. 38, No. 12, 2010, pp. 7960-7970.

[21] J. I. Peréz-Díaz, R. Millán, D. García, I. Guisández and J. R. Wilhelmi, "Contribution of Regulation Reservoirs Considering Pumping Capability to Environmentally Friendly Hydropower Operation," *Energy*, Vol. 48, No. 1, 2011, pp. 144-152.

[22] European Parliament and of the Council, "Establishing a Framework for Community Action in the Field of Water Policy," Directive 2000/60/EC of the European Parliament and of the Council, 2000. http://www.madrid.org/rlma_web/html/web/FichaNormativa.icm?ID=296

[23] Government of Spain, Ministry of the Presidency, Royal Legislative Decree 1/2001, Water Act, 2001. http://www.boe.es/boe/dias/2001/07/24/pdfs/A26791-26817.pdf

[24] Consejeria de Medio Ambiente, Junta de Andalucia, Royal Decree Law 1/2001, 2011. http://www.juntadeandalucia.es/medioambiente/

[25] CNE, Comisión Nacional de Energía, Royal Decree Law 54/1997, Law of the Electric System. http://www.cne.es/cne/doc/legislacion/NE_LSE.pdf

[26] Government of Spain, Ministry of the Presidency. Royal Decree 661/2007, Establishing the Regulation of the Ac-

tivity of the Electric Power Product gime. http://www.cne.es

[27] J. M. Y. Loyo, "The Electric Demand," 2011. http://www.unizar.es/jmyusta/wp-content/uploads/2011/01/CONTRATACION-SUMINISTRO-ELECTRICO-Enero-2011.pdf

[28] OMEL-OMIL http://www.omel.es/inicio/mercados-y-productos/mercado-electricidad/diario-e-intradiario/mercado-diario

[29] J. Martínez-Crespo, J. Usola and J. L. Fernández, "Security-Constrained Optimal Generation Scheduling in Large-Scale Power Systems," *IEEE Transactions on Power Systems*, Vol. 21, No. 1, 2006

[30] MATLAB, "The Languages of Technical Computing," Version 7.10.0.499, Math Works, 2010.

[31] Government of Spain, Ministry of Environment, "Guadalquivir's Description," 2011. http://www.chguadalquivir.es/opencms/portalchg/laDemarcacion/guadalquivir/breveDescripcion/

[32] Confederación Hidrográfica del Guadalquivir, "Ecological Flows," 2011. http://www.chguadalquivir.es/opencms/portalchg/marcoLegal/planHidrologicoCuenca/

Appendix

Table 3. Hydro plants data.

HPP	Type	Prev. HPP	Tv [h]	$V_i^{CT\min}\left[\text{Hm}^3\right]$	$h_i^{\max}[\text{m}]$
1	R	-	-	0.5	13
2	F	1	2	0.3	-
3	R	-	-	0.8	82
4	R	2, 3	6, 8	0.6	7

HPP	$V_i^{\max}\left[\text{Hm}^3\right]$	$V_i^{T\max}\left[\text{Hm}^3\right]$	$V_{i,1}^{SP}\left[\text{Hm}^3\right]$	$V_{i,T}^{SP}\left[\text{Hm}^3\right]$	$V_i^{U}\left[\text{Hm}^3\right]$
1	23	0.513	1.3	1.3	20
2	19	0.206	0	0	11
3	347	0.839	0.8	0.8	-
4	13	0.850	0	0	10

HPP	$V_i^{C\max}\left[\text{Hm}^3\right]$	η_i	$V_i^{EC\min}\left[\text{Hm}^3/\text{h}\right]$
1	0.3	0.7	0.0576
2	0.1	0.79	0
3	0.5	0.796	0.0576
4	0.4	0.7962	0

In **Table 3**, Prev. HPP is the number of the HPP upstream to the current HPP (i.e., upstream HPP 4 there are the HPP's 2 and 3).

Table 4. Coefficients volume-height of the hydro plants.

HPP	k_0 [m]	k_1 [m^{-2}]	k_2 [m^{-5}]	k_3 [m^{-8}]
1	−3.58E + 01	4.94E + 00	−1.07E − 01	1.02E − 07
2	25	0	0	0
3	2.53E + 00	4.75E − 01	−6.85E − 04	1.03E − 07
4	9.63E − 01	3.71E − 01	−3.90E − 04	1.05E − 07

Biographies

Gloria Hermida was born in Coruña (1973), received her B.S degree in Industrial Engineering from University of La Coruña (2007) and her Master degree in Electrical, Electronics and Automation Engineering (2011) from University Carlos III de Madrid. She is working as Assistant Professor in the Department of Electrical Engineering of University Carlos III de Madrid. Her research interests include the optimization of water resources and operation planning.

Edgardo D. Castronuovo received a B.S. degree (1995) in Electrical Engineering from the National University of La Plata, Argentina; both M.Sc. (1997) and Ph.D. (2001) degrees from the Federal University of Santa Catarina, Brazil, and performed Post-Doctorate (2005) at INESC-Porto, Portugal. He worked at the Power System areas of CEPEL, Brazil, and INESC-Porto, Portugal. Currently, Dr. Castronuovo is an Associate Professor at the Department of Electrical Engineering, University Carlos III of Madrid, Spain. His interests are in optimization methods applied to power system problems, renewable production, storage and deregulation of the electrical energy systems. Prof. Castronuovo is Senior Member of IEEE.

5

Considerations on Research of Environmental Risk Prevention and Control in "12th Five-Year Plan" Period

Pengli Xue[1], Xiaofeng Sun[1], Yun Song[1], Yanjun Cheng[1], Dezhi Sun[2]
[1]Environmental Protection Institute of Light Industry, Beijing, China
[2]Beijing Forestry University, Beijing, China

ABSTRACT

Along with the environmental pollution causes complexity and diversity increases ceaselessly, "national environmental protection" Twelfth Five "planning" (hereinafter referred to as "planning") will be the environmental risk prevention as the "12th Five-Year Plan" one of the important tasks, including advancing environmental risk management in the whole process, key areas the environmental risk prevention measures. The whole process environmental risk management covers a risk source recognition, receptor vulnerability assessment, environmental risk characterization, risk decision and risk assessment of accident loss. This article from the environmental risk source classification, environmental risk classification management, environmental emergency response and environmental risk and insurance environment four aspects put forward the "12th Five-Year Plan" whole process environmental risk management content, to further reduce our country environmental pollution accident risk and policy makers to provide some decision support.

Keywords: Whole-process Environmental Risk Assessment and Management; Environmental Risk Source Stage Division; Environmental Risk Regionalization; Emergency Management; Environment Insurance

1. Introduction

In recent years, along with the rapid advance of China's industrialization and urbanization, the density of population and economy increase sharply; meanwhile, the chemical enterprises are gathering towards the major riverside and seaside cities with the convenient transportation and resources, therefore leading to the enormous and sudden environmental risks caused by the improper distribution. The production, storage and transportation of dangerous materials of chemical enterprises are on the rise, which results in the frequent environmental pollution accidents of China. Based on the analysis of the dynamic change in time and space of environmental pollution accidents from 1990 to 2007, Xue (2011) points that China's sudden environmental pollution accidents decrease from 3.462 (1990) to 462 (2007) thanks to the release of a series of management rules on poisonous, harmful and dangerous chemicals, the implementation of evaluation system of safety production and environmental impact, and the issuance of technical guide for environmental risk evaluation of projects, especially the stricter environment supervision of risk enterprises. Seen from the spatial distribution characteristics, China's sudden environmental pollution accidents in the past ten years mainly came from the Yangtze River Delta and Liaodong Peninsula; however, with the implementation of the western development strategy, the exhaustion of eastern environmental capacity and the industrial upgrading, many heavily polluting enterprises begins moving to the west, and the enterprises specialized in paper making, coal chemicals and heavy metals smelting have become the major risk sources of western areas; the environmental risk caused by improper distribution and the soar of regional industrialization and urbanization result in the continuous increase in environmental pollution accident loss of China.

As the cause of environmental pollution is more and more complicated and diversified, the National 12th Five-Year Plan for Environmental Protection (hereinafter called the Plan), for the first time, lists the environmental risk prevention as an important task in the "12th Five-Year Plan" period which is parallel to the total emission reduction and quality control, and makes decisions to promote the whole-process risk management, improve the environmental risk management measures and establish the mechanism of environmental accident damage and compensation.

2. Risk Source Classification/Gradation Management

The classification of environmental risk source is the premise for risk identification; after the classification, it is necessary to establish the grading standard system for each kind of environmental risk source to assess the hazard of risk source and further ascertain the grade of environmental risk source, and then work out the monitoring and management measures against each grade of environmental risk source, especially the major ones. In this way, the environmental risk can be reduced from the source,

To comprehensively master the spatial distribution and hazard level of the fixed environmental risk sources of China, identify the major risk sources, and provide basic information for the establishment of China's regional environmental risk source files and database, the Ministry of Environmental Protection, in May 2009, carried out the environmental risk and chemicals inspection on the enterprises of 3 national key sectors (petroleum processing and coking, chemical material and product manufacture and medicine manufacturing), totaling 10 industries and 35 sub-industries. In a sense, the enterprises of 35 sub-industries involved in this inspection belongs to the environmental risk sources defined according to the industry characteristics classification, with each enterprise equivalent to one environmental risk source. However, the relevant research by Wei Keji et al. (2010) summarizes the classification of environmental risk source into the followings: the environmental sufferer-based classification, such as the water environment risk source, atmospheric environment risk source and soil environment risk source; the material state-based classification, such as the gaseous environment risk source, liquid environment risk source and solid environment risk source; the transmission route-based classification. Wen Lili et al. (2010) presented the design principle of major environmental risk source monitoring system based on the classification into liquid source, solid source and mobile source.

The environmental risk source gradation management aims to effectively avoid risk of each grade of environment risk of the same category; the risk source gradation management can save the management cost and shorten the emergency response time, therefore enhancing the efficiency of risk management. Wei Keji (2010) built the environment hazard index according to the hazard of risk source to water, atmosphere and soil to divide the risk sources into four grades, and then presented the identification standard system of major environmental risk source according to the said gradation.

Many industries in the various categories of national economy have the environmental risk and it is hard to make the uniform classification and assessment, therefore, the chemical raw materials/products manufacturing industry or the chemical industry park shall be chosen for the initial research on the environmental risk source assessment and gradation management standards, and we shall work out the prevention and early warning measures against the major risk source of such industry in aspect of technology transformation, equipment elimination, personnel training, etc. with a purpose to eliminate the major petrochemical environmental pollution accidents at their early stage.

As to the VOCs pollution risk, source tracing and characteristics of industrial VOCs emissions in China should pay more attention to the production of VOCs, the storage and transport, industrial processes using VOCs as raw materials and use of VOCs-containing products.

3. Research on Vulnerability of Environmental Risk Targets

From the point of region, the environmental risk sufferer is a compound system including the social economy, natural ecology and other factors, so the object suffering the sudden environmental pollution accident risk includes the social economy and ecosystem, and the integrated vulnerability of risk sufferer is composed of the vulnerability of social economy and that of ecosystem.

For the social economic system, we mainly consider the scale of regional population and economy, the regional accident prevention and emergency measures, the scale of other infrastructures and so on; for the natural ecosystem, we mainly consider the sensitive ecosystem, including the ecosystem of drinking water source area, natural reserve and wetland.

During the "12th Five-Year Plan" period, we shall research the vulnerability of social environmental risk sufferer through such aspects as population density, the density of hospital and urban lifeline, regional early-warning capacity, proportion of public education in GDP, number of hospital bed per 10,000 persons, per capita GDP, regional emergency evacuation capacity, sudden environmental pollution accident response capacity and prevention capacity, etc. and figure out the pertinent and practical measures to isolate risks or enhance the risk resistance capacity for the much vulnerable social risk sufferers; the research on vulnerability of natural ecological environmental risk suffer shall be carried out through the sensitivity and adaptability of regional ecosystem (including water area ratio, regional environmental improvement index, river regulation investment ratio, standard discharge rate of industrial wastewater) and related to the regional emergency monitoring and response to effectively reduce the damage caused by the ecosystem pollution accident.

4. Environmental Risk Zoning Management

At present, the Emergency Response Law of the People's Republic of China is lack of the relevant implementing rules, there is no adequate legal basis for implementation and the relevant supporting policies are not formulated yet, so there is no pertinent guide on the prevention and management of environment risk. During the "12th Five-Year Plan" period, we shall, based on the classification and gradation of environmental risk source, comprehensively consider the environmental risk sufferer to realize the regionalization for environmental risk in the major industries, such as the petrochemical industry, and regard it as the important basis for the environmental risk zoning management.

The environmental risk regionalization is the basis and prerequisite for the environmental risk zoning management. At present, the method and theoretical system of China's environmental risk regionalization are still in the research phase, and the environmental risk zoning management is not implemented in any specific administrative and regional system yet.

Yang Jie et al. (2006) took the economic zone along Yangtze River in Jiangsu Province as an example wherein a single development zone is regarded as the research unit to analyze the risk source, control mechanism and sufferer, and obtained the comprehensive risk index of each industrial park, which was then made into different grades, and finally they presented the corresponding risk management measures. Peng Wangminzi (2009) analyzed the characteristics of each risk source of a certain petrochemical industrial source in Guangzhou and then assessed the environmental risk though the information diffusion method; on the basis of this, he divided the environmental risk of industrial park into four grades according to the relevant grading standard. Xue Pengli (2011) adopted the top-down and bottom-up method to regionalize the sudden environment risk of Shanghai and got 2 risk regions, 5 risk sub-regions and 21 risk segments; meanwhile she presented accident prevention and early-warning measures for different types of risk areas, realizing the environmental risk zoning management in Shanghai.

In order to further prevent the environmental risk caused by the improper industrial distribution during the "12th Five-Year Plan" period, we shall, based on the environment function zoning, ecological zoning and main function zoning of our country, further consider the environmental risk source, the vulnerability of social environment and ecological environment sufferer in a comprehensive manner to assess the environmental risk and obtain the environmental risk zoning map, and then analyze this map together with the land use plan of China and the overall urban planning map to strengthen the space regulation and provide the objective and visual basis for the adjustment of industrial structure and layout, the reasonable layout and development of industries with high risk of serious pollution and the implementation of risk prevention measures, with a purpose to actually realize the environmental risk zoning management.

Based on the current and future industrial emissions of VOCs in China, the research on the risk zoning for the air pollution can provides scientific results for control the VOCs pollution.

5. Environmental Risk and Environmental Insurance

The environmental insurance is an effective green insurance system against the pollution issues which is widely adopted in the world and aims to compensate for the personal injury and properly loss of the third party caused by the environmental pollution in the insured site and assume the expenses incurred in the process of pollution cleanup and elimination; this system can not only help to reduce the environmental disasters and accidents but also alleviate the government's environmental protection burden and guarantee the rights and interests of the environmental pollution victims and their families.

During the "12th Five-Year Plan" period, we shall further establish and improve the laws and regulations system of environmental pollution responsibility insurance and specify the compulsory direction and transitional measures for the establishment of this insurance system. Additionally, the Ministry of Environmental Protection shall work together with the China Insurance Regulatory Commission and insurance companies to develop the reasonable and practical environmental insurance products for the major pollution industries, and establish a three-party interaction mechanism among the government department, insurance company and bank; improve the establishment of some environment insurance product service agencies and the environmental responsibility arbitration agencies; make more efforts to explore the implementation of the compulsory liability insurance in the industries with serious pollution and high environmental risk, make the environmental protecttion department to formulate the catalogue of such Industries (and their products) and relate the environmental pollution responsibility insurance with the credit qualifycation of the enterprises in this catalogue. At the same time, the government shall also provide the necessary policy support for the development of environment pollution responsibility insurance. Set up the compensation funds for environmental pollution damage to pay for the part beyond the insurance in case of major environment pollution accident and pay the emergency disposal expense for any other party. On this basis, during the "12th Five-Year Plan" period, we shall extensively carry out the pilot work in the high accident area and the special region with high environmental risk as much as possible, such as selecting the enterprises in the industry dealing with hazardous chemicals transportation, taking hazardous materials as raw materials and discharging hazardous pollutants or other hazardous wastes in the Yangtze River Delta and Zhujiang River Delta.

6. Conclusions

The frequent sudden environmental pollution events not only caused great loss to the social economy and affected the harmony and stability of society but also vanished the environment governance achievement of many years, with the damage caused to the ecological environment hard to be restored within short time. From the "12th Five-Year Plan" to 2020 is the mid-to-late period of industrialization, which will see the fast development and reformation of China's economy and society, and the stage characteristics determine the importance of environmental risk factors. The environmental risk factors, even reduced in amount and improved in quality, will also exert negative influence to the ecosystem and human health, and there are still some factors of environment safety and environment risk are beyond the control over amount and quality. the National 12th Five-Year Plan for Environmental Protection , for the first time, lists the environmental risk prevention as an important task in the "12th Five-Year Plan" period which is parallel to the total emission reduction and quality control, and make decisions to strengthen the environmental risk prevention and control over the key fields through implementing the whole-process management of environmental risk, strengthening the nuclear and radiation safety management, restraining the high-frequency state of heavy metal pollution events, promoting the safe treatment and disposal of solid wastes and improving the chemical-induced environmental risk prevention and control system.

This paper analyzed the classification and gradation of environmental risk source, environmental risk regionalization, environmental pollution emergency response at three levels (enterprise, park and region) and environmental risk & environmental insurance, and presented the emphases of the whole-process management of environmental risk of China during the "12th Five-Year Plan" period, and provided theimportant basis for the prevention of China's environmental pollution accident and the industrial distribution in the same period.

REFERENCES

[1] P. L. Xue and W.H. Zeng, "Trends of environmental accidents in China from 1990 to 2007", *Frontiers of Environmental Science and Engineering in China,* Vol. 5, No. 2, 2011, pp. 66-276.

[2] F. Y. Li, J. Bi, C. S. Qu et al. "Research and Application of Whole-process Environmental Risk Assessment and Management Mode," *China Environmental Science*, Vol. 30, No. 6, 2010, pp. 858-864.

[3] K. J. Wei, "Research on Identification of Major Environmental Risk Sources in Nanjing Chemical Industrial Park", Master Thesis of Beijing Forestry University, [D], 2009, pp. 49-52.

[4] L. L. Wen, Y. H. Song, B. F. Yue et al, "Research on Construction of Major Environmental Risk Source Monitoring System", *Chinese Perspective on Risk Analysis and Crisis Response*, 2010, pp. 154-197.

[5] P. L. Xue, "Research on Sudden Environmental Pollution Accident in Shanghai," Doctoral Theses of Beijing Normal University [D]. Beijing, 2010, pp.151-156.

[6] Q. Jia, "Research on Assessment and Grading Method of Sudden Environmental Risk of Petrochemical Enterprises," *Journal of Environmental Sciences*, Vol. 30, No. 7, 2011, pp. 1510-1517.

[7] J. Yang, J. Bi, and J. B. Zhou, "Development of Environmental Risk Monitoring and Early-warning System along the Yangtze River in Jiangsu Province," *Resource and Environment in the Yangtze Basin,* Vol. 15, No. 6, 2006, pp. 745-749.

[8] W. M. Peng and X. F. Shi, "Environmental Risk Regionalization Based on Information Diffusion Method," *Environmental Science and Technology*, Vol. 32, No. 9, 2006, pp. 191-193.

[9] P. L. Xue and W. H. Zeng, "Shanghai Environmental Pollution Emergency Risk Zoning along the Yangtze River," *China Environmental Scienc*e, Vol. 31, No. 10, 2011, pp. 1743- 1750.

6

Modeling and OLAP Cubes for Database of Ground and Municipal Water Supply

Taskeen Zaidi[1], Annapurna Singh[2], Vipin Saxena[1*]
[1]Department of Computer Science, B. B. Ambedkar University, Lucknow, India
[2]Department of Environmental Science, B. B. Ambedkar University, Lucknow, India

ABSTRACT

Modeling plays an important role for the solution of the complex research problems. When the database became large and complex then it is necessary to create a unified model for getting the desired information in the minimum time and to implement the model in a better way. The present paper deals with the modeling for searching of the desired information from a large database by storing the data inside the three dimensional data cubes. A sample case study is considered as a real data related to the ground water and municipal water supply, which contains the data from the various localities of a city. For the demonstration purpose, a sample size is taken as nine but when it becomes very large for number of localities of different cities then it is necessary to store the data inside data cubes. A well known object-oriented Unified Modeling Language (UML) is used to create Unified class and state models. For verification purpose, sample queries are also performed and corresponding results are depicted.

Keywords: Modeling; Database; Object-Oriented; Unified Modeling Language; OLAP Data Cubes; Water Supply

1. Introduction

In the current scenario, modeling becomes an integral part of the solution of any kind of research problems whether it is related to life sciences, medical sciences, engineering sciences, etc. An object-oriented modeling language is most popular language because of the evolution of the Graphical User Interface (GUI) applications in the computer science filed. On the basis of object-oriented technology, Object Management Group (OMG) [1] has launched the various version of one of the most powerful platform independent modeling language *i.e.* Unified Modeling Language (UML). It contains various kinds of symbols for drafting a design on a piece of paper. In this connection, Booch and Rambaugh [2] have designed various diagrams for Unified Modeling Language. By the use of UML, the present work is based on the On Line Analytical Processing (OLAP) by designing the OLAP data cubes. The research on the design of three dimensional the data cubes is available in [3]. By the use of the various tools, one can design the OLAP data cubes. In the data cubes information in the form of data is stored along the three axes *i.e.* x, y and z axis and it supports the object-oriented and structured technologies.

In this work, the data were taken from the observations related to the ground water and municipal water supply of different localities of a city. Let us describe some of the important references related to the status of supply of water. In the current scenario, surface water has polluted due to the various reasons overall known as the weather pollution. The different reasons and challenges in the water technology are described by Barua [4]. Drinking water standards are available in BIS [5], but in the different localities of different cities, the drinking water is going to be polluted daily. This is because of unbalanced chloride and nitrate concentration. The sample database was analyzed on the basis of physiochemical parameters based on standard methods for analysis of water described by Clesceri *et al.* [6]. World Health Organization [7] has released the various guidelines for the drinking water.

The present work deals with the storage of database related to ground and municipal water supply of drinking water. A concept of UML modeling is used for extracting the information called as Knowledge Discovery in Database (KDD) and it is useful when size of data becomes complex. A three dimensional representation of database in the form of data cubes is designed for storing the large

*Corresponding author.

database of ground water and municipal water supply. For demonstration purpose, a sample size of nine is considered for storing this database in data cubes and sample queries were performed for the verification purpose.

2. Background

2.1. Unified Modeling Language

UML is one of the powerful modeling languages and it is a platform independent. One can develop the code very easily by using the object-oriented programming language. It has two views of the problem called as the static and dynamic views. Both contains different types of the diagrams namely class, object diagrams represent the static view of the problem while activity, sequence and state diagrams show the dynamic view of the problem. This modeling language has been developed by Object Management Group.

2.2. Online Analytical Processing (OLAP) Cube

It is a multidimensional database which is used by software professionals for optimization of dataware houses. From the dataware houses, data cubes are designed and from the literature it is observed that three dimensional axes are used to design the data cubes and each cell represents the data which may be in the form of text, string or numerals forms. Multidimensional Expression (MDX) Language is used for representing the multidimensional database. The idea of OLAP cube is that it can work faster on the Local Area Network (LAN) or the distributed Wide Area Network (WAN) on which heterogeneous collection of devices can work together. One can get the desired information within fraction of seconds. Really this is a great achievement in the field of the large database and anyone can perform any query and can get result quickly.

2.3. A Sample Database

Database is a collection of the information and a real database is taken for ground and municipal water supply of the different localities of a city. For demonstration purpose, a sample size of nine is considered for the database related to ground and municipal water supply and observations [8] are based on the physical chemical characteristics of ground and municipal water quality of various localities of a city. The data is based upon the vicinity of the temperature 20°C.

PH was estimated using a potentiometer and it was calibrated using a buffer solution of PH 9.2, PH4 and PH7. Later about 100 ml of sample was taken in a 250 ml beaker and the electrode was dipped to get the PH value of the sample. For computation of total hardness as per following formula, about 25 ml sample was taken in a 100 ml flask and a pinch of Erichrome black-T was added to get a vine red colour. The sample was titrated with 0.01 M ethylene diamine tatra aceti Acid to a blue colour.

$$\text{Total hardness } (\text{mg/l}) = \text{T} \times 1000\text{xD/V}; \qquad (1)$$

where;

T = ml of EDTA used.

D = mg of $CaCo_3$ equivalent to 1 ml EDTA titrant (1 mg for 0.01 m EDTA used hear) therefore D = 1.

V = Volume of water sample.

For computation of alkalinity, 50 ml of sample was taken and 2 drops of phenolphthalein indicator was added. An absence of colour showed the presence total alkalinity but absence of phenolphthalein alkalinity. The sample was further titrated with 0.025% H_2SO_4 using mixed indicator. The colour of the solution became pinkat at end point.

$$\text{Total alkalinity as } (\text{mg/l}\,CaCO_3) = \text{T} \times \text{N} \times 50 \times 1000/\text{volume}; \qquad (2)$$

where;

T = Volume of titrant used in ml.

N = Normality of H_2SO_4.

For computation of chloride, 10 ml of sample was taken in a 100 ml flask and 3 drops of k_2CrO_4 solution was added to give a yellow colour. The sample was them titrated with (0.025 N) $AgNO_3$ to get a brick red colour at end point.

$$\text{Chloride mg/l} = \text{T} \times \text{N} \times 35.45 \times 1000/\text{V} \qquad (3)$$

where;

T = Volume of titrant used.

N = Normality of titrant ($AgNO_3$).

V = Volume of Sample in ml.

For estimation of Nitrate, 0.2 ml of clear sample was added with 0.5 ml 5% salicylic acid and 19 ml of 20% NaOH. A greenish yellow colour indicates the presence of nitrate which is estimated using a spectrophotometer at 410 nm wavelength.

On all these aspects the practical results are computed and given in **Tables 1** and **2**, respectively for the ground and municipal water supply.

3. UML Modeling for KDD

3.1. UML Class Diagram

In the object-oriented technology, UML class diagram shows the static behavior of the system. It can be drawn on a piece of paper and errors can be uncovered during the early stage of software development. Generally, software designer designs such type of diagram for implementation in the object-oriented programming style. A class is defined as group of attributes *i.e.* variables and

Table 1. Database of ground water supply (table name: tblgws.ldf).

Temperature (Kelvin)	PH	Total hardness (mg/lit)	Total suspended solids	Chloride (mg/lit)	Total alkalinity (mg/lit)	Nitrate (mg/lit)
293.15	7.77	180	0.035	42.5	190	33.4
294.15	7.76	180	0.032	28.4	200	16.4
294.65	7.7	200	0.035	85.1	214	18.1
294.95	7	180	0.03	28.4	192	53.2
295.15	7.05	220	0.035	70.9	234	37.9
295.15	7.3	140	0.28	121	198	26.6
294.15	7.6	180	0.36	21.3	228	12.7
294.15	7.5	160	0.34	35.5	180	31.1
294.44	7.46	180	0.1434	54.2	205	28.7

Table 2. Database of municipal water supply (table name: tblmws.ldf).

Temperature (Kelvin)	PH	Total hardness (mg/lit)	Total suspended solids	Chloride (mg/lit)	Total alkalinity (mg/lit)	Nitrate (mg/lit)
294.15	8.25	60	0.02	14.18	110	20.36
295.15	8.1	216	0.043	14.18	196	25.17
295.55	7.98	200	0.036	21.27	192	25.17
295.55	7.53	200	0.032	21.27	180	24.6
296.55	7.1	180	0.03	70.94	212	23.19
296.15	7.3	180	0.3	121.4	210	27.9
295.15	7.99	200	0.39	21.27	196	14.9
295.15	8.09	160	0.32	21.27	204	17.81
294.44	7.792	174.5	0.1464	38.22	184.5	22.39

the methods applied on the attributes. The accessing of the attributes and methods may be private, public or protected. AUML class diagram for accessing of the desired information is shown in the **Figure 1**. There are six classes namely User, Storage, Handheld devices, KDD, Data cubes and Search pattern. KDD class stands for the Knowledge Discovery Database. By the use of User class, user may login on the handheld device which may be laptop, I-pad, mobile, etc. The user desires to search knowledge database from a large database which is controlled by the class KDD. Search pattern class is responsible for the optimized search technique from the designed data cubes. The searching of the database is faster in comparison of the direct access from the database which is controlled by the Storage class as shown in the **Figure 1**.

3.2. UML State Diagram

The dynamic behavior of the system is represent by the state diagram and in the object-oriented technology, UML state diagram represents the functioning of the clock of handheld device in which the events are happening as per the forward clock of the device. **Figure 2** shows the state diagram for the display of the desired information by the use of data cubes. Initially user enters its id and password on the hand-held device. A large database is converted into the Knowledge Discovery Database (KDD) and thereafter data cubes are designed and user can found the display of the desired database on the device.

4. Design of OLAP Cube

The physiochemical characteristics of ground water and municipal water supply are stored in a three dimensional cubes. The three axis x, y, z are represented as hardness, chloride and nitrate, respectively. The database is recorded in each cell of the cube represented along x, y and z axes. The cells can be increased for the finite values of hardness, chloride and nitrate. The above database is represented in the following data cubes for the ground

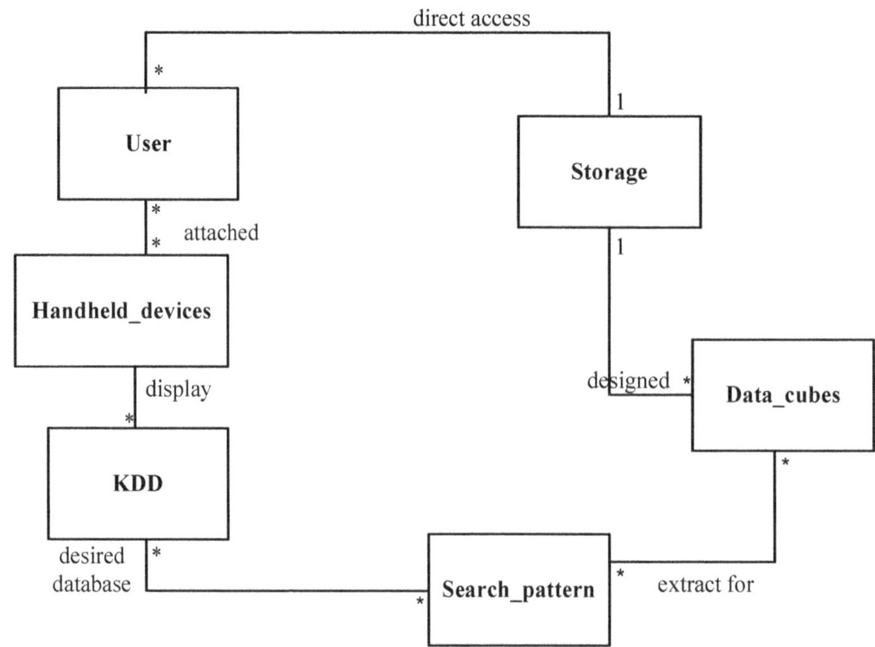

Figure 1. UML class diagram for storage of large database.

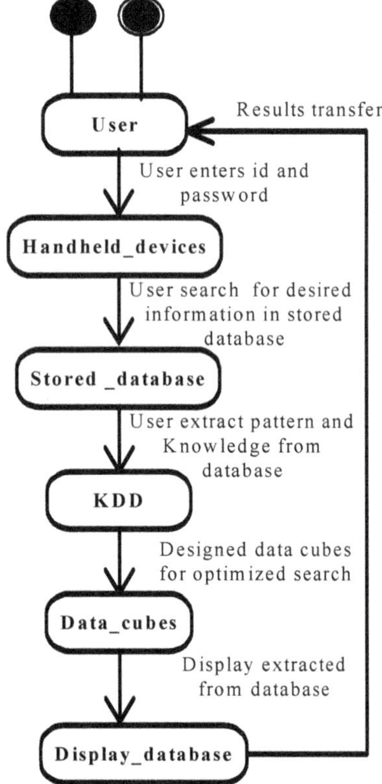

Figure 2. UML state diagram for KDD.

water and municipal water supply in **Figures 3** and **4**, respectively. These cubes support both kind of technology *i.e.* structured and object-oriented technologies. The implementation can be extracted by using object-oriented Unified Modeling Language.

On the above database, sample queries have been performed for the verification of the data from the database and these are described below briefly by the use of SQL Server:

Sample Query-I

Select Temperature, PH from tblgws where Totalhardness = "180".

The output of the above query is shown in **Table 3**.

Sample Query-II

Select Temperature tblmws where Totalhardness = "180" *having Chloride* = "70.94".

The output of the above query is shown in **Table 4**.

5. Conclusion and Future Scope of Work

From the above work, it is concluded that the modeling of the research problem is necessary for getting the solution of the problem in optimized way. UML is a powerful modeling language as shown above used to design the models for the ground and municipal water supply. If the database is large, then it can be stored inside the data cubes and user can extract the desired database within a fraction of seconds as shown above. Three dimensional storage of database is an excellent way for storing the large and complex database and one can extract the desired data in a few seconds. The data presented in the tables are real data which can be further extended for the number of localities and then transformed in the form of data cubes and one can get the desired information within a few seconds after executing the SQL queries. The other techniques like co-relation, entropy, Gini indexing can be applied for further interpretation of the results.

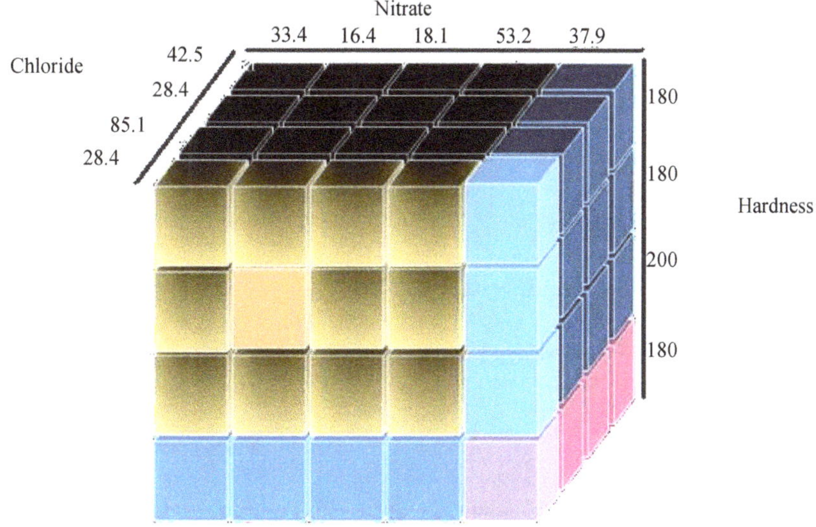

Figure 3. Data cube for ground water supply.

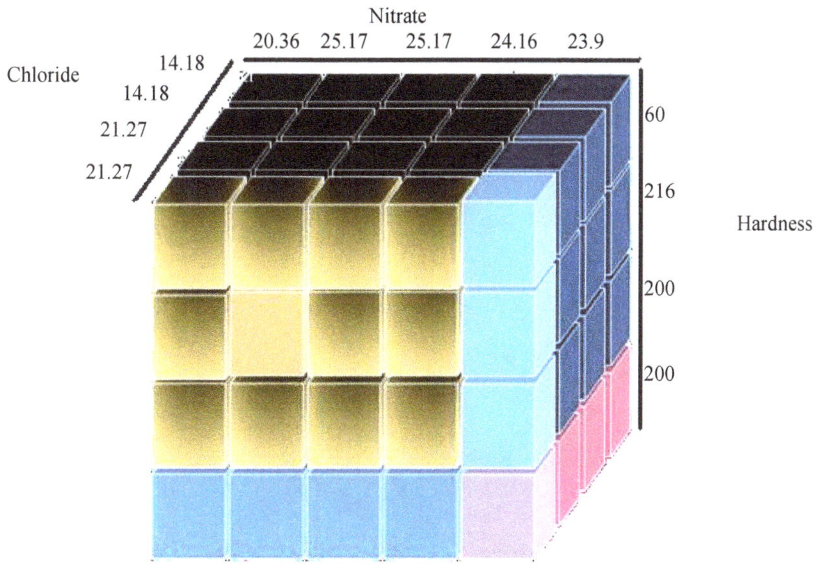

Figure 4. Data cube for municipal water supply.

Table 3. Result of sample query-I.

Temperature (Kelvin)	PH	Total hardness (mg/lit)	Total suspended solids	Chloride (mg/lit)	Total alkalinity (mg/lit)	Nitrate (mg/lit)
293.15	7.77	180	0.035	42.5	190	33.4
294.15	7.76	180	0.032	28.4	200	16.4
294.95	7	180	0.03	28.4	192	53.2
294.15	7.6	180	0.36	21.3	228	12.7
294.44	7.46	180	0.1434	54.2	205	28.7

Table 4. Result of sample query-II.

Temperature (Kelvin)	PH	Total hardness (mg/lit)	Total suspended solids	Chloride (mg/lit)	Total alkalinity (mg/lit)	Nitrate (mg/lit)
296.55	7.1	180	0.03	70.94	212	23.19

6. Acknowledgements

Thanks are due to University Grants Commission, New Delhi, India for providing financial assistance to carry out the above work.

REFERENCES

[1] Object Management Group, "Unified Modeling Language Specification," 2012. http://www.omg.org

[2] G. Booch and J. Rambaugh, "The Unified Modeling Language," User Guide Addison Wesley, Reading, 1999.

[3] Wikipedia, "OLAP Cube," 2013. http://en.wikipedia.org/wiki/OLAP_cube

[4] A K. Barua, "Water Technology Management Challenges and Choices in Sustainability of Water Use," Dominant Publishers and Distributors, New Delhi, 2001, pp. 2-3.

[5] Bureau of Indian Standards, "Indian Standard Specifications of Drinking Water," B.S. 2001.

[6] J. A. Camargo and A. Alonso, "Ecological and Toxicological Effects of Inorganic Nitrogen Pollution in Aquatic Ecosystems: A Global Assessment," *Environment International*, Vol. 32, No. 6, 2006, pp. 831-849.

[7] L. S. Clesceri, *et al.*, "Standard Methods for the Examination of Water and Waste Water," American Public Health Association, Washington DC, 1998, p. 19.

[8] A. Singh, *et al.*, "Status of Ground Water and Municipal Water Supply of Lucknow Region U.P.," *International Journal of Plant, Animal and Environmental Sciences*, Vol. 2, No. 4, 2012, 4 Pages.

Key Points

- Modeling is necessary to uncover the errors before implementing the solution.
- UML is an object-oriented technique which can be applied in the biological systems also.
- UML shows the static and dynamic behavior of the system.
- OLAP data cubes are used to store the large amount of database as it supports LxMxN data storage for finite values of L, M and N.

7

Adsorption Characteristics of Zinc (Zn^{2+}) from Aqueous Solution by Natural Bentonite and Kaolin Clay Minerals: A Comparative Study

Tushar Kanti Sen, Chi Khoo
Department of Chemical Engineering, GPO Box U1987, Curtin University, Perth, 6845 Western Australia, Australia

ABSTRACT

Clay minerals are one of the potential good adsorbent alternatives to activated carbon because of their large surface area and high cation exchange capacity. In this work the adsorptive properties of natural bentonite and kaolin clay minerals in the removal of zinc (Zn^{2+}) from aqueous solution have been studied by laboratory batch adsorption kinetic and equilibrium experiments. The result shows that the amount of adsorption of zinc metal ion increases with initial metal ion concentration, contact time, but decreases with the amount of adsorbent and temperature of the system for both the adsorbents. Kinetic experiments clearly indicate that adsorption of zinc metal ion (Zn^{2+}) on bentonite and kaolin is a two-step process: a very rapid adsorption of zinc metal ion to the external surface is followed by possible slow decreasing intraparticle diffusion in the interior of the adsorbent. This has also been confirmed by an intraparticle diffusion model. The equilibrium adsorption results are fitted better with the Langmuir isotherm compared to the Freundlich model. The value of separation factor, R_L from Langmuir equation give an indication of favourable adsorption. Finally from thermodynamic studies, it has been found that the adsorption process is exothermic due to negative ΔH^0 accompanied by decrease in entropy change and Gibbs free energy change (ΔG^0). Overall bentonite is a better adsorbent than kaolin in the the removal of Zn^{2+} from its aqueous solution.

Keywords: Metal Ion Adsorption; Clay Minerals; Kinetics; Isotherms

1. Introduction

Heavy metal ion pollution is currently of great concern due to the increased awareness of the potentially hazardous effects of elevated levels of these materials in the environment [1,2]. There is an increasing and alarming challenge to researchers and environmental control agencies from the indiscriminate disposal of metals in the environment. The main sources of zinc in the environment are the manufacturing of brass and bronze alloys and galvanization [3,4]. It is also utilized in paints, rubber, plastics, cosmetics and pharmaceuticals [4]. Zinc is an essential element for life and acts as micronutrient when present in trace amounts [3]. The WHO recommended maximum acceptable concentration of zinc in drinking water as 5.0 mg/L [5]. Beyond the permissible limits, Zn^{2+} is toxic [6]. Precipitation, ion exchange, filtration, solvent extraction and membrane technology and adsorption on activated carbon are the conventional method for the removal of heavy metal ions from aqueous solutions and all of which may be ineffective or extremely expensive, when the metals are dissolved in large volumes of solution at relatively low concentration [3]. Adsorption on activated carbon is the conventional methods for the removal of heavy metal ions from aqueous solutions but its high cost limits its use [7]. Therefore, adsorption is used especially in the water treatment field and the investigation has to be made to determine inexpensive and good adsorbent. Clay minerals such as kaolin, bentonite are the most wide-spread minerals of the earth crust which are known to be good adsorbents/sorbents of various metal ions, inorganic anions and organic ligands [8]. These clay minerals are good adsorbent alternatives to activated carbon because of their large surface area, high cation exchange capacity, chemical and mechanical stability and layered structure Moreover, oxides and clay minerals are important tropical soil secondary minerals, responsible for the low mobility and bioavailability of heavy metals [9]. Although, there are reported results on the adsorption capacity of clay minerals towards heavy metal ions [2] but a systematic studies on zinc (Zn^{2+}) adsorption characteristics on kaolin, bentonite under various physicochemical parameters are limited and also very scare. Therefore a study was conducted in order to determine the influence of initial metal ion concentration, adsorbent dosages and temperature changes on adsorption characteristics of natural bentonite and kaolin. It is also essential to understand the mechanism and kinetics of adsorption, because the studies of adsorption kinetics and mechanism are ultimately a prerequisite for designing an adsorption column [3]. Another reason for this study is the importance of adsorption on solid surfaces in many industrial applications in order to improve efficiency and economy. The kinetic adsorption results have been analysed using both pseudo-first-order and pseudo-second-order kinetics models. The mechanism of the adsorption process has been explained based on intra-particle diffusion model. The isotherm equilibrium results are better fitted with Langmuir model. Finally thermodynamic parameters are determined at three different temperatures and it has been found that the adsorption process is exothermic due to negative ΔH^0 accompanied by decrease in entropy change and Gibbs free energy change (ΔG^0).

2. Materials and Methods

2.1. Chemicals

All chemicals used were of analytical grade. Stock standard solution of Zn^{2+} has been prepared by dissolving the appropriate amount of its nitrate salt in deionised water, acidified with small amount of nitric acid. This stock solution was then diluting to specified concentrations. Kaolin BET surface area of 15.72 m^2/g, mean particle size of 17.94 μm) was obtained from Chem-Supply pty Ltd, Perth WA. Bentonite (BET surface area of 238.47 m^2/g and mean particle size of 7.49 μm) was obtained from Bronson & Jacobs Pty Ltd Australia. All plastic sample bottles and glassware were cleaned, then rinsed with deionised water and dried at 60°C in a temperature controlled oven. All measurements were conducted at the room temperature (28 ± 2, ℃). The concentration of Zn^{2+} was measured using a double beam flame atomic absorption spectrophotometer. Sizes of particles were measured by Malvern Master Seizer, Ver 1.2, UK. The pH was measured by Orion pH meter.

2.2. Adsorption Procedure

Adsorption measurements were determined by batch experiments of known amount of the sample with 40 mL of aqueous Zn^{2+} solutions as per Aries & Sen [3] in a series of 60 ml plastic bottles. The mixture were shaken in a constant temperature orbital shaker at 120 rpm at 30 ℃ for a given time and then the suspensions were filtered through a What man glass micro filter and the filtrates were analyzed using flame atomic absorption spectrophotometer with an air-acetylene flame. The experiments were carried out by varying concentration of initial Zn^{2+} solution, contact time, amount of adsorbent and temperature of the system. Adsorption mechanisms were studied according to predefined procedure with the Zn^{2+} concentration ranging from 1.0 to 40 mg /L. The Zn^{2+} concentration retained in the adsorbent phase was calculated according to Equation (1)

$$q_t = \frac{(C_0 - C_t)V}{m} \quad (1)$$

where C_0 (mg/L) and C_t (mg/L) are the concentration in the solution at time t = 0 and at time t, V is the volume of solution (L) and m is the amount of adsorbent (g) added.

The kinetics of adsorption of Zn (II) was carried out at low and high initial metal ion concentration using the same adsorption procedure started above. The only difference was that samples were collected and analyzed at regular time intervals during the adsorption process.

The transient behavior of the Zn (II) adsorption process was analyzed using two adsorption kinetic models; pseudo first and pseudo-second-order rate models. The rate constant of adsorption was determined from the pseudo-first-order rate model [10] as

$$\log(q_e - q_t) = \log q_e - \frac{K_1}{2.303}t \quad (2)$$

where q_t and q_e represents the amount of metal ion adsorbed (mg/g) at any time t and at equilibrium time respectively and K_1 represents the adsorption first-order rate constant (min^{-1}). Plot of Log ($q_e - q_t$) versus t gives a straight line for pseudo first-order adsorption kinetics which allow computation of the rate constant K_1.

The pseudo-second-order model [3,10] based on equilibrium adsorption is expressed as:

$$\frac{t}{q_t} = \frac{1}{K_2 q_e^2} + \frac{1}{q_e}t \quad (3)$$

A plot between t/q_t versus t gives the value of the constants K_2 (g/mg h) and also q_e (mg/g) can be calculated.

The Constant K_2 is used to calculate the initial sorption rate h, at t → 0, as follows

$$h = K_2 q_e^2 \quad (4)$$

Thus the rate constant K_2, initial adsorption rate h and predicted q_e can be calculated from the plot of t/q versus time t using Equation (3).

According to Weber & Morris (1963) [11] the intra-particle diffusion model for most the uptake varies almost proportionately with $t^{1/2}$ rather than with the contact time and can be represented as follows:

$$q_t = K_{id} = t^{0.5} \quad (5)$$

where q_t is the amount adsorbed at time t and $t^{0.5}$ is the square root of the time and K_{id} (mg/g.min$^{0.5}$) is the rate constant of intraparticle diffusion. When intra-particle diffusion plays a significant role in controlling the kinetics of the adsorption processes, the plots of q_t vs. $t^{0.5}$ yield straight lines passing through the origin and the slope gives the rate constant K_{id}.

Thermodynamic parameters such as Gibb's free energy (ΔG^0), enthalpy change (ΔH^0) and change in entropy (ΔS^0) for the adsorption of zinc on aluminum oxide has been determined by using the following equations [2]:

$$\Delta G^0 = \Delta H^0 - T\Delta S^0 \tag{6}$$

$$\log(\frac{q_e}{C_e}) = \frac{\Delta S^0}{2.303R} + \frac{-\Delta H^0}{2.303RT} \tag{7}$$

where q_e is the amount of zinc adsorbed per unit mass of aluminium oxide (mg/g), C_e is equilibrium concentration (mg/L) and T is temperature in K. q_e/C_e is called the adsorption affinity. The above equation is for unit mass of adsorbent dose.

3. Results and Discussions

3.1. Characterization of Adsorbenst

The characterization of the structure and surface chemistry of the adsorbent is of considerable interest for the development of adsorption and separation processes. FT-IR spectroscopy of bentonite which is not shown here indicated the presence of hydroxyl, carboxyl and Si-O which are important sorption sites. The particle size distribution of bentonite which is not shown here for which mean particle size was 7.49 μm, whereas mean particle size of kaolin was 17.95 μm. XRD analysis also indicates that the main mineral of kaolin was kaolinite with trace impurities of quartz, whereas in bentonite four different mineral phases were present: mainly quartz (SiO_2) and muscovite.

3.2. Effect of Contact Time and Initial Metal Ion Concentration on Zn(II) Metal Ion Adsorption Kinetics

Figure 1 represents a plot of the amount of zinc metal ion adsorbed (mg/g) versus contact time for Zn-kaolin and Zn-bentonite system. **Figure 2** represents a plot of the amount of Zn (II) adsorbed at different initial metal ion concentration range for both the system. From these plots, it is found that the amount of adsorption *i.e.* mg of adsorbate per gram of adsorbent increases with increasing contact time at all initial metal ion concentrations and equilibrium is attained within 80 minutes for both the systems Further it was observed that the amount of metal ion uptake, q_t (mg/g) is increased with increase in initial metal ion concentration (**Figure 2**). The increase in adsorption is more pronounced for the Zn-bentonite system compared to the Zn-Kaolin system. These kinetic experiments clearly indicate that the adsorption of Zn (II) on clay surface is a two-step process: a rapid adsorption of

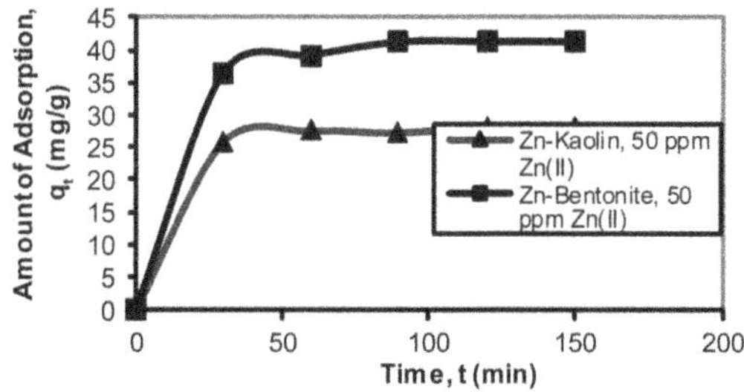

Figure 1. Effect of contact time on Zn (II) metal ion adsorption by kaolin and bentonite. Initial metal ion concentrations = 50 ppm, Initial solution pH = 6.65.

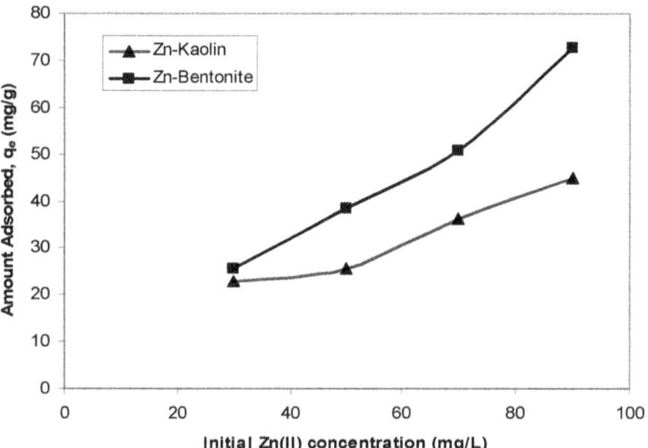

Figure 2. Effect of initial Zn (II) concentration on adsorption.

metal ions to the external surface is followed by possible slow intraparticle diffusion in the interior of the particles [10]. This has been confirmed by fitting experimental data with diffusion model which is presented latter section. This two-stage process is also due to presence of two different types of binding sites on the adsorbents. Moreover, the amount of Zn(II) ions adsorbed per unit mass of the biosorbent increases with the initial metal ion concentration (**Figure 2**) which might be due to the higher availability of Zn(II) ions in solution. Further a higher initial concentration provides increased driving force to overcome all mass transfer resistance of metal ions between the aqueous and solid phases resulting in higher probability of collision between Zn(II) ions and sorbents.

3.3. Effect of Adsorbent Dose on Zn(II) Adsorption Kinetics

The results of the kinetic experiments with varying adsorbent concentrations for which plots are not presented here. It has been found that the amount of Zn (II) adsorbed per unit mass of adsorbent decreases as the adsorbent mass increase for both systems. Several other investigators have also reported the same trend of adsorbent concentration effect on metal ion adsorption [3]. Although the number of adsorption sites per unit mass of an adsorbent should remain constant, independent of the total adsorbent mass, increasing the adsorbent amount in a fixed volume reduces the number of available sites as the effective surface area is likely to decrease

3.4. Effect of Temperature on Zn(II) Adsorption Kinetics and Thermodynamics Study

To observe the effect of temperature on the adsorption capacity, experiments were carried out in three different temperatures of 30℃, 50℃ and 70℃ for a fixed initial metal ion concentration of 50 ppm for which plots are not presented here. It was found that with the increased in temperature, adsorption capacity decreased for both the systems. This is mainly because of decreased surface activity suggesting that adsorption between metal ion and clay minerals was an exothermic process. With increasing temperature, the attractive forces between the clay surfaces and metal ion are weakened and then sorption decreases. This may be due to a tendency for metal ion to excape from the solid phase of clay to the liquid phase with an increase in temperature of the solution. The values of Gibbs free energy (ΔG^0) have been calculated by knowing the value of the enthalpy of adsorption (ΔH^0) and the entropy of adsorption (ΔS^0) which are obtained from the slope and intercept of a plot of log (q_e/C_e) versus 1/T (not shown here).

All these thermodynamic parameters are presented in **Table 1**.

3.5. Zn(II) Adsorption Kinetic Models & Isotherm

In this study, the two most widely used kinetic models; pseudo-first-order and pseudo-second-order were employed which are described in earlier section. In the pseudo-first-order model, the rate constant k_1 and correlation coefficient, R^2, were determined by plotting log ($q_e - q_t$) against time, t for both systems (not presented) with very poor regression coefficient, R^2 with range of 0.15 to 0.74 for various physicochemical parameters. Moreover, the pseudo-first-order kinetic model predicts a much lower value of the equilibrium adsorption capacity than the experimental value for this system and hence it gives the inapplicability of this model. But the pseudo-second-order kinetic model is fitted very well with very high regression coefficient (R^2) which is shown in **Table 2**. The pseudo-second-order rate constant, k_2, equilibrium sorption capacity, q_e and initial rate constant, h were calculated for these systems from the fitted model equations which are tabulated in **Table 2**. Higher correlation coefficients (R^2) with respect to fitted pseudo-first-order model suggest that adsorption of zinc metal ion on clay minerals follow Pseudo-second-order.

Table 1. Thermodynamic parameters for zinc adsorption at different temperatures.

System	Temperature (℃)	ΔG^0 (kJ·mol^{-1})	ΔH^0 (kJ·mol^{-1})	ΔS^0 (j·mol^{-1}·K^{-1})
Zn-Kaolin	30	2.18	−0.228	−0.0072
	50	2.10	−0.228	−0.0072
	70	2.24	−0.228	−0.0072
Zn-Bentonite	30	0.065	−0.268	−0.0011
	50	0.087	−0.268	−0.0011
	70	0.109	−0.268	−0.0011

The most commonly used technique for identifying the mechanism involved in the sorption process is by fitting the experimental kinetic data with intraparticle diffusion plot (Equation (5)). It has been found that the adsorption plots (Which are not shown here) are not linear over the whole time range and can be separated into two-three linear regions which confirm the multi stages of adsorption.

The adsorption equilibrium data was fitted with Langmuir and Frenundlich isotherms within the metal ion concentration range of 30 - 90 ppm respectively. Linear regression was used to determine the most fitted isotherm. The Freundlich adsorption isotherm can be expressed as [10]

$$\ln q_e = \ln K_f + 1/n\,(\ln C_e) \quad (8)$$

where q_e is the amount of metal ion adsorbed atequili-

Table 2. Pseudo-second-order kinetic parameters.

System	System parameters	k_2 (g/mg.min)	q_e (mg/g)	H (mg/g.min)	R^2
Zn-kaolin	Initial Zn(II) concentration				
	30.00 ppm	0.0143	25.4	9.25	0.999
	50.00 ppm	0.0250	28.4	20.0	0.999
	70.00 ppm	0.0071	37.3	9.80	0.998
	Kaolin dosages				
	0.01 g	0.0329	28.4	26.0	0.998
	0.02 g	0.0535	13.0	9.61	0.996
	0.03 g	0.0253	9.09	2.05	0.9
	Temperature				
	30°C	0.0358	27.5	27.7	0.999
	50°C	0.0271	24.9	16.2	0.999
	70°C	0.0066	21.3	2.88	0.9
Zn-bentonite	Initial Zn (II) concentration				
	30.00 ppm	0.0090	30.9	8.69	0.9
	50.00 ppm	0.010	41.6	18.3	0.995
	70.00 ppm	0.002	57.1	7.63	0.999
	Bentonite dosages				
	0.01 g	0.013	45.2	27.7	0.9
	0.02 g	0.440	19.4	166.62	0.9
	0.03 g	0.022	9.85	2.19	0.990
	Temperature				
	30°C	0.013	36.2	17.1	0.990
	50°C	0.0197	35.9	24.9	0.997
	70°C	0.0194	30.0	17.5	0.9

brium time, C_e is equilibrium concentration of nickel metal ion in solution. K_f and n are isotherm constants which indicates the capacity and the intensity of the adsorption respectively and can be calculated from the intercept and slope between ln q_e and lnC_e which are shown in **Table 3** for both the systems. The Freundlich plots are not shown here.

Also Langmuir isotherm equation was also fitted for both the system with this same metal ion concentration range. The linearized form of Langmuir can be written

$$C_e/q_e = (1/K_a\, q_m) + C_e/q_m \quad (9)$$

The Langmuir constants, q_m (maximum adsorption capacity) and K_a (values for Langmuir-2) can be obtained from plots between C_e/q_e versus C_e which are shown in **Figure 3** & **Figure 4** respectively with fixed initial conditions. The maximum adsorption capacity of Zn (II), q_m and constant related to the binding energy of the sorption systems, K_a is calculated which are 56.49 mg/g and 0.0437 for Zn-kaolin and 62.5 mg/g and 0.0648 for Zn-bentonite respectively. Overall Langmuir isotherm model had higher regression coefficient (R^2) compared to the Freundlich isotherm model for both the systems.

A further analysis of the Langmuir equation can be made on the basis of a dimensionless equilibrium parameter, R_L, also known as the separation factor, given by (Sen & Gomez, 2011)

$$R_L = \frac{1}{1 + K_a\, C_0} \quad (10)$$

where K_a is the Langmuir constant and C_0 is the initial metal ion concentration (mg/L). The separation factor, R_L has been calculated from Langmuir plot. It has been found that the calculated range of RL values from 0.432 to 0.202 for Zn-kaolin and 0.339 to 0.146 for Zn-bentonite system with the initial metal ion range of 30 to 90 ppm. These R_L values indicates favourable adsorption as it lie in the range $< 0 < R_L < 1$ [2]. The maximum Langmuir adsorption capacity of bentonite was more than kaolin.

Table 3. Freundlich parameters obtained from Freundlich plots. Amount of iron oxide and also kaolin added = 8 mg; pH 4.5; Temperature = 28℃; Shaker speed = 120 rpm.

Adsorbent	Freundlich constants		
	K_F	n	R^2
Kaolin	9.58	2.92	0.874
Bentonite	7.69	1.75	0.994

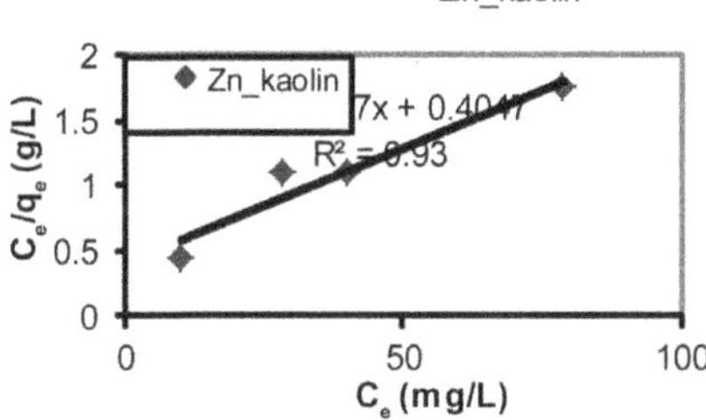

Figure 3. Langmuir plot Zn-Kaolin system.

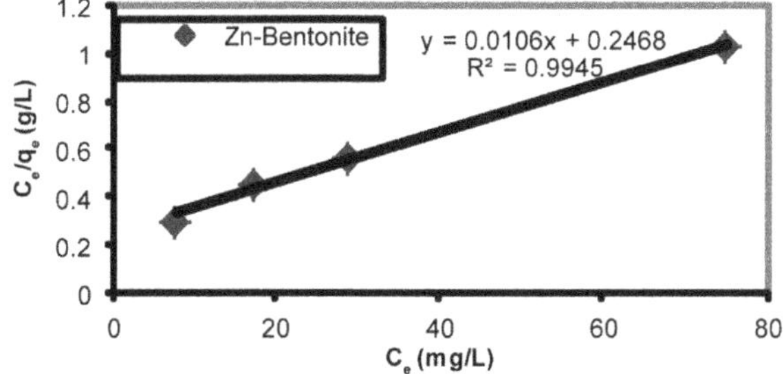

Figure 4. Langmuir plot for Zn-bentonite system.

4. Conclusions

The results obtained in this study demonstrated that kaolin and natural bentonite both can be used as an excellent natural adsorbent to remove Zn (II) from wastewaters with good efficiency and low cost. The amount of metal ion Zn (II) adsorption on both clay minerals was found to increase with increase in initial metal ion concentration and contact time but found to decreases with an increase in amount of adsorbent and temperature. The maximum adsorption capacity of bentonite was found to 62.5 mg/g with an initial Zn(II) concentration range of 30 to 90 ppm, whereas for kaolin it was 56.49 mg/g with the same metal ion concentration range. Kinetic experiments clearly indicated that sorption of Zn (II) on both kaolin and bentonite is a two steps process: a rapid adsorption of metal ion to the external surface followed by intraparticle diffusion into the interior of adsorbent which has also been confirmed by intraparticle diffusion model. Overall the kinetic studies revealed that adsorption process followed the pseudo-second-order kinetics model. The Langmuir isotherm model was applicable for both systems. The constant value, R_L (low separation factor) in Langmuir isotherms indicated that there was favourable adsorption for both systems. Finally thermodynamic parameters were determined at three different temperatures.

REFERENCES

[1] S. SenGupta and K. G. Bhattacharyya, "Adsorption of Ni (II) on Clays," *Journal of Colloid and Interface Science* Vol. 295, No. 1, 2006, pp. 21-32.

[2] T. K. Sen, and G. Dustin, "Adsorption of Zinc (Zn^{2+}) from Aqueous Solution on Natural Bentonite," Desalination, Vol. 267, No. 2-3, 2011, pp. 286-294.

[3] F. Arias and T. K. Sen, "Removal of Zinc Metal Ion (Zn^{2+}) from Its Aqueous Solution by Kaolin Clay Mineral: A

Kinetic and Equilibrium Study," *Colloids and Surfaces A*: Vol. 348, 2009, pp. 100-108.

[4] C. H. Weng and C. P. Huang, "Adsorption Characteristics of Zn (II) from Dilute Aqueous Solutions by Fly Ash," *Colloids Surf A*, Vol. 247, 2004, pp. 137-143.

[5] D. Mohan and K. P. Singh, "Single and Multi-component Adsorption of Cadmium and Zinc Using Activated Carbon Derived from Bagasse-an Agricultural Waste," *Water Research*, Vol. 36, 2002, pp. 2304-2318.

[6] A. K. Bhattacharya, S. N. Mandal and S. K. Das, "Adsorption of Zn (II) from Aqueous Solution by Using Different Adsorbents," *Chemical Engineering Journal* Vol. 123, 2006, pp. 43-51.

[7] M. Mohammad, S. Maitra, N. Ahmad, A. Bustam, T. K., Sen and B. K. Dutta, "Metal Ion Removal from Aqueous Solution Using Physic Seed Hull," *Journal of Hazardous Materials*, Vol. 179, 2010, pp. 363-372.

[8] C. Gurses, M. Dogan and M. Yalcin, "Adsorption Kinetics of the Cationic Dye Methylene Blue onto Clay," *Journal of Hazardous Materials*, Vol. 131, No. 1-3, 2006, pp. 217-228.

[9] E. V. Mellis, M. C. P. Cruz and J. C. Casagrande, "Nickel Adsorption by Soils in Relation to pH, Organic Matter and Iron Oxide," *Scientia Agricola,* Vol. 61, No. 2, 2004, pp. 190-195.

[10] T. K. Sen and M. V. Sarzali, "Removal of Cadmium Metal Ion (Cd^{2+}) from its Aqueous Solution by Aluminium Oxide: A Kinetic and Equilibrium Study," *Chemical Engineering Journal*, Vol. 142, 2008, pp. 256-262.

[11] W. Weber and J. Morris, "Kinetics of Adsorption on Carbon from Solution," *American Society of Civil Engineers*, Vol. 89, 1963, pp. 31-60.

Digital Interactive Kanban Advertisement System Using Face Recognition Methodology

Feng-Yi Cheng, Chu-Ja Chang, Gwo-Jia Jong
Department of Electronics Engineering, National Kaohsiung University of Applied Sciences, Kaohsiung.

ABSTRACT

Most of advertisement systems are presently still launch the publicity content by the static words and pictures. Recently, this static advertisement model will not be able to attract people's attention more and more. Moreover, the static information content of advertisement system is limited because of the layout shown size. It can not also fully demonstrate the information content of advertisement system. In this paper, we develop a digital interactive kanban advertisement system using face recognition methodology to solve these problems. The system captures the person's face through the camera. The digital advertisement content size is relevant by the person and camera observation locations. In this paper, we adopt the Adaboost algorithm to judge people face, and the system only need to grab the position of the face. The system doesn't built expensive and complex equipment to reduce the system cost and enhance the system performance. This system can also achieve the same similar digital interactive advertising effectiveness.

Keywords: Face Recognition; Kanban Advertisement; Adaboost; Interactive

1. Introduction

There are various kinds of advertising media now. For example: televisions, broadcastings, magazines, newspapers, outdoor advertising and transit advertising...etc. Beside above media, the appearance of the web advertising makes advertising having more and more development space. An advantage of web advertising is that consumers can select advertising freely and obtain messages advertising transmit immediately. Advertising can interact with consumers and transmit feedback instantly, but traditional advertising can't [1,2]. In this paper, interactive multimedia as the theme to explore the visual interface design and interactive multimedia development, according to the popular trend of today's interactive multimedia authoring the common interactive multimedia such as: interactive web pages, DVD movies menu, web advertising, teaching CD-ROM...etc., because of the rapid growth of technology and the technology matures, the traditional static advertising, web, gradually interactive multimedia replaced [3].

In recent years, face detection technology tends to mature, the technology has been widely used in cameras, computer identification system and interactive advertising. In [4], they research proposes a system that engages audience to the advertisement through interactive applications and provides data to the advertiser/producer about their audience, but we think that is too complex. Therefore, we propose an interactive system, using text and pictures to do the interactive display, it can allow users to quickly understand the contents of advertisements.

2. Interactivity Kanban Advertisement

The **Figure 1** is the configuration of interactive kanban advertisement system. At first, we capture the camera image, then color segmentation extracts the skin color of a face from a cluttered image; then, binary imaging further forms a more complete region. Next, morphological erosion eliminates some of the small spots in a tested image. Contrary to erosion, dilation enlarges and connects a small and disconnected, but marked, facial region. Subsequently, connected component labeling is employed to mark multiple faces in the image. Finally, an area threshold and an aspect ratio are used to validate the corrected facial region. After then we use the Adaboost algorithm [7-10] to make face recognition. At last, we can judge by the distance between the captured face and advertising, when the person is closer and closer, the words will accord to the distance for scaling to achieve

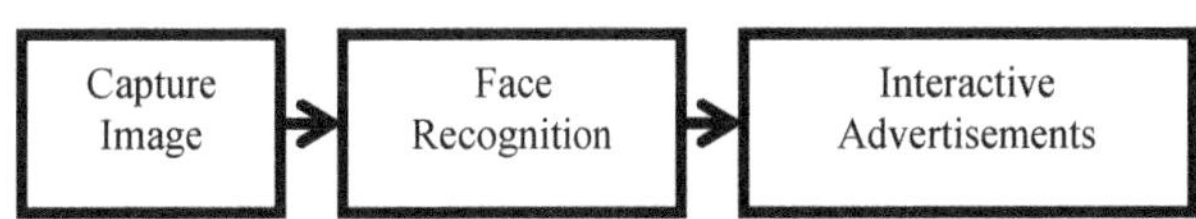

Figure 1. Interactivity kanban ad system block diagram.

the purpose of attracting users.

The **Figure 2** is our hardware configuration of capture image, the camera mounts on the Kanban Advertisement above.

3. Face Recognition

The **Figure 3** is our face recognition system, the face recognition has three processing steps: Skin color detection, Detect face region, facial feature points and finally, through Adaboost screening the most right face of people. This part is mainly to capture each person's face form the image.

3.1. Skin Color Detection

The color segmentation is an important pre-processing step in the face recognition methods. We are used HIS method [5] to detection face skin color. Frist, we transform the image of RGB three color changes to H, S. It can be written as

$$H = \begin{cases} \theta, & if\ B \le G \\ 360-\theta & if\ B > G \end{cases} \quad (1)$$

where

$$\theta = \cos^{-1}\left\{ \frac{\frac{1}{2}\left[(R-G)+(R-B)\right]}{\left[(R-G)^2+(R-B)(G-B)\right]^{1/2}} \right\} \quad (2)$$

$$S = 1 - \frac{3}{(R+G+B)}\left[\min(R,G,B)\right] \quad (3)$$

Then we can follow the rule to find skin color f_c:

Figure 2. Interactivity kanban ad system configuration.

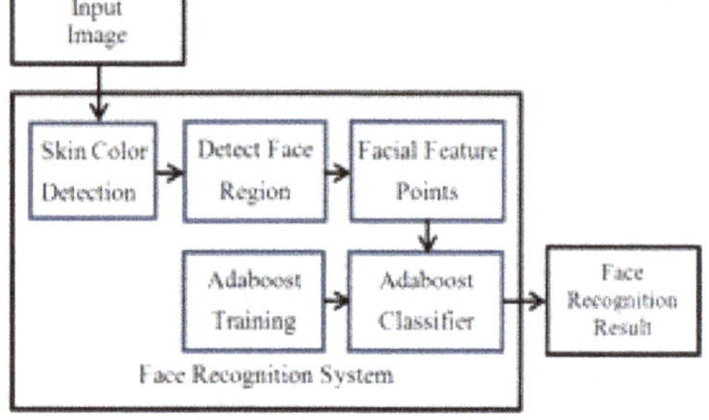

Figure 3. Face recognition block diagram.

$$f_c = \begin{cases} 1, if\ 8 \le H \le 32\ and\ 30 \le S \le 163 \\ 0, otherwise \end{cases} \quad (4)$$

After the skin color detection, we only see the portion of skin color, as shown in **Figure 4**.

3.2. Face Region Detection Method

After the Skin color detection, we change the RBG to binary used thresholding. Then, we consulted the paper's method [6] to the binary image is sub-divided into blocks. Then, the total skin area within a block is computed, and if this is greater than or equal to 40% of the block area, the block label is assigned to be skin. A connected region step is then performed by examining the 8 neighbourhood connectivity among the blocks to create a set of candidate regions. Face regions are selected amongst the candidate regions as the regions having an aspect ratio corresponding to the 1.2 - 2.0 ratio, than use the rate value of the image to define the threshold. It is showed in **Figure 5**.

3.3. Capture Facial Feature Points

3.3.1. Eyes Detection

It is obvious that eyes are non-skin color regions, the C_r and C_b component of eyes and skin contains bigger difference in the YC_rC_b space, and the C_b is higher than the C_r in the eyes region. It can detection location and size by above information.

3.3.2. Mouths Detection

Mouth is also non-skin color region, in the mouth the C_r is much higher and the C_b much lower. Increasing the difference between the C_b and C_r can accurately detected size and location.

3.4. Adaboost Algorithm

Machine learning algorithm is flourishing in recent year, widely used at various levels. Face Detection this issue in order to obtain better characteristics also introduces a machine learning concepts, these studies are a breakthrough in the past to the face detection frame, most notably the 2004 study is presented using the integral image Viola for the characteristic value of the AdaBoost face detection method. AdaBoost is an algorithm for constructing a "strong" classifier as linear combination of "weak" classifiers [7-10].

3.4.1. Haar-Like Features and Integral Image

A set of Haar-like features, used as the input features to the cascaded classifiers, are shown in **Figure 6**. In our work, Haar-like features consideration is using integral image to improve computation efficiency.

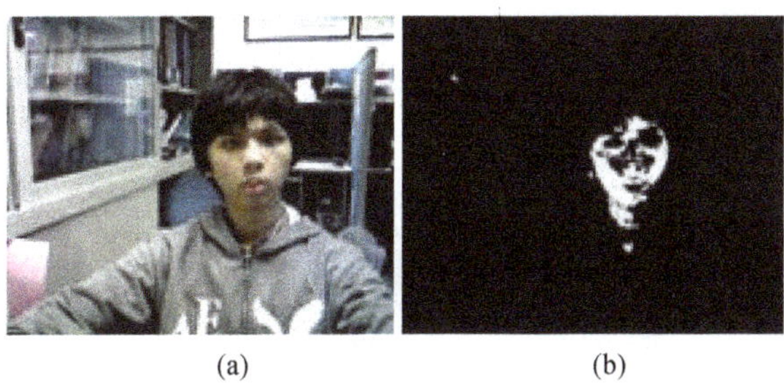

(a) (b)

Figure 4. (a) Original image; (b) Skin color detection.

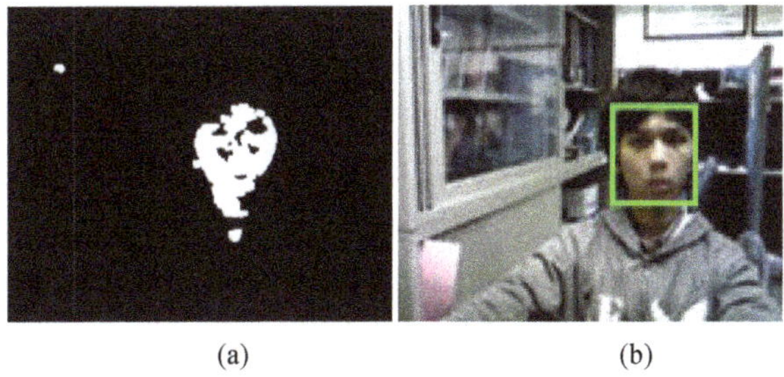

(a) (b)

Figure 5. (a) Morphological process; (b) Face region detection.

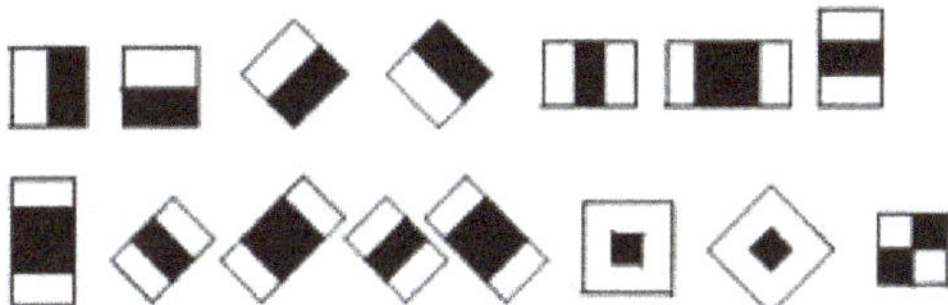

Figure 6. The Haar-like features for AdaBoost algorithm.

The Haar-like features that shown in **Figure 6**. It used in our face detection system. The features can be rapidly computed at different scales by introducing "Integral Image".

3.4.2. AdaBoost Algorithm

In fact AdaBoost is a classification of concepts, for example, In order we pick a better than normal a little bit (>= 50%) of the algorithm, it can again and again use update weighting approach to reduce error rate, The process is as follows

1) Input M sample of the target image and N sample not of the target image, and I search the number of features.

2) Initialize weights

Target image samples weights

$$WP_m = \frac{1}{2M} \tag{5}$$

Non-target image samples weights

$$WN_n = \frac{1}{2N} \tag{6}$$

3) For each feature j, train a weak classifier T, and evaluate its error E with respect to W

$$E = \sum_{m=1}^{M} WP_m(1-T_m) + \sum_{n=1}^{N} WN_n \cdot T_n \tag{7}$$

In this derivation when $T = 1$ consistent with the image features, $T = 0$ does not meet.

4) Using step 3. Add the choose features to the stage and determine the corresponding weights

$$W_i = \log\left(\frac{1-E}{E}\right) \tag{8}$$

5) For step 3. Searching the better features to updates image sample weights. Updated target image sample weights.

$$WP_m \leftarrow WP_m \cdot \left(\frac{E}{1-E}\right)^{1-E} \tag{9}$$

Updated non-target image samples weights.

$$WN_n \leftarrow WN_n \cdot \left(\frac{E}{1-E}\right)^{1-E} \tag{10}$$

6) Normalize the weights

$$WP_m \leftarrow \left(\frac{WP_m}{\sum_{m=1}^{M} WP_m}\right) \tag{11}$$

$$WN_n \leftarrow \left(\frac{WN_n}{\sum_{n=1}^{n} WN_n}\right) \tag{12}$$

7) Check whether the number of the current search features to meet the demand, if the lack of jump back to step 3, otherwise the end of.

4. Interactive Advertisement System

The **Figure 7** is our interactive advertisement system flow process. At first, we capture the camera's image,

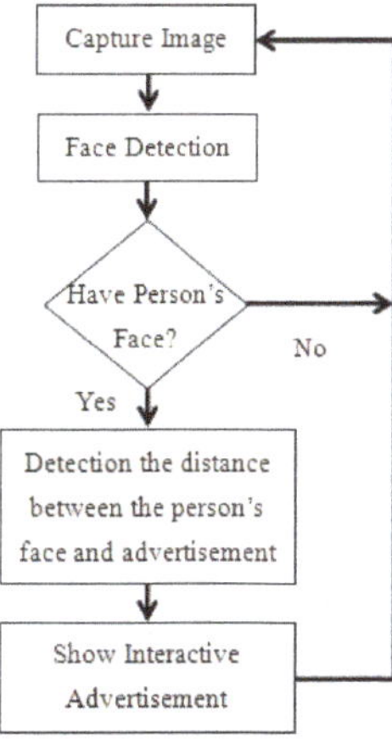

Figure 7. Interactive ad system flow process.

then thought the image to do face recognition processing, to get the information of people who watch the Advertisement, and then the system further determine whether capture face or not, in other words, to determine if someone is watch the kanban advertisement system, after then to detection the distance between the person's face and advertisement, when people are approached the advertisement, the advertisement will also show more message telling the people, let people can learn more about the details of the advertisement, to increase the impression of people watch advertisement.

5. Result and Discuss

We use a 20 million pixels webcam and a 36-inch TV to achieve the Interactive Kanban Advertisement System. The webcam is set in place of 130cm high and angle of 90 degrees. The **Figure 8** shows the information that the relationship between pixel size of face and the distance from person to camera. When the system captures the person's face, we can use the pixel size of face to determine where the person.

The **Figure 9** is the user interface of program, this program of interface can divided into two parts, the camera of image is on the left, the interactive kanban advertisement is on the right. If someone walks past in front of the kanban advertisement, the system will catch person's face, and the kanban advertisement also shows some words to attract people's attention, it is like **Figure 9**.

When person is closer and closer, the system will calculate the face of pixels to detect the distance, if people rely on close enough, the system will change the advertising content, display more information attract people to continue to watch, it is like **Figure 10**.

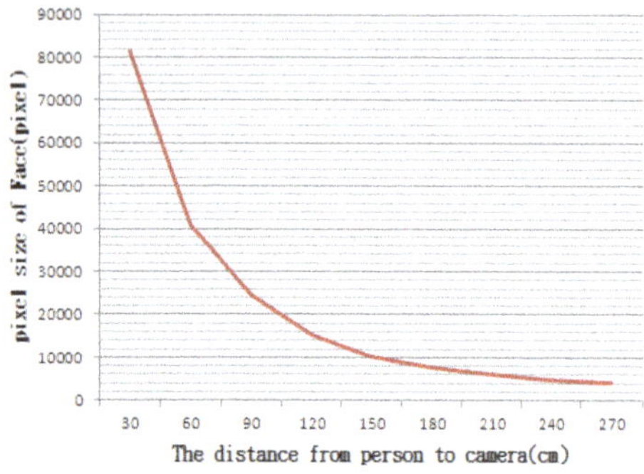

Figure 8. Graph of ratio between person and camera.

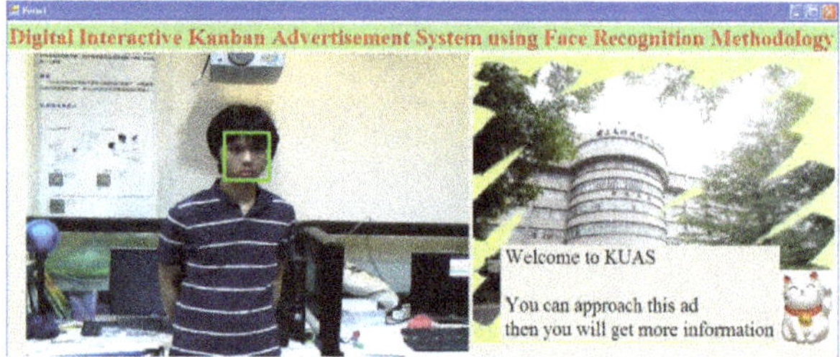

Figure 9. The user interface of program.

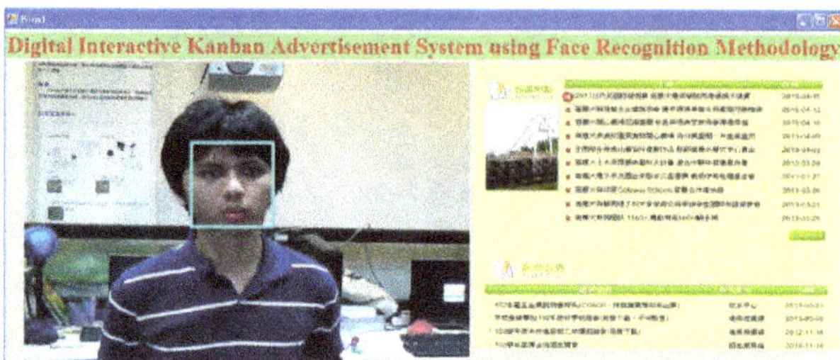

Figure 10. It change the ad content when people rely on close enough.

6. Conclusions

In this paper, we propose an Adaboost algorithm approach to the face recognition for applications of interactive advertisement system. Although it has a lot of people to research for the subject so far, however, the proposed approach are more complex, build may more cost or need to take the time to interact with advertisement, caused people inconvenience. Our system only through the distance between the face and advertising to interact, through words and pictures to attract people, reduce the complexity of the system also allows people quickly to understand more information of advertisements.

REFERENCES

[1] J. H. Cho, Y. J. Sah and J. Ryu, "A New Content-related Advertising Model for Interactive Television," Broadband Multimedia Systems and Broadcasting 2008, March 31 2008-April 2 2008, pp. 1-9.

[2] M.-H. Hsieh, D.-L. Yang and J.-Y. Dai, "A Face Recognition System Prototype to Evaluate the Effectiveness of Digital Advertisement," 2010 Conference on Computer Vision, Image Processing and Information Technology, 2010-06. Zhongli, Taiwan, pp. 283-289.

[3] J. Kim and S. Kang, "An Ontology-Based Personalized Target Advertisement System on Interactive TV," Consumer Electronics (ICCE), *2011 IEEE International Conference,* 9-12 Jan. 2011, pp. 895 - 896.

[4] M. Taspinar, A. T. Naskali, M. Kurt and G. Eren, "The Importance of Customized Advertisement Delivery Using 3D Tracking and Facial Recognition," in Proc. The Second International Conference on Digital Information and Communication Technology and its Applications (DICTAP), 2012, pp. 526-530.

[5] S. Guerfi, J.-P. Gambotto and S. Lelandais, "Implementation of the Watershed Method in the HSI Color Space for the Face Extraction," Advanced Video and Signal Based Surveillance, Sept. 2005, pp. 282-286.

[6] M. Rahman and N. Kehtarnavaz, "Real-Time Face-Priorit y Auto Focus for Digital and Cell-Phone Cameras," *IEEE Transactions on Consumer Electronics*, Vol. 54, No. 4, 2008, pp. 1506-1513.

[7] J. X. Ruan and J. X. Yin "Multi-Pose Face Detection Using Facial Features and AdaBoost Algorithm," Second International Workshop on Computer Science and Engineering, 2009, pp. 31-34.

[8] Y.-W. Wu and X.-Y. Ai, "Face Detection in Color Images Using AdaBoost Algorithm Based on Skin Color Information," Workshop on Knowledge Discovery and Data Mining, 2008, pp. 339-342.

[9] S. A. Inalou and S. Kasaei "AdaBoost-Based Face Detection in Color Images with Low False Alarm," Second International Conference on Computer Modeling and Simulation, 2010, pp.101-111.

[10] Y. C. Xing, Z. Z. Wang and W. P. Qiang, "Face Tracking Based Advertisement Effect Evaluation," Image and Signal Processing, 2009. CISP '09. 2nd, 2009, pp. 1-4.

Adsorption of Methylene Blue by NaOH-modified Dead Leaves of Plane Trees

Lei Gong, Wei Sun, Lingying Kong
College of Environment and Safety Engineering, Qingdao University of Science and Technology, Qingdao, China

ABSTRACT

NaOH-modified dead leaves of plane trees were used as bioadsorbent to remove methylene blue (MB) from aqueous solution. Variable influencing factors, including contact time, temperature, initial MB concentration and pH were studied through single-factor experiments. The results showed that the initial concentration 100 mg/L, bioadsorbent of 2.5 g/L, pH of 7, room temperature were the best adsorption conditions. The NaOH-modified bioadsorbent had a high adsorption capacity for MB, and its saturated extent of adsorption was 203.28 mg/g, which was higher than the un-modified dead leaves (145.62 mg/g) and some other bioadsorbents. Finally, adsorption kinetics and isotherms were discussed, suggesting that the Langmuir isotherm model and Pseudo-second order kinetics were fitted well with the adsorption process.

Keywords: Modified; Bioadsorbent; Methylene Blue; Kinetics; Adsorption Isotherm

1. Introduction

Dyes are extensively used in many industries such as leather, paper, textiles, printing and cosmetics. Low or high concentration of dye wastes can be often detected in the effluents arising from these industries. However, dye wastes are difficult to be removed from the effluents due to their complex composition and bad biodegradation [1].Therefore, removal of dyes is an essential procedure of wastewater t treatment before discharge.

The methods to treat dyeing wastewater can be classified into two types: physical and chemical processes. Among all these methods, activated carbon is the most popular used as solid adsorbent because of its high adsorption capability[2,3]. However, widely utilized activated carbon in commerce is vastly limited with its initial cost and the high expense of regeneration after exhausted. For these reasons, many researchers are aiming at developing cheaper substitutes to replace activated carbon, such as fly ash[4], cellulose-based waste[5], rice husk[6], tea waste[7], dehydrated wheat bran[8], bagasse fly ash[9], ect.

In many areas, plane trees with dense leaves and fast-growing character are often planted as urban greening tree species. Lots of plane tree leaves can be available in autumn when they fall, so they are obtained easily and low-cost. So in the present study, dead leaves of plane trees have been used as bioadsorbent for the removal of MB from aqueous solution, and relatively simple pretreatment with NaOH was used to modify the properties of them. The purpose of this study is to find the feasibility of plane tree leaves as a bioadsorbent.

2. Materials and Methods

2.1. Preparation of Adsorbents

The dead leaves used in the study were collected from Qingdao University of Science and Technology in autumn. The collected leaves were washed with deionized water several times to remove the dust and other water-soluble materials. The process couldn't stop until the washing water was colorless. The washed leaves were dried at 333K in a hot air oven for one day, and then shattered and sieved to get different geometrical sizes (0.45 - 0.9 mm).The sieved particles were washed repeatedly with deionized water till the UV absorbency of washing water was close to zero, and the final pH kept constant. The washed materials were then dried at 333K in the hot air oven for one day. Finally, the obtained particles were modified with NaOH (0.5 mol/L) [10] that was the bioadsorbent the experiments needed.

2.2. Adsorbate

The stock solution (1 g/L) of MB (methylene blue, $C_{16}H_{18}ClN_3S\cdot 3H_2O$, 82% pure, a molecular weight of 373.9, supplied by Shanghai reagent Co., Ltd), was prepared in distilled water. Calibration curve for MB was prepared by measuring the absorbance of different concentrations of MB at 665 nm (UV-VIS spectrophotometer, TU-1901, Beijing PGENERAL Co., Ltd).

2.3. Sorption Experiments

The experiments were conducted in 250 mL Erlenmeyer flasks containing different amount of bioadsorbent and 200 mL of MB solution at desired concentrations. The flasks were agitated in a Water Baths Shaker at constant shaking rate. Samples were pipetted from the mixed solution in flasks at prearranged time intervals to determine the residual MB concentration. The samples were centrifuged and the supernatant liquid was determined its absorbance after dilution. For isotherm studies, a series of flasks containing 200 mL MB solution in the range of 100 - 500 mg/L were prepared. Bioadsorbent of 0.5 g was added to each flask and then the mixtures were agitated at 293.15K.

The adsorption amount of MB adsorbed onto the bioadsorbent at equilibrium was calculated with the following equation:

$$q_{eq} = \frac{(C_0 - C_{eq})\,V}{X} \quad (1)$$

where C_0 (mg/L) and C_{eq} (mg/L) are the initial and equilibrium concentration of MB, V(L) is the volume of solution, X(g) is the weight of adsorbent in one container, C(mg/L is the solution concentration after reaction for a period of time.

2.4. Sorption Kinetics

For kinetic studies, the effect of initial MB concentration on the kinetics of MB adsorption was studied. Three adsorption kinetic models are used in this study to describe the adsorption characteristics.

The Pseudo-first order equation can be expressed in Equation (2):

$$\frac{dq}{dt} = K_1 (q_e - q) \quad (2)$$

where K_1(min^{-1}) is the rate constant of the pseudo-first order adsorption, q_e(mg/g) is the amount of equilibrium adsorption, q(mg/g) is the amount of dye adsorbed at any time.

Equation (2) can be integrated to the following form by applying the initial conditions $q = 0$ at $t = 0$:

$$\log(q_e - q) = \log q_e - \frac{K_1}{2.303} t \quad (3)$$

The value of K_1 can be determined by the slope of plot of log(q_e-q) versus t in Equation (3).

The Pseudo-second order equation can be expressed in Equation (4):

$$\frac{dq}{dt} = K_2 (q_e - q)^2 \quad (4)$$

where K_2($g \cdot mg^{-1} \cdot min^{-1}$) is the rate constant pseudo-second order, q_e and q are the same with Equation (2) .

Considering that q = 0 when t = 0 and that q = q when t = t, integrating Equation (5), result is as follows:

$$\frac{1}{q_e - q} = \frac{1}{q_e} + K_2 t \quad (5)$$

The value of K2 can be determined by the slope and intercept of the plot 1/(q_e-q) versus t or by t/q versus t.

Intraparticle diffusion process can be expressed in Equation (6):

$$q = K_i t^{0.5} + C \quad (6)$$

where K_i ($mg \cdot g^{-1} \cdot min^{0.5}$) is the Intraparticle diffusion constant,C is the interept.

Among the three steps—film diffusion, pore diffusion and intraparticle transport, of the adsorption mechanism, the slowest step usually controls the whole rate of the process. However, in a batch reactor, the intraparticle diffusion is often a rate-limiting step. Therefore, the rate constant of intraparticle transport (K_i) is estimated from the slope of the linear portion of the plot q against square root of t. Values of C represent the thickness of boundary layer, and that they are positive correlation[11,12].

2.5. Sorption Equilibrium Isotherm Models

The Langmuir adsorption model [13] is based on the assumption that maximum adsorption corresponds to a saturated monolayer of solute molecules on the adsorbent surface. The Langmuir equation may be written as

$$1/q_e = 1/q_m C_{eq} b + 1/q_m \quad (7)$$

where q_e(mg/g) is the amount of per unit weight of adsorbent, q_m(mg/g) is the monolayer adsorption capacity, b(L/mg) is the constant related to the free energy of adsorption, C_{eq}(mg/L) is the equilibrium concentration of the solute in the bulk solution.

The Freundlich model [14] is an empirical equation that assumes heterogeneous adsorption due to the diversity adsorption sites. Its equation may be written as:

$$q_e = K_F \cdot C_{eq}^{\ 1/n} \quad (8)$$

$$\ln q_e = \ln K_F + \frac{1}{n} \ln \rho_e \quad (9)$$

where q_e and C_{eq} are the same with the Langmuir's, ρ_e (mg/L) is the equilibrium concentration of the solute in the bulk solution, K_F ($mg^{1-1/n} L^{1/n}/g$) is a constant indicative of the relative adsorption capacity of the adsorbent, n is a constant indicative of the intensity of the adsorption.

3. Results and Discussion

3.1. Effect of Contact Time

The effect of contact time on the sorption of MB was

studied for an initial MB concentration of 100 mg/L. The result is shown in **Figure 1**. The obtained result reveals that the speed of MB uptake onto bioadsorbent is fast at the initial stage (before 50 min), and then it decreases gradually and becomes near the equilibrium in about 150min. This is obvious from the fact that there are abundant vacant surface sites available for adsorption during the initial stage, and with the lapse of time, the remaining vacant surface sites are difficult to be occupied as a result of repulsive forces between the dye molecules on the solid and bulk phases [15]. Besides, at the beginning, the dye ions are adsorbed onto the exterior surface of the bioadsorbent till reaching saturation at a fast rate, and after that, the dye ions enters the pores of the bioadsorbent and are adsorbed by the interior surface of the particles, which carries on at a relatively slow rate[16,17].

3.2. Effect of Adsorbent Dosage

To determine the effect of adsorbent dosage on the sorption of MB, the bioadsorbent concentration was set from 2.5 to 10.0 g/L. As shown in **Figure 2**, the percentage of MB uptake increases with increasing adsorbent dosage at the same initial MB concentration. As observed from **Figure 2(a)**, the removal rate increases from 71.5% (2.5 g/L) to 90.2%(10.0 g/L) due to the increased available surface area and more sorption sites. Though the range of particle size is constant, the surface area will enlarge along with the increase of the adsorbent dosage in the solution. However, as seen from **Figure 2(b)**, the more the adsorbent dosage, the lesser the adsorption capacity (from 42.09 mg/g to 11.87 mg/g), which could be explained with the reduction of the effective surface area.

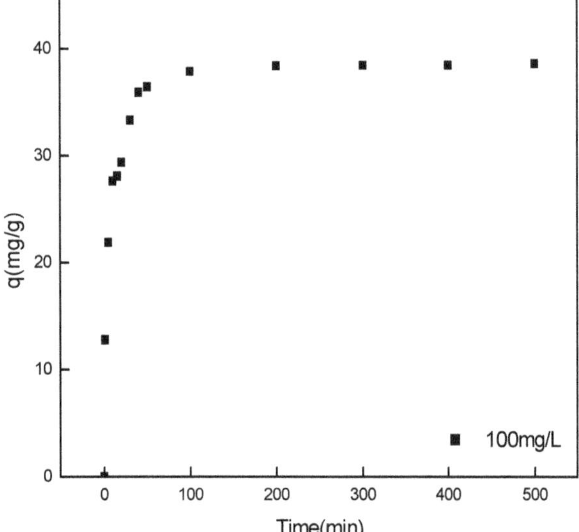

Figure 1. Effect of contact time on the sorption of MB onto dead leaves.

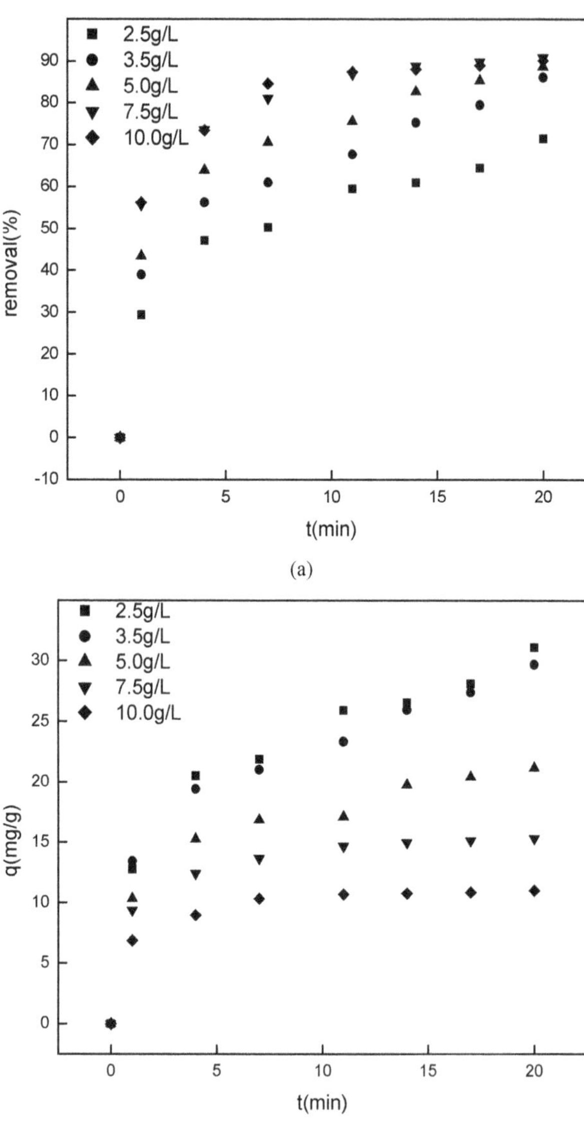

Figure 2. Effect of adsorbent dosage on the sorption of MB onto dead leaves.

3.3. Effect of Initial MB Concentration

The effect of different initial MB concentration ranging from 100 to 500 mg/L at 298.15K on dye adsorption is shown in **Figure 3**, which is shown that the initial MB concentration influences the contact time necessary to reach equilibrium: the shorter the contact time, the lower the initial MB concentration. 100 minutes are sufficient to reach the adsorption equilibrium in the range of 100 to 300 mg/L, whereas the equilibrium time for 500 mg/L increases to 400 min. It is reported that the adsorption spectrum shape of MB is dependent on its concentration. Higher concentration causes a decrease of the monomer band (Here MB has a long-wavelength maximum-665 nm) and an increase of the more energetic band (Here MB has a short wavelength shoulder-608 nm). In lower concentration, of about 1×10^{-5} mol/L, MB doesn't display obvious dimerization or aggregation. So the longer equilibrium time for the higher initial concentration can be accounted for the obvious dimerization or aggregation because of the lower intraparticle diffusion rate of aggregate forms[4]. It also can be seen that the saturated extent of adsorption of the adsorbent increases for the higher initial MB concentration (44.44 to 203.28 mg/g) as a result of stronger driving force. And that, the saturated extent of adsorption of un-modified adsorbent is only 145.62 mg/g in the same condition. This result is also obtained by other authors for the adsorption of dyes on various adsorbents [1,18].

3.4. Effect of Temperature

Figure 4 illustrates the effect of temperature on MB adsorption for different initial MB concentrations. It is shown that higher temperature can help the adsorbent acquire higher adsorption capacity for the same concentration, especially for higher initial concentration. This reveals that the adsorption of MB onto the bioadsorbent is an endothermic process. The temperature has several effects on the adsorption process: i)High temperature increases the rate of diffusion of MB molecules across the external boundary layer and into the internal pores of the adsorbent particles; ii)Rising temperature may increase the deaggregate tendency and the uptake of monomers of MB[4]; iii)High temperature could enlarge the pore size or create some new active sites on the adsorbent surface due to rupture of bond[19].

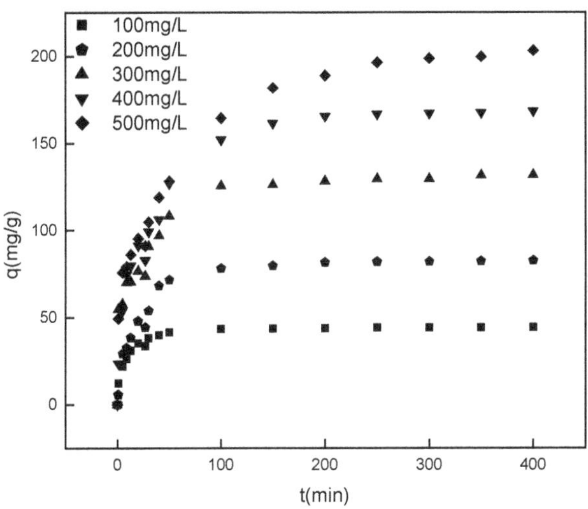

Figure 3. Effect of initial MB concentration on the sorption of MB onto dead leaves.

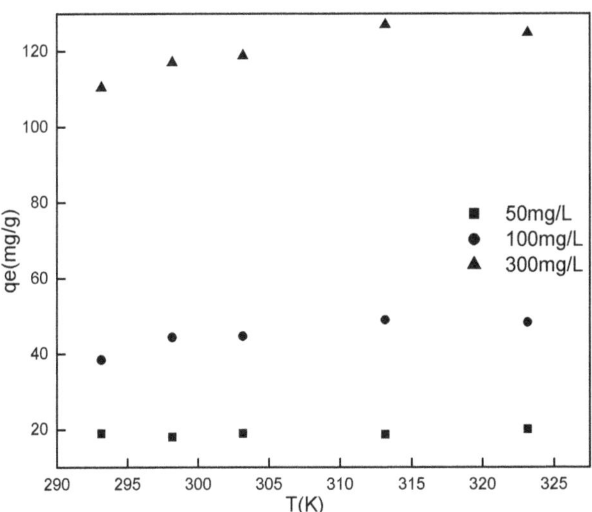

Figure 4. Effect of temperature on the sorption of MB onto dead leaves.

3.5. Effect of pH

To determine the effect of initial pH on MB adsorption, the pH of MB solution is varied from 3 to 13. As seen from **Figure 5**, the adsorption capacity of bioadsorbent increases from 12.8 mg/g to 25.8 mg/g in 1min at pH 13, which is obviously higher than other pH values. MB belongs to one of cationic dyes, which exists in aqueous solution in form of positively charged ions. Therefore, higher pH value leads to more electrostatic forces of attraction. In addition, at lower pH, H^+ could fight for the vacant adsorption sites with the positively charged MB cations, which also causes lower adsorption capacity.

However, the adsorption capacity of dead leaves at pH 7 comes to be close to that pH 13 after 30 minutes, so pH 7 is fine for the adsorption in the practical wastewater treatment.

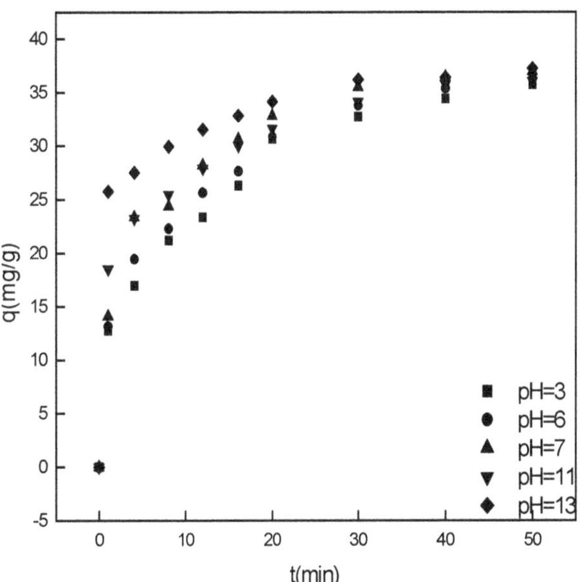

Figure 5 Effect of pH on the sorption of MB onto dead leaves.

3.6. Sorption Kinetic Models

The sorption kinetics could provide valuable insights into the reaction pathways and the mechanism of a sorption reaction, so they are worth paying attention to. The constants and correlation coefficient of three kinetics modesls are summarized in **Table 1**. The value of R^2 of pseudo-second order model is proved to be best among the three models. In addition, its calculated $q_{e(calc.)}$ is very close to the experimental $q_{e(exp.)}$. In the view of the result, it indicates that the sorption of MB onto adsorbent is more appropriate for the pseudo-second order model in contrast to the pseudo-first order model and intrapaticle diffusion model, suggesting that the adsorption process is controlled by chemisorption [20,21]. A similar result is obtained for the adsorption of MB from aqueous solution onto spent tea leaves [22].

From **Table 1**, the value of C increases with higher initial MB concentration, leading to a greater boundary layer effect. when $C = 0$, the intraparticle diffusion is thought to be the rate limiting step; If $C > 0$, both intraparticle diffusion and external mass transfer are considered as rate limiting steps[23]. Therefore, the adsorption of MB onto dead leaves is controlled by both external mass transfer and intraparticle diffusion.

3.7. Equilibrium Isotherms

Adsorption isotherms are necessary to design adsorption systems. Here Langmuir and Freundlich equations are used to model the adsorption data. To determine the equilibrium isotherms for each equation, initial MB concentrations are designed from 50 to 800 mg/L while the adsorbent concentration was kept constant (2.5 g/L) at 293.15K. The isotherm constants and correlation coefficients are shown in **Table 2**. The Langmuir isotherm is represented by equation (7). The Freundlich isotherm is represented by equation (9). As seen from **Table 2**, the R^2 of Langmuir model is better than the Freundlich model's, illustrating that the adsorption equilibrium can be best represented by the Langmuir isotherm model, with maximum monolayer adsorption capacity of 243.902 mg/g onto the adsorbent at 293.15K.

The characteristics of Langmuir isotherm can be expressed in terms of a dimensionless constant separation factor R_L that is given by equation(10) [24]:

$$R_L = \frac{1}{1+bC_0} \tag{10}$$

where C_0 (mg/L) is the highest initial concentration of adsorbate and b is Langmuir constant. The parameter R_L indicates the nature of the shape of the isotherm accordingly: unfavorable adsorption of $R_L > 1$, favorable adsorption of $0 < R_L < 1$, irreversible adsorption of $R_L = 0$, linear adsorption of $R_L = 1$.

The value of R_L in this study found to be 0.025 shows that the adsorption of MB onto adsorbent is favorable. And in the Freundlich model, $1/n$ is the adsorption intensity. It also gives an indication of the favorability of adsorption. Value of $n > 1$ means favorable adsorption conditions [25].

4. Conclusions

The results obtained show that the dead leaves of plane trees can be used for the removal of MB from aqueous solution. Kinetic data were tested using the pseudo-first, pseudo-second and intrapaticle diffusion kinetic models. The kinetics of the adsorption process was found to follow the pseudo-second order kinetic model, suggesting that the adsorption process was controlled by chemisorption. The plots of intraparticle diffusion didn't pass the origin, implying that there were other processes affected the adsorption except intraparticle diffusion. The equilibrium data were fitted to models of Langmuir and Freundlich,

Table 1. The constants and R^2 of three kinetic models for the adsorption of mb onto desd leaves.

MB concentration (mg/L)	$q_{e(exp.)}$	Pseudo-first order model			Pseudo-second order model			Intrapaticle diffusion model		
		$q_{e(calc.)}$	k_1	R^2	$q_{e(calc.)}$	k_2	R^2	C	k_i	R^2
100	44.38	27.38	4.72	0.9889	45.44	0.0059	0.9953	10.83	4.69	0.9664
200	82.61	72.86	3.77	0.9886	83.26	0.0015	0.9934	11.11	8.54	0.9744
300	131.77	82.20	2.33	0.9880	132.83	0.0006	0.9963	19.70	10.02	0.9695
400	168.35	126.01	2.11	0.9846	165.56	0.0003	0.9999	41.26	11.63	0.9617
500	203.28	142.45	1.31	0.9836	199.45	0.0002	0.9994	42.50	14.63	0.9717

q_e=mg/g; k_1=×10^{-2} min^{-1}; k_2=g mg^{-1} min^{-1}; k_i=mg · g^{-1} $min^{-0.5}$

Table 2. Langmuir and freundlich isotherm constants and correlation coefficients.

Isotherm	Langmuir			Freundlich		
Parameters	q_m (mg/g)	b (L/mg)	R^2	K_F	n	R^2
Values	243.902	0.046	0.996	16.592	1.677	0.963

and the adsorption equilibrium can be best represented by the Langmuir isotherm model, with maximum monolayer adsorption capacity of 243.902 mg/g of the adsorbent at 293.15K. The dead leaves of plane trees used to be adsorbent in this study are freely and abundantly available, and have high adsorption capacity for MB. Therefore, the prospect of dead leaves of plane trees used as adsorbent for removal of MB dye from aqueous solution is very promising.

REFERENCES

[1] D. Özer, G. Dursun and A. Özer, "Methylene Blue Adsorption from Aqueous Solution by Dehydrated Peanut Hull," *Journal of Hazardous Materials*, Vol. 144, No. 1-2, 2007, pp. 171-179.

[2] R. Perrich, "Activated Carbon Adsorption for Wastewater Treatment," CRC Press, Boca Raton, FL, 1981.

[3] K. Rastogi, J. N. Sahu, B. C. Meikap and M. N. Biswas, "Removal of Methylene Blue from Wastewater Using Fly Ash as an Adsorbent by Hydrocyclone," *Journal of Hazardous Materials*, Vol. 158, No. 2, 2008, pp. 531-540.

[4] S. Karaca, A. Gürse, M. Açıkyıldız and M. Ejder (Korucu), "Adsorption of Cationic Dye from Aqueous Solutions by Activated Carbon," *Microporous and Mesoporous Materials*, Vol. 115, No. 3, 2008, pp. 376-382.

[5] G. Annadurai, R. S. Juang and D. J. Lee, "Use of Cellu-Lose-Based Wastes for Adsorption of Dyes from Aqueous Solutions," *Journal of Hazardous Materials*, B, Vol. 92, No. 3, 2002, pp. 263-274.

[6] K. Mohanty, J. T. Naidu, B. C. Meikap and M. N. Biswas, "Removal of Crystal Violet from Wastewater by Activated Carbons Prepared from Rice Husk," *Industrial & Engineering Chemistry Research,* Vol. 45, No. 14, 2006, pp. 5165-5171.

[7] Md. Tamez Uddin, Md. Akhtarul Islam, Shaheen Mahmud and Md. Rukanuzzaman, "Adsorptive Removal of Methylene Blue by Tea Waste," *Journal of Hazardous Materials*, Vol. 164, No. 1, 2009, pp. 53-60.

[8] A. Özer and G. Dursun, "Removal of Methylene Blue from Aqueous Solution by Dehydrated Wheat Bran Carbon," *Journal of Hazardous Materials,* Vol. 146, No. 1-2, 2007, pp. 262-269.

[9] V. K. Gupta, D. Mohan, S. Sharma and M. Sharma, "Removal of Basic Dyes (Rhodamine-B and Methylene Blue) from Aqueous Solutions Using Bagasse Fly Ash," *Separation Science and Technology*, Vol. 35, No. 13, 2000, pp. 2097-2113.

[10] P. Janoš, S. Coskun, V. Pilařová and J. Rej-nek, "Removal of Basic (Methylene Blue) and Acid (Egacid Orange) Dyes from Waters by Sorption on Chemically Treated Wood Shavings," *Bioresource Technology*, Vol. 100, No. 3, 2009, pp. 1450-1453.

[11] K. Kannan and M. M. Sundaram, "Kinetics and Mechanism of Removal of Methylene Blue by Adsorption on Various Carbons—A Comparative Study," *Dyes Pigments*, Vol. 51, No. 1, 2001, pp. 25-40.

[12] I. D. Mall, V. C. Srivastqava and N. K. Agarwal, "Removal of Orange-G and Methyl Violet Dyes by Adsorption Onto Bagasse Fly Ash-Kinetic Study and Equilibrium Isotherm Analyses," *Dyes Pigments*, Vol. 69, No. 3, 2006, pp. 210-223.

[13] I. Langmuir, "The Adsorption of Gases on Plane Surfaces of Glass, Mica and Platinum," *Journal of the American Chemical Society*, Vol. 40, No. 9, 1918, pp. 1316-1403.

[14] H. M. F. Freundlich, "Über Die Adsorption in Lösungen (Adsorption in solution)," *Zeitschrift Fur Physikalische Chemie*, Vol. 57A, 1906, pp. 212-223..

[15] O. Hamdaoui, F. Saoudi, M. Chiha and E. Naffrechoux, "Sorption of Malachite Green by a Novel Sorbent, Dead Leaves of Plane Tree: Equilibrium and Kinetic Modeling," *Chemical Engineering Journal*, Vol. 143, No. 1-3, 2008, pp. 73-84.

[16] P. K. Malik, "Use of Activated Carbons Prepared from Sawdust and Rice-Husk for Adsorption of Acid Dyes: A Case Study of Acid Yellow36," *Dyes Pigments*, Vol. 56, No. 3, 2003, pp. 239-249.

[17] I. D. Mall, V. C. Srivastava, N. K. Agarwal and I. M. Mishra, "Removal of Congo Red from Aqueous Solution by Bagasse Fly Ash and Activated Carbon: Kinetic Study and Equilibrium Isotherm Analyses," *Chemosphere*, Vol. 61, No. 4, 2005, pp. 492-501.

[18] N. K. Amin, "Removal of Direct Blue-106 Dye from Aqueous Solution Using New Activated Carbons Developed from Pomegranate Peel: Adsorption Equilibrium and Kinetics," *Journal of Hazardous Materials,* Vol. 165, No. 1-3, 2009, pp. 52-62.

[19] F. Wu, R. Tseng and R. Juang, "Pore Structure and Adsorption Performance of the Activated Carbons Prepared from Plum Kernels," *Journal of Hazardous Materials*, B, Vol. 69, No. 3, 1999, pp. 287-302.

[20] Y. Nuhoglu and E. Oguz, "Removal of Copper(II) from Aqueous Solutions by Biosorption on the Cone Biomass of *Thuja orientalis*," *Process Biochemistry,* Vol. 38, No.

11, 2003, pp. 1627-1638.

[21] I. D. Mall and S. N. Upadhyay, "Treatment of Methyl Violet Bearing Wastewater from Paper Mill Effluent Using Low Cost Adsorbents," *Journal of Indian Pulp Paper Technology Association,* Vol. 7, No. 1, 1995, pp. 51-57.

[22] B. H. Hameed, "Spent Tea Leaves: A New Non-Conventional and Low-Cost Adsorbent for Removal of Basic Dye from Aqueous Solutions," *Journal of Hazardous Materials*, Vol. 161, No. 2-3, 2009, pp. 753-759.

[23] A. R. and P. P., "Batch and Column Studies of Biosorption of Heavy Metals by Caulerpa Lentillifera," *Bioresource Technology*, Vol. 99, No. 8, 2008, pp. 2766-2777.

[24] K. R. Hall, L. C. Eagleton, A. Acrivos and T. Vermeulen, "Pore-and Solid-Diffusion Kinetics in Fixed-Bed Adsorption under Constant-Pattern Conditions," *IEC Fundam*, Vol. 5, No. 2, 1966, pp. 212-223.

[25] Y. S. HO and G. McKay, "Sorption of Dye from Aqueous Solution by Peat," *Chemical Engineering Journal*, Vol. 70, No. 2, 1998, pp. 115-124.

10

Production of Hydrogen by Electrolysis of Water: Effects of the Electrolyte Type on the Electrolysis Performances

Romdhane Ben Slama
Higher Institute of Applied Sciences and Technology, University of Gabes, Gabes, Tunisia

ABSTRACT

The production of hydrogen, vector of energy, by electrolysis way and by using photovoltaic solar energy can be optimized by suitable choice of electrolytes. Distilled water, usually used, due to membrane presence may be substituted by wastewaters, which enters more in their treatment. Waste water such as those of the Cleansing National Office, and also of the factories such as those referring with ammonia, the margines, and even urines that make it possible to produce much more hydrogen as distilled or salted water, more especially as they do not even require an additive or membranes: conventional electrolysers with two electrodes. This study seeks to optimize the choice among waste water and this, by electrolysis in laboratory or over the sun according to produced hydrogen flow criteria, electrolysis efficiency and electric power consumption. The additive used is NaCl. The most significant results are on the one hand the significant increase in the produced hydrogen flow by the addition of the additive; on the other hand the advantage of gas liquor and urine compared to the others tested electrolytes.

Keywords: Hydrogen Production; Electrolysis; Electrolyte; Photovoltaic

1. Introduction

Water electrolysis has long been known to produce hydrogen. However, for membrane electrolysers, used water must be pure.

We show here that wastewater electrolysis gives the same performance or even better performance, because they contain bacteria that produce hydrogen.

Among wastewater include those of ONAS (municipal wastewater), the margine, ammonia water of ammonia production plants, but water with vinegar and urine deemed rich in nitrogenous matter (ammonia) according to recent research from the University of Ohio in the United States.

Hydrogen production by water electrolysis can be economically viable by using electrical energy from renewable sources such as photovoltaic solar energy [1-4].

Our previous studies have relied on the use of salt water as electrolyte [5-8]. In this article, we will vary the nature of the electrolyte leaning towards wastewater deemed by their richness in bacteria which are the basis for hydrogen production. Human urine will also be also used in reference to the work of Boot [9,10].

2. Experimental Set-Up

2.1. Parameters of Calculation

- Hydrogen production flow rate: $Q_v = V/t$
  With:
- Absorptive power by the electrolyser: Pa = U.I
- Useful power of the electrolyser: Pu = PCI.Q.r
  With PCI: lower thermal value of hydrogen (119.910^6 J/Kg)
  ρ: density of hydrogen (0.09 Kg/m^3)
- Consumed electric power: W = Pa.t (J)
- Useful efficiency: h = PCI. (V/(Pa.t)).r
- Consumed electric power per unit of volume: W/V = Pa.t/V (J/cm^3)

A photovoltaic module with its panel, electrolysers and their electrodes made in various materials, are represented by the following photographs (**Figure 1**).

2.2. Chemical Characteristics of Waste Water Used

Table 1 represents the characteristics of the waste water used, such as the pH, the resistivity and initial salinity,

Figure 1. Photographs of the photovoltaic model and the electrolysers with their corroded electrodes.

Table 1. Characteristics of the liquids before and after the addition of NaCl.

Liquids	Liquids in initial state characteristics (before the addition of NaCl)			Salted liquids characteristics (after the addition of NaCl)	
	pH	Resistivity (m·S)	Initial salinity (g/l)	pH	Conductivity ($S \cdot m^{-1}$)
Water of tap (Gabès, Tunisia)	6.97	4.67	2.5	6.9	-
Olive pucker	4.56	45.13	8.7	-	-
Margine	4.97	46.3	9.5	4.91	17.4
Urine	5.61	49.6	33.5	5.03	4.7
Vinegar of pink	5.33	4.8	2.6	4.62	3
Waste water	6.98	8.31	4.6	7.32	5.2
Gas liquor (NH_4OH)	7.9	120.3	-	7.37	7.9
Water of kitchen	4.18	8.5	4.7	8.34	10.3
Milk water	4.01	6.25	3.4	3.78	6.2

before and after the sodium chloride addition.

We can notice that the sodium chloride addition has a weak effect on the pH, however the resistivity is strongly reduced.

3. Influence of NaCl Additive on the Electrolysis of Various Waste Water

NaCl is used as additive for handling reasons facility, abundance on the market and low cost.

3.1. First Parameter: Released Hydrogen Flow

These experiments are carried out under a value of 6 V and an ambient temperature of 21°C.

In the initial state of each liquid (waste water) used, we can notice that the gas liquor and the urine are distinguished because they have release hydrogen flows higher than the other liquids. With the addition of the NaCl additive, all salted liquids will have their hydrogen flow increasing in a significant way. Urine and gas liquor remain among the best as well with as without additive salt.

3.2. Second Parameter: Electric Power Consumed

The electric power is mainly consumed to produce hydrogen. Thus these two parameters are almost proportional according to **Figures 2** and **3**. However the variation remains enormous between the two configurations with and without additive.

This variation will be reduced in **Figures 4** and **5** concerning, respectively, the energetic efficiency and the specific energy consumed for hydrogen production.

The NaCl addition increases the electrolyte conductivity and consequently the current passage and thus the consumed electric power and the produced hydrogen flow and this, whatever the type of electrolyte.

3.3. Third Parameter: Electrolyser Efficiency

The calculated efficiency is proportional to the quotient flow/electric power (**Figure 4**).

We can only notice a small difference on the results indicated in this **Figure 3**. In fact, it is noted that the efficiency of the liquids without sodium chloride is larger than that of the liquids with NaCl. This will be explained by the low electric power dissipated for the liquids without NaCl that for the liquids salted even those which have significant flows.

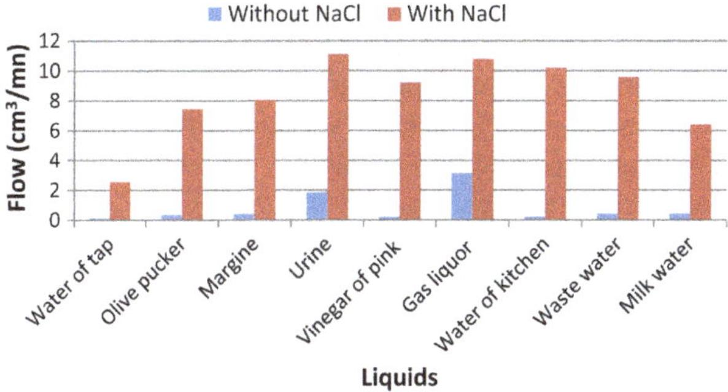

Figure 2. Salinity effect on hydrogen flow produced by electrolyse waste water.

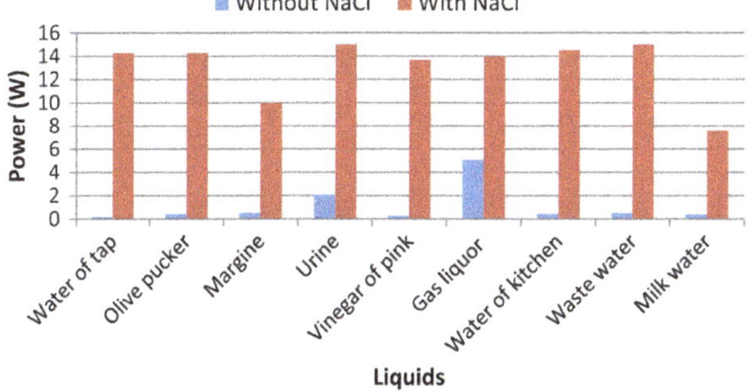

Figure 3. Salinity effect on the electric power consumed to produce hydrogen by waste water electrolysis.

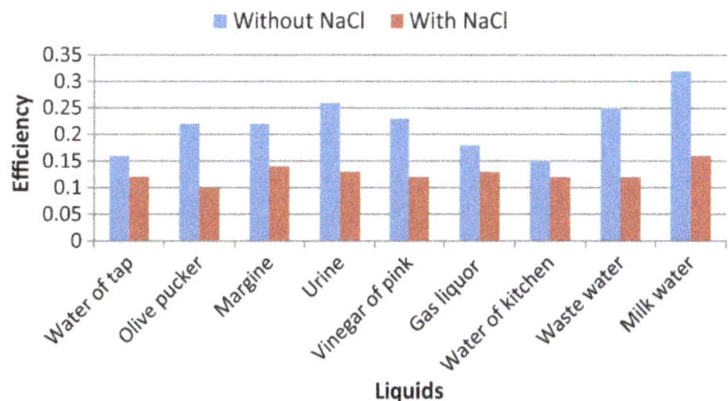

Figure 4. Salinity effect on the produced hydrogen efficiency by waste water electrolyse.

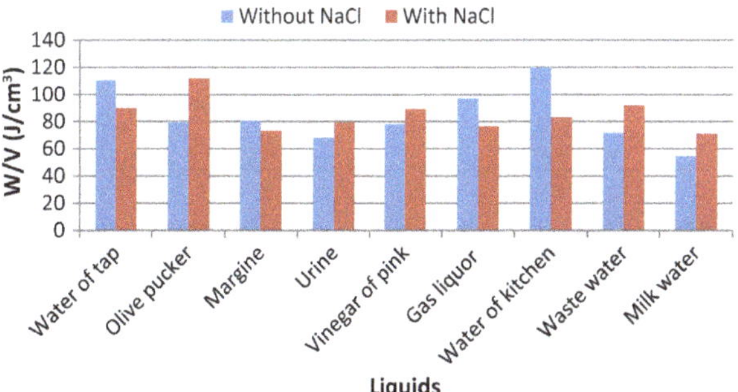

Figure 5. Salinity effect on the consumed specific electric power to produce hydrogen by waste water electrolysis.

3.4. Fourth Parameter: Consumed Specific Electric Power

By this measured parameter of consumed specific electric power, we wanted to alleviate the great variations which exist between the two configurations with and without additive, observed in **Figures 2** and **3**.

If it is true that the addition of NaCl does not have spectacular effect for all the electrolytes as in the case of flow and power consumption, we can however note that the addition of NaCl generates a decrease in the specific energy consumed in the case of the tap water, margine, gas liquor, cooking waste water and an increase in this energy to let us pucker olive, urinates, vinegar of pink, cleansing water and milk water.

4. Optimizing the Choice between Electrolytes, with 200 g/l NaCl

By generating hydrogen by solar water electrolysis, we seeks to maximize the produced hydrogen flow rate Q_v and to minimize the specific energy consumption while taking account of the energetic efficiency, so the ratio:

$$r = \frac{Q_v \cdot \eta}{\frac{W}{V} \cdot P}$$

Table 2. Electrolysis performance comparison.

Electrolyte	Q_v (cm^3/mn)	W/V (J/cm^3)	P (W)	Efficiency	$r = \frac{Q_v \cdot \eta}{\frac{W}{V} \cdot P}$	Rang
Tap water	9.8	95	14.2	0.115	0.000835	7
Olive pucker	7.5	118	14.2	0.1	0.0004476	9
Margine	8.3	76	10.1	0.14	0.00151	2
Urine	11.5	82.5	15	0.125	0.00116	4
Vinegar of pink	11.2	95	14	0.12	0.00101	6
Gas liquor	11.2	80	14.2	0.127	0.00125	3
Water of cooking	10.5	85	14.5	0.12	0.001022	5
Waste water	10	98	15	0.12	0.000816	8
Milk of water	6.7	75	7.9	0.161	0.001809	1

Hence the electrolysis performance comparison chart for different electrolyte containing NaCl at 200 g/l (**Table 2**).

Tested electrolytes can be classified into five families, in descending order of interest:

- First Family: milk of water, margine
- Second family: gas liquor, urine, water of cooking, vinegar of pink
- Third family: tap water, waste water
- Fourth family: olive pucker

Most hydrogen producers electrolytes, while minimizing energy consumption, are based on ammonia.

Their abundance varies: Ammonia is industrially produced, vinegar of pink can be synthesized. As for urine, and feed its human or animal production not requires no worries about its abundance and renewable character, throughout the human life.

5. Interpretation of Results

Various measurements give sometimes conclusions that can appear contradictory as for the interest of the NaCl additive. However, our purpose is to reach the best production flows with less power consumption; it is with the manner of the vehicles that must drive fasr without consuming too much!

Thus while using NaCl like additive, rank first the ammonia water and urine, to the second rank waste water of cooking and from the cleansing water and the margine, finally the water out of the tap and the others.

If theoretical energy to break the water molecule is of 285 kJ/mol or 12.7 J/cm^3, our found values are thus high and electrolysis still requires to be optimized for example by reducing the supply voltage by the parallel assembly of the electrolysers on the photovoltaic module.

The tests results over the sun show that the electric power consumed by the electrolyser follows that of the daily sunning; it is the reverse for the efficiency. We observed the same paces of efficiency for the case of the solar water heaters.

6. Conclusions

The addition of NaCl in the electrolytes had activated the electrochemical reactions and produced more hydrogen and this is the goal required in this work.

According to the type of used electrolyte (tap water, margine, gas liquor, waste water from cooking, puckered olive, urine, vinegar of pink, municipal waste water and finally milk water), there is variation of the hydrogen flow rate produced by supplying the electrolysers in electrical current by the photovoltaic module.

As the energetic efficiency does not change often in the same direction as the produced hydrogen flow, it is for that we use also the specific energy consumption parameter "W/V" which can be considered as being the most characteristic parameter in the comparisons between the various types of electrolytes.

REFERENCES

[1] E. Bilgen, "Solar Hydrogen from Photovoltaic Eletrolizer Systems," *Energy Conversion and Management*, Vol. 42, No. 9, 2001, pp. 1047-1057.

[2] S. H. Jensen, *et al.*, "Hydrogen and Synthetic Fuel Production from Renewable Energy Sources," *International Journal of Hydrogen Energy*, Vol. 32, No. 15, 2007, pp. 3253-3257.

[3] P.-H. Floch, *et al.*, "On the Production of Hydrogen via Alkaline Electrolysis during Off-Peak Periods," *International Journal of Hydrogen Energy*, Vol. 32, No. 18, 2007, pp. 4641-4647.

[4] L. Solera, J. Macanása, M. Muñoza and J. Casado, "Electrocatalytic Production of Hydrogen Boosted by Organic Pollutants and Visible Light," *International Journal of Hydrogen Energy*, Vol. 31, No. 1, 2006, pp. 129-139.

[5] R. Ben Slama, "Production of Hydrogen by Electrolyse of Water and Photovoltaic Energy," *Proceeding of the 3rd International Congress on Renewable Energies and Environment CERE*, Tunis, 6-8 November 2006.

[6] R. Ben Slama, "Tests on the Solar Hydrogen Production by Water Electrolysis," *Proceeding of the JITH*, Albi, 28-30 August 2007.

[7] R. Ben Slama, "Solar Hydrogen Generation by Water Electrolysis," *Proceeding of the First Francophone Conference on Hydrogen: Energy Vector*, Sousse, 9-11 May 2008.

[8] R. Ben Slama, "Génération d'Hydrogène par Electrolyse Solaire de l'eau," *Proceeding des Journées Annuelles* 2008 *Société Française de Métallurgie et de Matériaux*, Paris, 4-6 June 2008.

[9] M. Cooper and G. G. Botte, "Hydrogen Production from the Electro-Oxidation of Ammonia Catalyzed by Platinum and Rhodium on Raney Nickel Substrate," *Journal of the Electrochemical Society*, Vol. 153, No. 10, pp. A1894-A1901.

[10] F. Vitse, M. Cooper and G. G. Botte, "On the Use of Ammonia Electrolysis for Hydrogen Production," *Journal of Power Sources*, Vol. 142, No. 1-2, 2005, pp. 18-26.

Nomenclature

I	Electrical current (A)
U	Voltage (V)
P	Power (W)
PCI	Lower heating value (J/kg)
Q	Flow rate (m^3/s)
V	Volume of the test tube (m^3)
t	Tube filling time (s)
W	Electrical energy (J)
r	Density of hydrogen (Kg/m^3)

Indices

a	Absorbed
u	Useful

11

Refrigerator Coupling to a Water-Heater and Heating Floor to Save Energy and to Reduce Carbon Emissions

Romdhane Ben Slama
ISSAT Gabes Rue Omar Ibn Khattab, Gabes, Tunisia

ABSTRACT

With an aim of rationing use of energy, energy safety, and to reduce carbon emission, our interest was geared towards the refrigerators and all the refrigerating machines. Indeed the heat yielded by the exchanger condenser can be developed for the water heating, floors heating etc. After an encouraging theoretical study, two prototypes were produced in order to validate the theoretical results. A first refrigerator was coupled with a water-heater and another with a heating floor. The water temperature reached, in one day, is of 60˚C; which makes it possible to predict better results with a continuously used refrigerator. In the same way for the heating floor coupled with the second refrigerator, the temperature reached high values because the surface is reduced; however for the heating floors the standard fixes the temperature between 28˚C and 30˚C.

Keywords: Refrigerator; Recuperator; Heating Water; Heating Floor; Heat Pump

1. Introduction

Energy consumption in the world is significant and in continuous growth. It is fundamentally linked to the life level. It promotes the thermal comfort of the citizen, through the heating and cooling systems (air-conditioners, water-heaters, refrigerators…). With the aim of saving energy and to reduce carbon emission, it is necessary to find some innovating solutions in this field and to create negawatts.

For this reason, we propose to recover the waste heat on the refrigerator condenser level, and this by two methods:

- The first method is to enhance the heat lost during condensation of Freon, and this, recovering it in a water cumulus to heat it and to use in the domestics or other needs.
- the second method consists in heating a floor to replace the traditional heating radiators.

Finally, the heat recovery system does not have any modification on the basic refrigeration cycle. The tests and experiments show the effectiveness of this heat recovery project [1-6].

2. Bibliographical Study

The publications evoking the use of heat pumps for water heating are relatively very few. Let us quote Grazini, Skrivan, Huang, Jie [7-10] which extracts heat from the ambient air like cold source of the heat pump. **Figure 1** is given as an example.

Others authors use solar energy like evaporator such as Hawlader, Li, Chyng, Borges, Huang [11-16].

Among the few publication jointly using the condenser and the evaporator of a heat pump (or refrigerator), we even quote our work concerning a solar sea water still developed by our self, at the National School Engineers of Gabes, Tunisia [17-23].

Figure 2 shows our system of solar desalination provided with the heat pump.

The condenser (6) heats water to be evaporated and the evaporator (2) makes it possible to condense instantaneously the vapour and thus to form the condensate collected in the gutter (3).

For the principle of refrigerator operation (**Figure 3**), this system can be considered as an insulated cupboard whose interior temperature is lower than the ambient temperature. To obtain cold inside the refrigerator, heat is extracted with the air and food and then is rejected by the condenser in the kitchen.

3. Theoretical Considerations of the Coupling of Refrigerator to a Water-Heater and a Heating Floor

We will begin by determining the intervals of real operating time of the refrigerator and the heat transfer between the condenser and water or floor to be heated.

3.1. Preliminary Study of Dimensioning

The tracing of the histogram of **Figure 4** allows to know

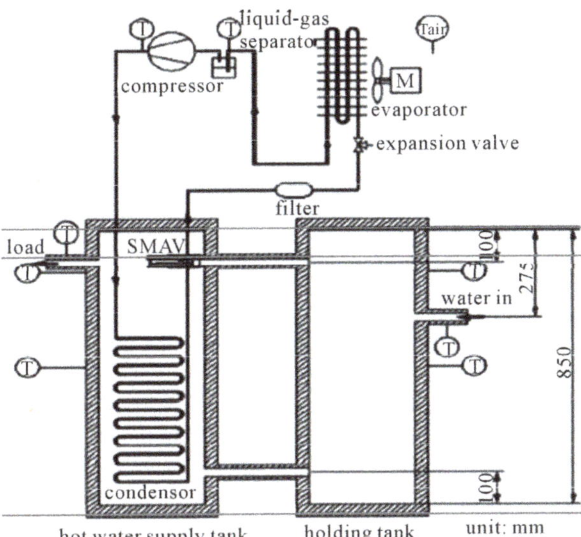

Figure 1. Water heating by heat pump.

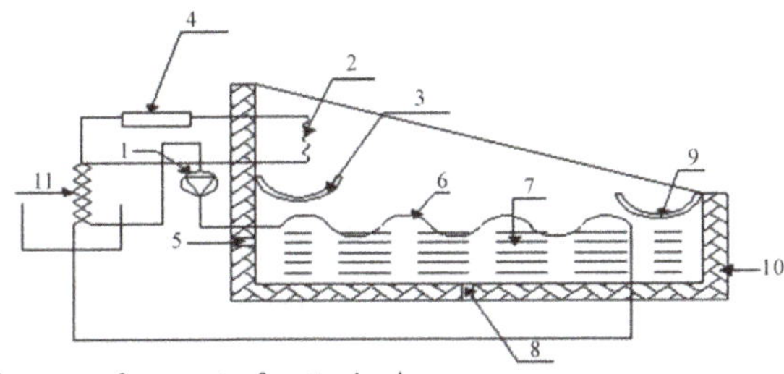

Figure 2. Solar distiller provided with heat pump.

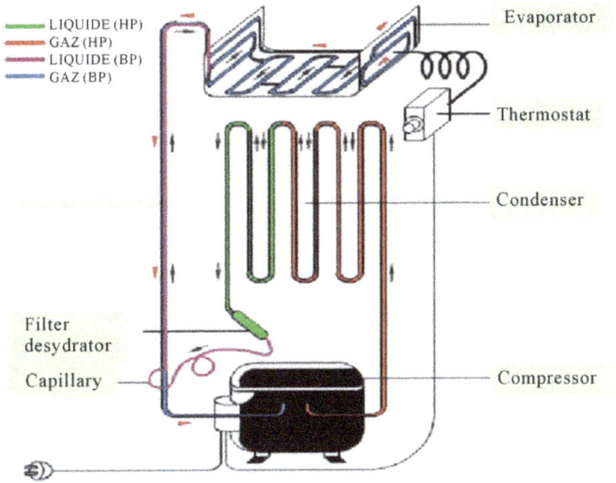

Figure 3. Refrigerator operation cycle [2].

the operating time of refrigerator' compressor during 24 hours. The operating time varies from one moment to another as required by the refrigerator.

Operation of a refrigerator during 24 hours.

$$\sum t_{compressor} = 9\text{ h }43\text{ min/day}$$

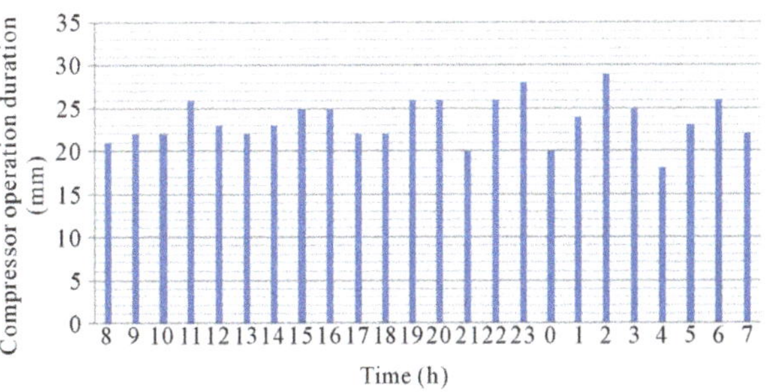

Figure 4. Time operation life of a refrigerator lasting 24 h. Compressor power of 140 W and using R13a freon. Room temperature = 19°C (Tozeur, Tunisia, on March 20, 2009).

3.2. Theoretical Study of a Refrigerator to a Water-Heater Coupling

The energy yielded by the condenser is transferred to water to be heated. However, when water heats in the storage tank, thermals loss will appear (coefficient U = 4.92 W/°C).

3.2.1. Modeling

The refrigerating machine can be modeled by considering energies on the level of the compressor, the condenser and the evaporator, as follows:

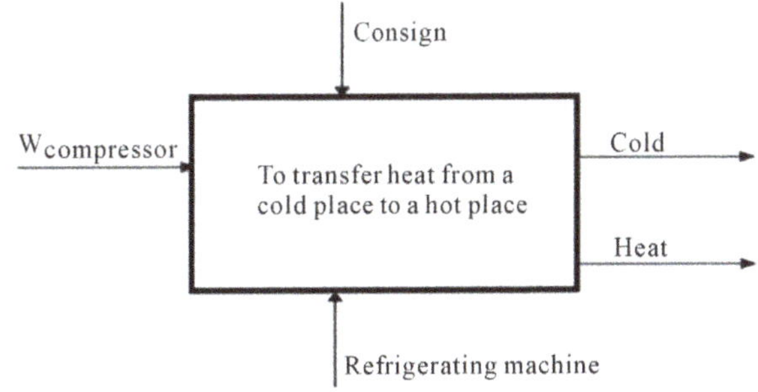

$$P_{comp} - Q_{condenser} + Q_{evaporator} = 0$$

$$P_{comp} - P_{comp} \cdot COP_{th} + P_{comp} \cdot COP_{fr} = 0$$

$$1 - COP_{th} + COP_{fr} = 0$$

$$COP_{total} = COP_{th} + COP_{fr}$$

The total COP can thus reach raised values, which confers on our system a great energetic effectiveness, because, usually the refrigerating machine is used either like refrigerator to produce the cold, or like heat pump to heat.

3.2.2. Thermal Transfer

For the coupling of a refrigerator to a water cumulus study, we make the calculation of the water mass which can be heated with different temperatures. The efficiency is then evaluated.

It is supposed that the heat yielded by the condenser is

recovered by water to heat.

$$Q_{condenser} = Q_{water}$$

$$COP \cdot P_{comp} \cdot t \cdot \eta = M_{water} \cdot Cp \cdot \Delta T \tag{1}$$

ΔT is the water temperature increase from initial to final state.

It is possible to determine the variation of ΔT according to time, in the same way for the efficiency.

Determination of heat water storage efficiency

$$\eta = Q_{useful} / Q_{absorbed} \tag{2}$$

With:

$$Q_{absorbed} = P_{comp} \cdot COP \cdot t$$
$$Q_{useful} = Q_{abs} - Q_{loss}$$
$$Q_{loss} = U \cdot \Delta T \cdot t$$

P_{comp}: Compressor power.

t: summon operating time of compressor.

U: thermal loss ratio.

$$U = \frac{\rho \cdot Cp \cdot V}{\Delta t} \cdot \ln\left(\frac{T_i - T_{am}}{T_f - T_{am}}\right) \tag{3}$$

Therefore we have:

$$\eta = \frac{P_{comp} \cdot COP \cdot t - U \cdot \Delta T \cdot t}{P_{comp} \cdot COP \cdot t} \tag{4}$$

Temperature variation

We replace (4) in (1), then:

$$\Delta T = \frac{P_{comp} \cdot COP \cdot t}{U \cdot t + M_{water} \cdot Cp} \tag{5}$$

Mass of water to heat

Replacing (5) in (4), so:

$$M_{water} = \frac{P_{comp} \cdot COP \cdot t}{\Delta T \cdot \left(\frac{Cp}{\Delta t} \cdot \ln\left(\frac{T_i - T_{amb}}{T_f - T_{amb}}\right) \cdot t + Cp\right)} \tag{6}$$

The calculation of the water mass to heat, during 24 hours, can be carried out for different water temperature rise.

According to expression (6) we give **Table 1**.

1) ΔT variation and efficiency according to time for water capacity of 50 liters which is the experimentally capacity used.

Thermal loss coefficient

With:

$$Cp_{water} = 4185 \text{ J}/(\text{kg}^\circ\text{C})$$

$$\rho_{water} = 1000 \text{ kg/m}^3$$

$$\Delta t = 11 \text{ hours}; 18:00 \rightarrow 5:00$$

$$V = 0.05 \text{ m}^3$$

$$T_i = 59^\circ C;\ T_{ami} = 26^\circ C$$

$$T_f = 35^\circ C;\ T_{amf} = 22^\circ C$$

$$U = 4.92 \text{ W}/^\circ C$$

According to the expressions (2) and (3) we calculated the temperature rise, the efficiency, the useful and absorptive energies, gathered in the following **Table 2** and **Figure 5**.

With:

$$P_{comp} = 140 \text{ W};\ COP = 3;\ M_{water} = 50 \text{ Kg}$$

With time, water warms up by the heat yielded by the condenser; then, the quantity of heat contained in water increases. However, the energetic efficiency decreases following the increase in the losses by convection with the ambient air (**Figure 5**).

2) Checking length of the condenser immersed in the hot water cumulus

In our case, the flow of the refrigerant (R134a freon) in the condenser immersed in the cumulus is done from top to bottom, *i.e.* in against current with the circulation of the water to be heated which is done upwards.

Table 1. Value of water mass for different rise in temperature.

ΔT (°C)	20	25	30	35	40
M (kg)	90	72	60	51	45

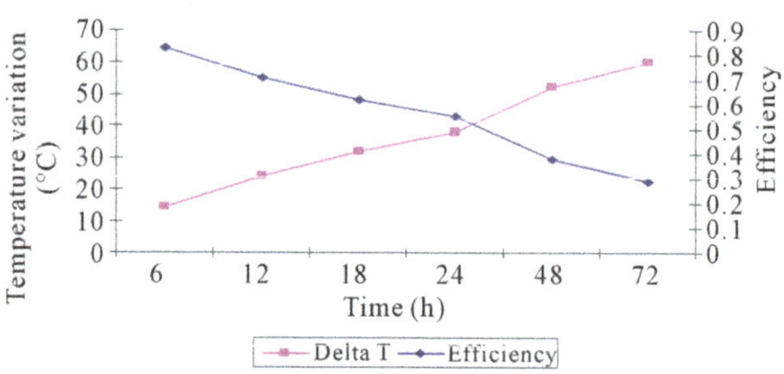

Figure 5. Rise variation in water temperature according to time.

Table 2. Variation of ΔT and efficiency according to time.

Time (hours)	ΔT (°C)	efficiency	Q_{useful} (J)	Q_{abs} (J)
6	14.20	0.83	2972654	3565800
12	24.35	0.71	509736	7131600
18	32	0.62	6691162	10697400
24	38	0.55	7930746	14263200
48	52.50	0.38	10986005	28526400
72	60.22	0.29	12604349	42789600

Therefore the difference in logarithmic average temperature curve will be carried out in an against current cycle (**Figure 6**).

According to **Table 2** of comparison during 12 hours and according to measurements:

- Hot source: (condenser)
  $T_e = 62^{\circ}C$ and $T_s = 43^{\circ}C$
- Cold source: (water to be heated); with ($\Delta T = 24.35^{\circ}C$)
  $t_e = 20^{\circ}C$ and $t_s = 44.35^{\circ}C$
- Determination of logarithmic average temperature difference ΔTmL

The expression of ΔTmL is:

$$\Delta TmL = \frac{\Delta T0 - \Delta TL}{\ln\left(\frac{\Delta T0}{\Delta TL}\right)} \quad (7)$$

$$\Delta T0 = Te - ts = 62 - 44.35 = 17.65^{\circ}C$$

$$\Delta TL = Ts - te = 43 - 20 = 23^{\circ}C$$

$$\Delta TmL = 20.16^{\circ}C$$

- Heat exchange coefficient:

The condenser exchanger contains freon which passes from vapor state to liquid state by yielding its heat to water to be heated (**Figure 7**).

$$Ui = \frac{1}{\frac{1}{h\ freon} + di \cdot \frac{\ln \ln\left(\frac{de}{di}\right)}{2 \cdot \lambda_{copper}} + \frac{1}{h_{water}}} \quad (8)$$

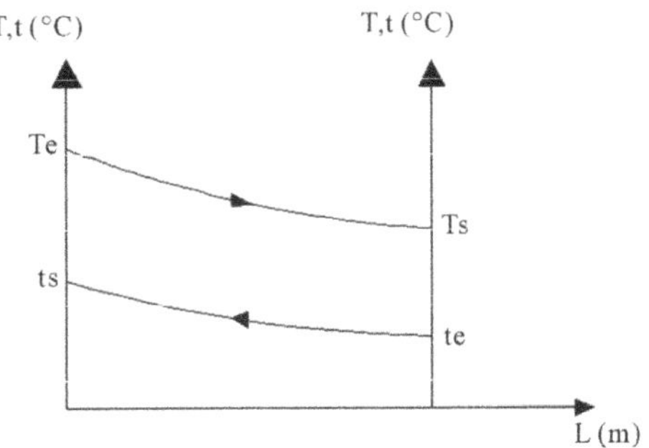

Figure 6. Cycle of against current.

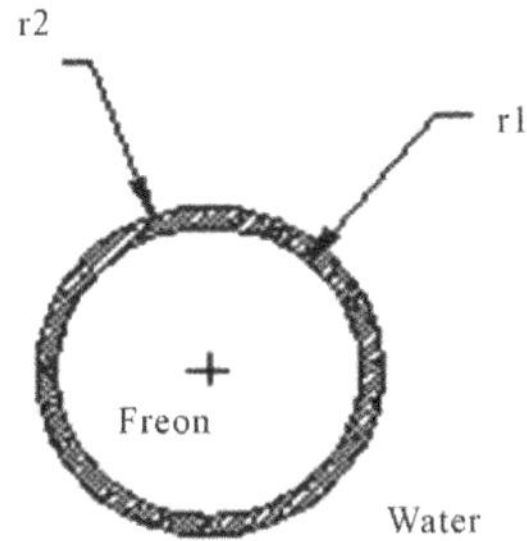

Figure 7. Heat exchanger (water condenser).

With:

$h\ water = 50\ W/(m^{2^{\circ}}C)$; $h\ freon = 4200\ W/(m^{2^{\circ}}C)$;

$\lambda\ copper = 390 W/(m^{\circ}C)$

$$di = 0.004\ m;\ de = 0.006\ m$$

Then:

$$Ui = 50\ W/(m^{\circ}C)$$

3.3. Theoretical Study of the Coupling of a Refrigerator to a Heating Floor

In this section we will make the study of a refrigerator to a heating floor coupling, and the determination of the temperature variation for each surface for different periods and for different surfaces to be heated.

Calculation is carried out on concrete surfaces height h = 0.1 m.

There is a thermal contact between two bodies thus there is a heat transfer between the two bodies in contact (condenser/concrete).

$$Q_{condenser} = Q_{concrete}$$
$$COP \cdot P_{comp} \cdot t \cdot \eta = M_{concrete} \cdot Cp \cdot \Delta T \quad (9)$$

ΔT: variation between the final and the initial tem- perature of the floor.

- Determination of the temperature variation ΔT

We have:

$$\eta = Q_{useful}/Q_{absorbed} = Q_{absorbed} - P_{loss}/Q_{absorbed}$$

With:

- $Q_{absorbed} = P_{comp} \cdot COP \cdot t$
- $P_{loss} = U \cdot S \cdot \Delta T \cdot t$
- $Q_{useful} = Q_{absorbed} - P_{loss}$

The initial temperature is equal to the ambient temperature.

Thus:

$$\eta = \frac{P_{comp} \cdot COP \cdot t - U \cdot S \cdot \Delta T \cdot t}{P_{comp} \cdot COP \cdot t} \quad (10)$$

We replace (10) in (11):

$$\Delta T = \frac{P_{comp} \cdot COP \cdot t}{U \cdot S \cdot t + M_{concrete} \cdot Cp} \quad (11)$$

With:

$$M_{concrete} = \rho \cdot S \cdot H$$

Finally:

$$\Delta T = \frac{P_{comp} \cdot COP \cdot t}{S \cdot (\rho \cdot H \cdot Cp + U \cdot t)} \quad (12)$$

Table 3 and **Figure 8** indicate the variations of ΔT according to time and the floor surface.

The reached temperature on the floor level is a function of the exchange surface with the condenser; in fact also with the compressor power.

In practice, a prototype of heating sand floor of surface S = 0.43 m^2 was produced and in which a condenser of 6 m length is immersed there.

4. Experimental Study

First, we make the water cumulus then its coupling with the refrigerator and the heating floor.

4.1. Water Cumulus Realization

We have a cylindrical tank in plastic, diameter 0.38 m and height 0.48 m, which is isolated by five cm glass woo. The unit is then covered by a thin stainless steel jacket (**Figure 9**).

4.2. Coupling between the Refrigerator and the Water Cumulus

We replaced the refrigerator condenser by a copper spiral serpentine placed in the water cumulus, thus the freon condensation will allow to yield the latent heat to water and thus to constitute a hot water storage (**Figures 10, 11**).

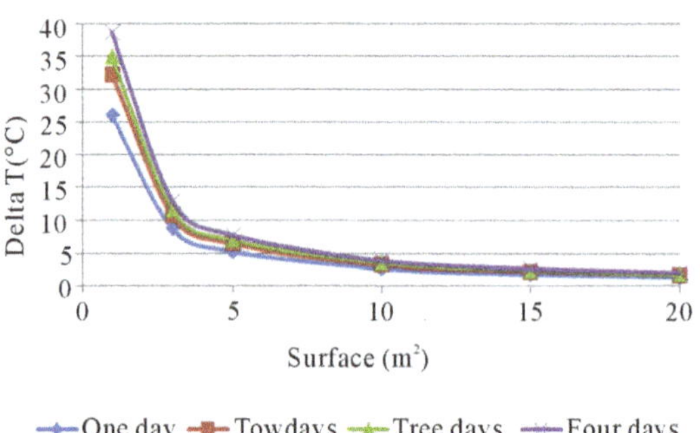

Figure 8. Floor temperature rise according to surface and a number of day.

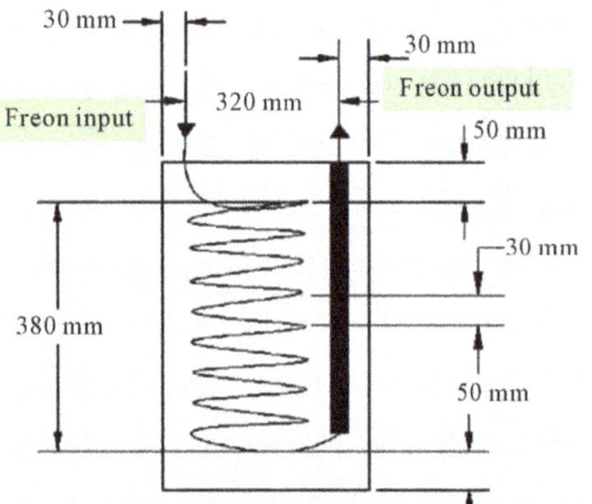

Figure 9. Dimensions of the copper condenser immersed in the water cumulus.

Table 3. Variation of ΔT according to time and of surface.

Surface (m^2)	ΔT (°C)			
	1 day	2 days	3 days	7 days
1	26	32	35	38
3	8.77	10	11	13
5	5.26	6	7	7
10	2.63	3.2	3.5	3.8
15	1.75	2.15	2.33	2.6
20	1.31	1.61	1.75	2

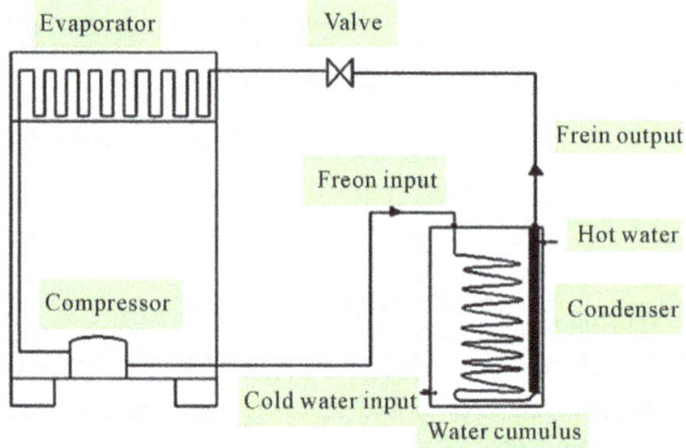

Figure 10. Refrigerating circuit binding the refrigerator and the water-heater.

Figure 11. Coupling of the refrigerator to a water-heater photograph.

4.3. Heating Floor Realization

The goal for the moment is to construct a transportable prototype. Thus the floor is carried out by some sand filling of a wood box of dimensions 0.66 m × 0.66 m, that is to say a surface of 0.43 m^2. As the selected height is 0.1 m, then the sand mass is:

$$M = \rho \cdot S \cdot H = 66.6 \text{ Kg}$$

The condenser of refrigerator is then placed in the wood box before filling this latter by dry sand (**Figure 12**).

4.4. Refrigerator and Heating Floor Coupling

For this system, the refrigerator' condenser is placed in the box filled with sand, thus, the foran condensation does not directly heat the air but the sand, and stores its heat in the heating floor (**Figures 13-15**).

Notice:

It is also possible to couple the water cumulus and the floor heating with only one refrigerator using an electromagnetic sluice gate which makes it possible to ensure the commutation of the refrigerating circuit of the refrigerator either towards the hot water cumulus, or towards the heating floor.

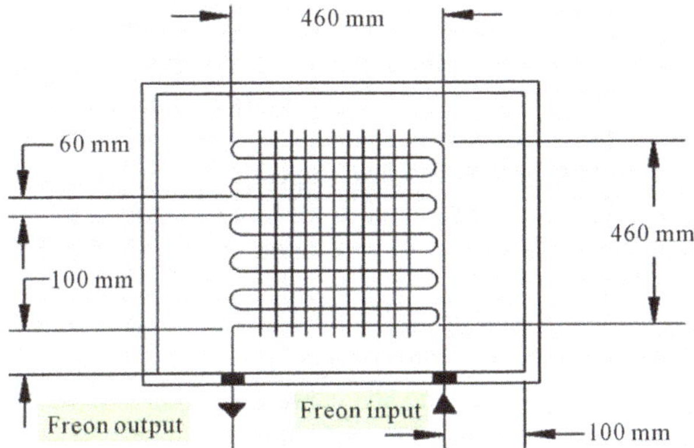

Figure 12. Dimensions of the refrigerator condenser immersed in the sand floor.

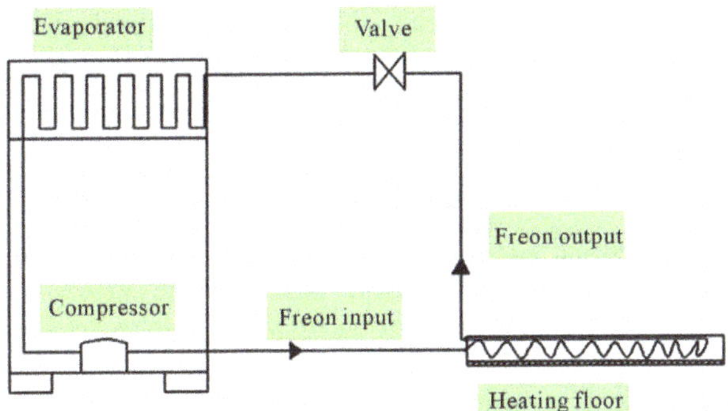

Figure 13. Installation circuit of refrigerator/heating floor coupling.

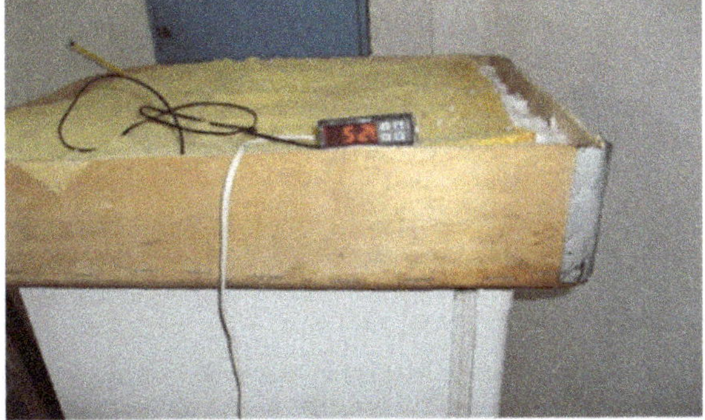

Figure 14. Small-scale model of the heating sand floor photographs.

Figure 15. Photographs of the two refrigerators: One coupled with the water-heater, the other with the heating floor.

It is also to be noted that the floor can be heated in a "conventional" way by the water from big tank storage, heated by the condenser of fridges which has restaurants and hotels or refrigeration rooms of foods conservation, or other.

4.5. Experimental Results

The experiments make it possible to bind the refrigerator to the system to be heated (water and heating floor) and to record the heating, refrigeration and ambient temperatures.

4.5.1. Evolution of Water-Heater Temperature

We study here the coupling of the refrigerator to a water heating cumulus.

1) First test

Figure 16 shows that in the morning, with the refrigerator starting, the rise in the storage water temperature is slower than that of the interior of refrigerator (approximately 8 h 30 against 2 h), this is because of the difference in heat mass capacity between water and air (4185 against 1000 J/kg˚C).

With an ambient temperature of 26˚C, water temperature reached is 59˚C, that is to say a rise in water temperature of 33˚C.

2) Second test

At the end of the one day operation, the refrigerator is voluntarily stopped for reasons of night safety. The following day at 5 am, the tests began again by the refrigerator restarting. The water temperature was 35˚C, that is to say 15˚C higher than for the previous day; what shows that the cumulus stores well heat thanks to its heat insulation with glass wool.

According to **Figure 17**, the temperature increases in the same manner than the first test, it reaches the permanent mode (59˚C) at the end of eight hours. A racking of ten liters hot water was carried out from the cumulus, followed by a filling of the same cold water capacity. Quickly the water temperature decreased up to 45˚C then it increased gradually to reach again the permanent mode at the end of four hours, against five hours and half for the preceding heating and the same variation in temperature. This is to show that hot water can be used several times per day.

4.5.2. Heating Floor Temperature Evolution

Is studied here the coupling of the refrigerator to a heated floor consisting of a layer of sand.

For the input/output curves of temperatures in the floor (**Figure 18**), we distinguish three zones: Zones 1 and 2 without insulation and zone 3 with insulation.

- Zone 1 [5 h – 10 h]: Fast rise in floor temperatures according to time.

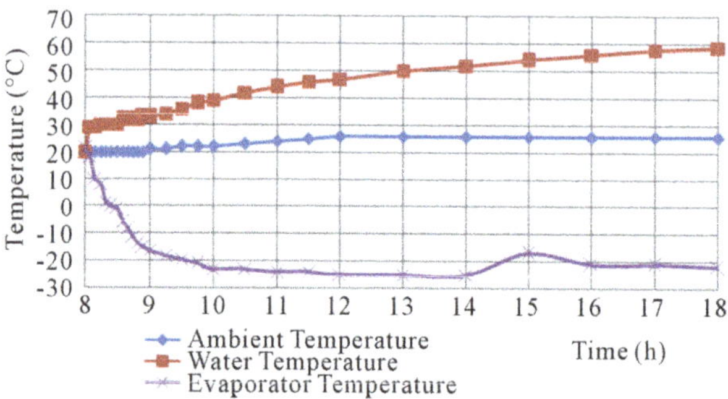

Figure 16. Variation of the temperatures according to time. Water heating case.

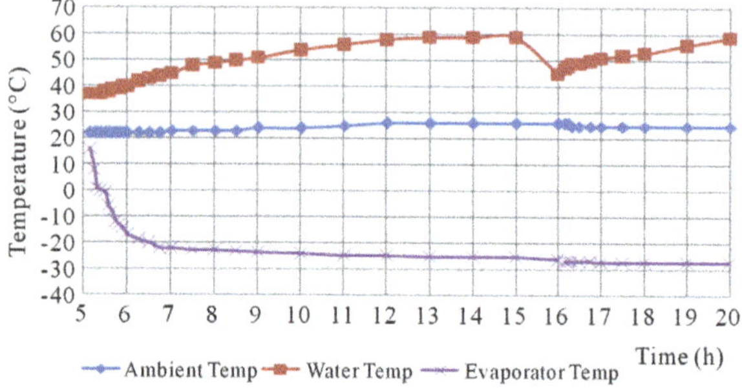

Figure 17. Variation of water, evaporator and ambient temperatures before and after draining of 10 liters water.

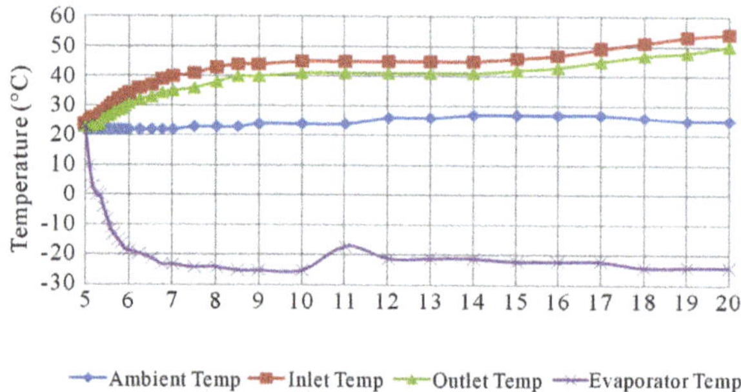

Figure 18. Temperature variation according to time. Heating floor case.

- Zone 2 [10 h - 14 h]: The temperatures reach their maximum values (permanent mode with 45˚C to 43˚C).
- Zone 3 [14 h - 20 h]: Floor temperatures rise compared to Zone 2 up to 54˚C/52˚C thanks to the heat insulation.

If the heating floor is insulated, then the heat quantity stored in sand increases in a significant way.

5. Conclusions

By this project, we tried to widen the use of a refrigerating system for the water heating, spaces or buildings, and this by the exploitation of the energy previously rejected by the condenser. With this manner the refrigerator can contribute to heat water and/or floor-heating; while keeping its principal function to cool.

The temperatures reached respectively, by water and floor, are approximately 60˚C and 50˚C, without modifying the temperature of the evaporator, which is at approximately −20˚C.

The elimination of the grid of the usual condenser located at the back of the refrigerator creates an advantage: to avoid its heating and consequently up time of the compressor and thus its electric consumption.

Finally, the results obtained as well theoretically as in experiments show clearly that the heat withdraws coming from the condenser immersed in water or sand, is a reliable source of heat, being able to be useful at least for pre-heating of water or room. The recovered heat enters in the négawatts production concept.

REFERENCES

[1] B. Slama and R. Thermodynamic, "Heat Water by the Condenser of Refrigerator," *International Symposium on Convective Heat and Mass Transfer in Sustainable Energy*, Hammamet, April 26 to May 1, 2009.

[2] S. R. Ben, "Water-Heater Coupled with the Refrigerator to Develop the Heat of the Condenser," *International Renewable Energy Congress IREC*, Sousse, 5-7 November 2009, pp. 12-18.

[3] S. R. Ben, "Thermodynamic Heat Water by the Condenser of Refrigerator," 1ère *Conférence Maghrébine Sur les Matériaux et l'Energie*, Gafsa, 26-28 Mai 2010.

[4] S. R. Ben, "Coupling of a Refrigerator to a Water Heater and Heating Floor," GRETH Heat Set, Opatija, 18-22 October 2010.

[5] S. R. Ben, "Production de Negawatts Par Couplage d'Un Refrigerateur à Un Chauffe-Eau," *Congres de la Société Française de Thermique*: *Energie Solaire et Thermique*, Perpignan, 24-27 Mai 2011.

[6] S. R. Ben, "Water Heating by Recovery of Heat Released by the Refrigerator," *23rd IIR International Congress of Refrigeration*, Praha, 21-26 August 2011.

[7] G. Grazzini and R. Rinaldi, "Thermodynamic Optimal

Design of Heat Exchangers for an Irreversible Refrigerator," *International Journal of Thermal Sciences*, Vol. 40, No. 2, 2001, pp. 173-180.

[8] V. Skrivan, "L'Utilisation de La Chaleur de Condensation des Groupes Frigorifiques de Moyenne Puissance Pour le Chauffage de L'Eau," *Revue Internationale du Froid*, 1984, pp. 14-20.

[9] B. J. Huang, J. H. Wang, J. H. Wu and P. E. Yang, "A Fast Response Heat Pump Water Heater Using Thermostat Made from Shape Memory Alloy," *Applied Thermal Engineering*, Vol. 29, No. 1, 2009, pp. 56-63.

[10] J. Ji, K. Liu, T.-T. Chow, G. Pei, W. He and H. He, "Performance Analysis of a Photovoltaic Heat Pump," *Applied Energy*, Vol. 85, 2008, pp. 680-693.

[11] M. N. A. Hawlader and K. A. Jahangeer, "Solar Heat Pump Drying and Water Heating in the Tropics," *Solar Energy*, Vol. 80, 2006, pp. 492-499.

[12] M. N. A. Hawlader, S. K. Chou, K. A. Jahangeer, S. M. A. Rahman and L. K. W. Eugene, "Solar-Assisted Heat-Pump Dryer and Water Heater," *Applied Energy*, Vol. 74, 2003, pp. 185-193.

[13] Y. W. Li, R. Z. Wang, J. Y. Wu and Y. X. Xu, "Experimental Performance Analysis on a Direct Expansion Solar-Assisted Heat Pump Water Heater," *Applied Thermal Engineering*, Vol. 27, 2007, pp. 2858-2868.

[14] J. P. Chyng, C. P. Lee and B. J. Huang, "Performance Analysis of a Solar-Assisted Heat Pump Water Heater," *Solar Energy*, Vol. 74, 2003, pp. 33-44.

[15] B. J. Huang and J. P. Chyng, "Performance Characteristics of Integral Type Solar Assisted Heat Pump," *Solar Energy*, Vol. 71, 2001, pp. 403-414.

[16] B. J. Huang, J. P. Lee and J. P. Chyng, "Heat-Pipe Enhanced Solar-Assisted Heat Pump Water Heater," *Solar Energy*, Vol. 78, 2005, pp. 375-381.

[17] S. R. Ben and S. Gabsi, "Distillateur Solaire Activé Par Une Pompe à Chaleur à Compression," *Journées Tunisiennes des Ecoulements et des Transferts JTET*, Monastir, 19-21 Mars 2006.

[18] S. R. Ben and S. Gabsi, "Conception et Expérimentation d'un Distillateur d'Eau de Mer Solaire/Pompe à Chaleur," *Premières Journées Tunisiennes sur le Traitement et le Dessalement de l'Eau JNTDE*, Hammamet, 4-7 Novembre 2006.

[19] S. R. Ben, K. Hidouri and S. Gabsi, "Study of An Hybrid Water Solar Distiller Coupled with Heat Pump," *GREITH Heat Set* 2007, Chambery, 18-20 Avril 2007, pp. 677-684.

[20] S. R. Ben, K. Hidouri and S. Gabsi, "Performance of an Hybrid Sea Water Distiller Solar/Heat Pump," *ICAMEM*, Hammamet, 19 December 2006.

[21] S. R. Ben and S. Gabsi, "Design and Experimentation of a Solar Sea Water Distiller/Heat Pump," *TIWATMED*, Djerba, 24-26 Mai 2007.

[22] S. R. Ben, K. Hidouri and S. Gabsi, "Distillateur Solaire Hybride à Pompe à Chaleur à Compression," 1er *Colloque Maghrebin sur le Traitement et le Dessalement des Eaux CMTDE*, Hammamet, 7-10 Décembre 2007.

[23] K. Hidouri, S. R. Ben and S. Gabsi, "Hybrid Solar Still by Heat Pump Compression," *Desalination*, Vol. 250, 2010, pp. 444-449.

Nomenclature

COP	coefficient of performance
Cp	heat mass ($J \cdot kg^{-1} {}^{\circ}C^{-1}$)
h	heat exchange cefficient ($Wm^{-2} \cdot K^{-1}$)
M	Mass (kg)
P_{comp}	power of the compressor (W)
Q	Heat quantity (J)
S	surface (m^2)
t	time (h)
U	convection losses coefficient ($Wm^{-2} \cdot K^{-1}$)
S	surface (m^2)
T	Temperature (°C)
ΔT	Temperature variation (°C)
Δt	time variation (s)
η	efficiency
ρ	density (kg/m^3)
λ	conductivity transfer coefficient ($Wm^{-1} \cdot K^{-1}$)
Signes	
amb	ambient
i	initial
f	final
th	thermal
fr	refrigerator

12

Time-Effect Relationship of Toxicity Induced by Roundup® and Its Main Constituents in Liver of Carassius Auratus

Jinyu Fan, Jinju Geng, Hongqiang Ren, Xiaorong Wang
State Key Laboratory of Pollution Control and Resource Reuse, School of the Environment, Nanjing University, Nanjing, China

ABSTRACT

In order to evaluate the eco-toxicological effects of Roundup® on *Carassius auratus* (*C. auratus*), fish were exposed to 32 μg/L Roundup®, isopropylamine salt of glyphosate (G.I.S) and polyoxyethylene amine (POEA) over different periods (0.5, 1, 3, 7 and 14 d). Hydroxyl radical (·OH), malondialdehyde (MDA) and acetylcholinesterase (AChE) in liver were detected in this study. Results showed that the generation of ·OH increased before 7 d, but without significantly difference. ·OH was induced at 1 d for POEA group, 3 d for Roundup® group and 7 d for G.I.S group. At 14 d, ·OH generation returned to normal levels. MDA contents all increased significantly ($p < 0.01$) during 7 days and then reached a normal level at 14 d. AChE activity in all group tests revealed a significant inhibition ($p < 0.01$) after 7 days exposure and then rebounded a little, but remained below the control after 14 days exposure. The rate of AChE inhibition range from 13% - 42% in Roundup®, 6% - 40% in G.I.S, and 21% - 54% in POEA, suggesting that POEA was more toxic compared to Roundup® and G.I.S. 32 μg/L Roundup® exposure led to the change of physiological and biochemical indexes in *C. auratus*, which was a reversible process in the long run.

Keywords: *Carassius Auratus*; Roundup®; Hydroxyl Radical (·OH); Toxicity

1. Introduction

Roundup®, the main glyphosate formulations, is already mixtures of glyphosate and various adjuvants at different concentrations [1]. The original formulation of Roundup® contains isopropylamine salt of glyphosate (G.I.S) as the active ingredient and polyoxyethylene amine (POEA) as the surfactant agent [2]. Since Roundup® can easily reach the aquatic systems by runoff, drainage, leaching or inadvertent aerial overspray, the herbicide represents a dangerous and widely spread group of environmental contaminants [3]. However, the knowledge on the time–effect of toxicity induced by environmental concentration of Roundup® and its main constituents to fish is still limited.

Researches suggest that reactive oxygen species (ROS), can be induced in organisms exposed to some environmental contaminants [4-6]. Recent study suggested that Roundup® might induce oxidative stress in aquatic organisms through the increased levels of tissue lipid hydroperoxides [7]. Few direct evidences can prove ROS generation and oxidative stress in aquatic organisms exposed to Roundup® and its main constituents.

The measurement of acetylcholinesterase (AChE) activity in different fish tissues has also proved to be a sensitive method for detecting the presence of several herbicides [8,9]. The inhibition of AChE causes an accumulation of acetylcholine in the synapse, which therefore AChE cannot function in a normal way [10]. Glusczak et al. [11] reported that Rhamdia quelen showed significant reduction in AChE activity after exposed to Roundup®.

In this study, *Carassius auratus* (*C. auratus*), commonly found in China, is chosen as the testing aquatic organism. The aim of the study is to investigate the time-effect of environmental concentration glyphosate herbicide on oxidative stress and AChE activity of native freshwater fishes, and to compare the toxicity difference of Roundup® and its main constituents (G.I.S and POEA) to *C. auratus*.

2. Experimental Methods

2.1. Chemicals

Roundup® solution (41% purity, containing 41% G.I.S and 18% POEA), was obtained from Monsanto Company (St. Louis, MO, USA). G.I.S (41% purity) was purchased from Sigma Chemical (St. Louis, MO, USA). POEA was purchased from Haian petrochemical complex (Jiangsu, China). The other reagents used were analytically pure, such as the spin-trapping agent, a-phenyl-N-tert-butylnitrone (PBN), 2-thiobarbituric acid (TBA), and bovine serum albumin (BSA), which were purchased from Sigma Chemical.

2.2. Experimental Fish and Pollutants Treatment

Gold fish (*C. auratus*) with 9.9 ± 0.10 cm body length and 20.2 ± 0.75 g body weight were purchased from Fuzimiao aquatic breeding base (aquaculture facility, Nanjing, China). All fish were acclimatized to water dechlorinated with activated carbon for two weeks before the experiment. The total mortality of fish was below 3%.

In the Canadian Water Quality Guideline, the safety exposure concentration of glyphosate was 65 μg/L considered protective of aquatic life [12]. So in this experiment, we set half of 65 μg/L as the sole concentration. After acclimatization, fish were randomly divided into sixteen groups and kept in glass aquaria. One group was designated for control and the other groups were employed as experimental groups that received concentrations 32 μg/L of Roundup® (containing 41% G.I.S and 18% POEA), G.I.S (41% purity), and 18% POEA (equal to the concentration in Roundup®), respectively, for 0.5, 1, 3, 7 and 14 d. During the experiment, 50% water was replaced daily by adding fresh Roundup®, G.I.S and POEA solution to minimize contamination from metabolic waste. Artificial dry food was provided once a day. Fish were sampled after exposure. Then the fish were dissected to obtain some fresh livers for the determination of hydroxyl radical. The rest of the livers were homogenized at 4°C for other experiments. During the experiment, the water conditions were as follows: Keep the dissolved oxygen levels at 5 mg/L by continuous aeration, temperature at 20°C ± 1°C, pH 7.0 ± 0.3.

2.3. PBN Adduct Extraction and Electron Paramagnetic Resonance (EPR) Analysis

ROS production in livers of *C. auratus* were measured using PBN as the spin-trapping agent [13]. After being rinsed with ice-cold physiological salt water, 0.1g of an fish liver sample was removed and homogenized quickly in 1.0 mL 50 mmol/L PBN (dissolved in dimethylsulfoxide (DMSO)) using a Teflon pestle in a potter homogenizer. 0.1mL supernatants was transferred to a capillary tube with a diameter of 0.9 mm, and placed in liquid nitrogen for EPR meaurements. The whole operation was carried out in an incubation system with continuous N_2 puring. The EPR spectra were recorded on a Bruker EMX 10/12 X-band spectrometer (Bruker, Germany) at room temperature (25°C), with the operation conditions of magnetic field center 3470 G, scan range 200 G, modulation frequency 100 kHz; modulation amplitude 0.5 G, microwave frequency 9.751 GHz, incident microwave power 20 mW, and sweep time 84 s for 5 scans.

2.4. Degree of Lipid Peroxidation Determined Using Malondialdehyde (MDA)

MDA content was measured by previous published thiobarbituric acid assay of Miller and Aust [14] with some modification. The reaction mixture containing 0.2 mL of tissue homogenate, 0.2 mL 8.1% sodium dodecyl sulfate (SDS), 1.5 mL 20% acetic acid buffer (pH 3.5), 1.5 mL 1% TBA, and 1 mL distilled water was heated at 90°C for 90 min, then cooled at room temperature and centrifuged for 15 min at 3,000 rpm/min. The absorbance of the supernatant was determined at 532 nm using a UV-220 spectrophotometer (Shimadzu, Japan). The amount of MDA formed was calculated by measuring the absorbance at 532 nm using a molar extinction coefficient of $1.56 \times 105\ M^{-1}\cdot cm^{-1}$.

The total protein content from enzyme extraction was measured using BSA as a standard [15]. All the determination assays were performed in triplicate, at a minimum.

2.5. AChE Activity Assay

Fish liver samples were weighed and homogenized in normal saline. The homogenate was centrifuged for 10 min at 4°C at 3,500 rpm and the supernatant was used as the enzyme source. AChE activity was determined with a AChE Detection kit from Nanjing Jian Cheng Biology Company (Nanjing, China).

2.6. Statistical Analysis

Statistical analyses were performed using SPSS statistical package version 16.0. Data were expressed as mean values±standard deviation (SD). The differences between experimental groups and the control were compared by a one-way analysis of variance (ANOVA), and significantly different treatments were identified by Dunnett's test. The level of statistical significance was set at significantly different from control $p < 0.05$ (*) and highly significantly different from control $p<0.01$ (**).

3. Results and Discussion

3.1. Free Radical Production by Induction of Roundup®, G.I.S and POEA

Signals of PBN adducts of fish hepatic after Roundup®, G.I.S and POEA exposure could be detected using EPR (**Figure 1**). A six line spectrum of three groups with two hyperfine coupling splitting peaks was observed. According to previous literature [4,5,13,16], the trapped ROS was likely to be the hydroxyl radical (·OH) and the levels could be expressed with the second couplet intensity of the triplets in the EPR spectra.

Figure 2 showed the kinetics of ·OH generation over

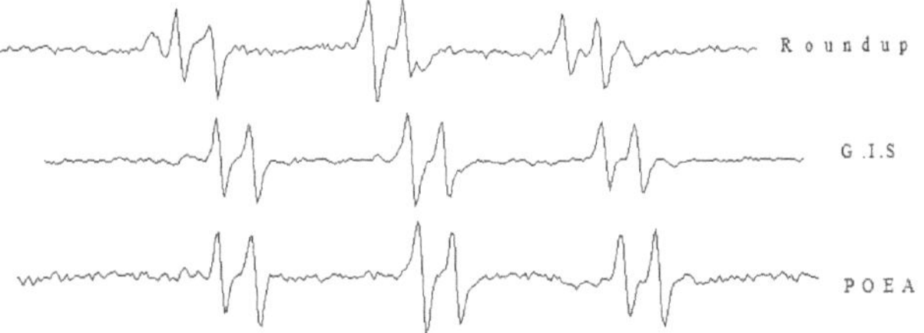

Figure 1. EPR was used to detect PBN-radical adducts in the liver of *C.auratus* with Roundup®, G.I.S and POEA.

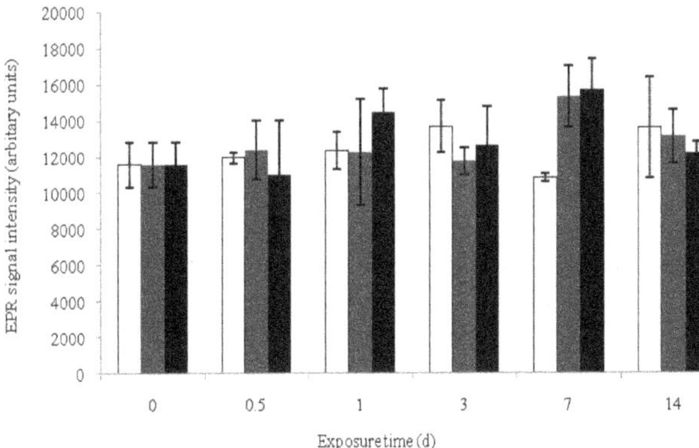

Figure 2. ·OH signal intensity in fish liver after exposure to 32 μg/L Roundup® (blank bars), G.I.S (grey bars) and POEA (black bars), for different experimental periods.

different periods (0.5, 1, 3, 7 and 14 d) of 32 μg/L Roundup®, G.I.S and POEA exposure. ·OH generation increased first and then decreased nearly with the control group, but without significantly difference. ·OH was induced at 1 d for POEA group, 3 d for Roundup® group and 7 d for G.I.S group. ·OH accumulation was earlier under POEA group conditions than under Roundup® group and G.I.S group, indicating that the speed of oxidant stress of POEA is faster than Roundup® and G.I.S in *C. auratus*. The maximum accumulation of ·OH generation was 118% (of the control) in Roundup®, 132% (of the control) in G.I.S and 135% (of the control) in POEA. Considering the speed and the maximum of ·OH accumulation, POEA is supposed to be more toxic than the other two substances. Howe et al. [17] found that for *Rana clamitans*, acute toxicity values in order of decreasing toxicity were POEA > Roundup®.

It was reported that the ·OH can be significantly induced by phenanthrene, pentachlorophenol, pyrene and 2-Chlorophenol in fish liver [18-21]. In the present study, the ·OH generation induced by Roundup®, G.I.S and POEA, but without obviously accumulated. It might be explained that the activities of antioxidant enzymes were activated and induced to remove ·OH to protect organisms from oxidative stress. After 14 days, the ·OH signal intensity returned to normal levels. This could be explained as follows: First, the ·OH had been reduced by antioxidant defense systems, or metabolized to less harmful radicals. Second, the fish had adjusted itself to combat the oxidative stress. Third, glyphosate formulations does not accumulation in vivo [22] and protect the fish from many harmful effects.

3.2. Changes in MDA

One of the most damaging effects ROS and their products in cells is the peroxidation of membrane lipids, which can be indicated by MDA detection [16]. MDA contents in fish livers after exposure to Roundup®, G.I.S and POEA were illustrated in **Figure 3**. Roundup®, G.I.S and POEA group elevated ($p < 0.01$) the MDA contents during 7 days and returned to normal levels at 14 d. The MDA contents increased at 0.5 d after Roundup®, G.I.S and POEA exposure, and they kept increasing until a maximum level were reached at 3 d in Roundup® (148% of the control) and G.I.S (170% of the control), at 1 d in POEA (180% of the control). The speed of lipid peroxidation may be further confirmed that POEA was more toxic than the other two pollutants. Obviously, Roundup® and its main constituents exposure resulted in an accumulation of lipid peroxidation in *C. auratus* during 7 days. But at 14 d, MDA contents were the same as the control group, indicating that exposure of low concentration of Roundup® to fish is a reversible process. This maybe explained by an adaptive response takes place in the cells or might be due to the activities of various damage removal and repair enzymes, to minimize the concentration of ·OH to the basal level and blocked lipid peroxidation in the cell.

Detailed studies have provided evidence that some xenobiotics induced MDA contents following stress on many species [5,21,24]. In the present study, ·OH generation induced without significantly difference, which suggested the self-adjusting of fish by activating antioxidant capacity to combat the cellular excess ROS generation. However, the increased MDA contents during 7 days proved that the lipid peroxidation in fish liver was promoted and further revealed that the fish was already

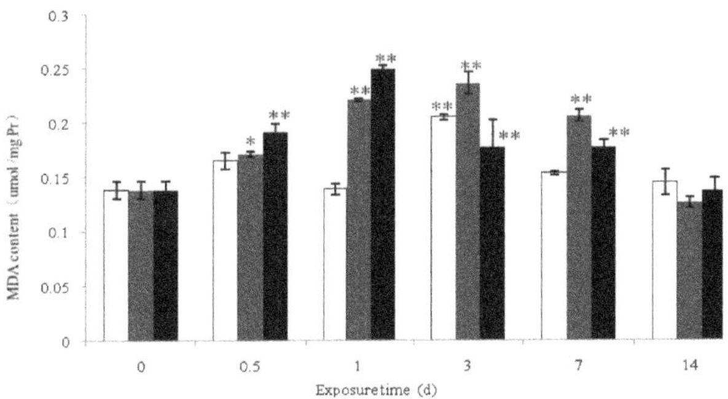

Figure 3. MDA content in fish liver after exposure to 32 μg/L Roundup® (blank bars), G.I.S (grey bars) and POEA (black bars), for different experimental periods.

in the status of oxidative stress although the results did not showed the accumulate of ·OH. According to Luo et al. [20], the higher MDA level by the end of the exposure to 2-chlorophenol in *C. auratus* suggested the oxidative damage occurred although hydroxyl radical returning to the normal level. In the study, the MDA content did not have a time-response pattern consistent with ·OH generation at all exposure time in Roundup®, G.I.S and POEA, which might be due to the effectiveness of the antioxidant system in providing protection. Sun et al. [21] found the similar change pattern of MDA in liver after different doses of pyrene were exposed to the fish (*C. auratus*).

3.3. Changes in AChE Activity

AChE activity in the liver of *C. auratus* exposed to Roundup®, G.I.S and POEA, for different experimental periods was displayed in **Figure 4**. All tests revealed a significant inhibition of AChE activity during 7 days exposure, then restored to the level of control group after 14 d. The inhibition percentages in Roundup®, G.I.S and POEA of the control fish was 13% - 42%, 6% - 40% and 21% - 54%, respectively. The degree of AChE reduction showed the toxicity order of the chemicals was: POEA > Roundup® > G.I.S. On the basis of Giesy et al. [24], the LC_{50} values (mg/L) for rainbow trout were between 8.2 and 27 for Roundup®, between 0.65 and 7.4 for POEA.

Organophosphorus pesticides have several toxic properties, the most prominent effect of which is AChE inhibition. AChE activity is therefore widely used in biomonitoring studies as a biomarker of organopho- sphorus pesticide exposure [25]. In this study, the reduction of AChE activity is assumed to have been resulted from the direct action of Roundup®, G.I.S and POEA exposure on active site of this enzyme. Glusczak et al. [26] reported the inhibition of this enzyme in the brain of *Leporinus obtusidens* exposed to 3, 6, 10 and 20 mg/L glyphosate for 96 h. Modesto and Martinez [27] also reported a decreasing AChE activity in the brain of *Prochilodus*

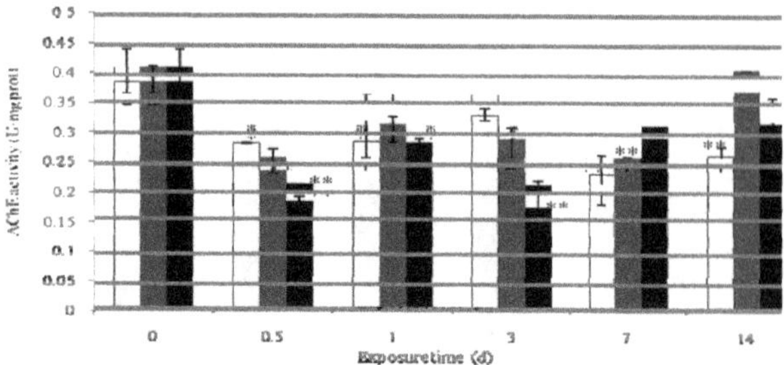

Figure 4. AChE activity in fish liver after exposure to 32 μg/L Roundup® (blank bars), G.I.S (grey bars) and POEA (black bars), for different experimental periods.

lineatus after exposure to 1 and 5 mg/L Roundup® for 96 h. The exposure of the three pollutants led to the maximum inhibition of AChE in POEA by 54%. The rate of AChE inhibition may not be considered a life-threatening situation since fish are capable of tolerating over 90% AChE inhibition [28]. After 14 days exposed to Roundup®, G.I.S and POEA, AChE activity rebounded a little, which were consistant with the alteration of ·OH generation and MDA contents, indicating that long-term exposure of Roundup®, G.I.S and POEA with low concentration in *C. auratus* is a reversible process.

4. Conclusions

The present study showed that Roundup®, G.I.S and POEA may cause changes in the metabolic and enzymatic parameters of fish during 7 days, such as ·OH generation addition, lipid peroxidation and AChE inhibition, implying that the fish was already in the status of oxidative stress. At 14 d, ·OH generation, MDA contents and AChE activity all returned to the normal level, indicating that the exposure of Roundup®, G.I.S and POEA with 32 μg/L in *C. auratus* is a reversible process in a long time. According to the toxic effects of Roundup®, G.I.S and POEA on *C. auratus*, POEA may be the most toxic pollutant. Roundup®, G.I.S and POEA should be distinguished when assessing the toxicity of this pesticide.

5. Acknowledgements

The research was funded through the National Science Foundation of China (No. 21077051, 51278241) and the Jiangsu Natural Science Foundation (No. BK2011057).

REFERENCES

[1] N. Benachour and G. E. Seralini, "Glyphosate Formulations Induce Apoptosis and Necrosis in Human Umbilical, Embryonic, and Placental Cells," *Chemical Research in Toxicology*, Vol. 22, No. 1, 2009, pp. 97-105.

[2] R. A. Releya, "The Impact of Insecticides and Herbicides on the Biodiversity and Productivity of Aquatic Communities," *Ecological Applications*, Vol. 15, No. 2, 2005, pp. 618–627.

[3] S. Guiherme, M. A. Santos, C. Barroso, I. Gaivao and M. Pacheco, "Differential Genotoxicity of Roundup Formulation and Its Constituents in Blood Cells of Fish (Anguilla Anguilla): Considerations on Chemical Interactions and DNA Damaging Mechanisms," *Ecotoxicology*, Vol. 21, No. 5, 2012, pp. 1381-1390.

[4] Y. Luo, X. R. Wang, H. H. Shi, D. Q. Ma, Y. X. Sui and L. L. Jin, " Electron Paramagnetic Resonance Investigation of in Vivo Free Radical Formation and Oxidative Stress Induced by 2,4-Dichlorophenol in the Freshwater Fish Carassius Auratus," *Environmental Toxicology and*

Chemistry, Vol. 24, No. 9, 2005, pp. 2145–2153.

[5] Y. Luo, X. R. Wang, L. L. Ji and Y. Su, " EPR Detection of Hydroxyl Radical Generation and Its Interaction with Antioxidant System in Carassius Auratus Exposed to Pentachlorophenol," *Journal of Hazardous Materials*, Vol. 171, No. 1-3, 2009, pp. 1096-1102.

[6] X. C. Xie, Y. X. Wu, M. Y. Zhu, Y. K. Zhang and X. R. Wang, "Hydroxyl Radical Generation and Oxidative Stress in Earthworms (Eisenia Fetida) Exposed to Decabromodiphenyl Ether (BDE-209) ," *Ecotoxicology*, Vol. 20, No. 5, 2011, pp. 993-999.

[7] M. J. Costa, D. A. Monteiro, A. L. Oliveira-Neto, F. T. Rantin and A. L. Kalinin, "Oxidative Stress Biomarkers and Heart Function in Bullfrog Tadpoles Exposed to Roundup Original®," *Ecotoxicology*, Vol. 17, No. 3, 2008, pp. 153–163.

[8] E. Sancho, J. J. Cerón and M. D. Ferrando, "Cholinesterase Activity and Hematological Parameters as Biomarkers of Sublethal Molinate Exposure in Anguilla Anguilla," *Ecotoxicology and Environmental Safety*, Vol. 46, No. 1, 2000, pp. 81–86.

[9] A. Ferrari, A. Venturino and A. M. P. de D'Angelo, "Muscular and Brain Cholinesterase Sensitivities to Azinphos Methyl and Carbaryl in the Juvenile Rainbow Trout Oncorhynchus Mykiss," *Comparative Biochemistry and Physiology C-Toxicology & Pharamacology*, Vol. 146, No. 3, 2007, pp. 308–313.

[10] H. M. Dutta and D. A. Arends, "Effects of Endosulfan on Brain Acetylcholinesterase Activity in Juvenile Bluegill Sunfish," *Environmental Research*, Vol. 91, No. 3, 2003, pp. 157–162.

[11] L. Glusczak, D. S. Miron, B. S. Moraes, R. R. Simoes, M. R. C. Schetinger, V. M. Morsch and V. L. Loro, "Acute Effects of Glyphosate Herbicide on Metabolic and Enzymatic Parameters of Silver Catfish (Rhamdia quelen)," *Comparative Biochemistry and Physiology C-Toxicology & Pha-ramacology*, Vol. 146, No. 4, 2007, pp. 519–524.

[12] J. Struger, D. Thomspon, B. Staznik, P. Martin, T. McDaniel and C. Marvin, "Occurrence of Glyphosate in Surface Waters of Southern Ontario," *Bulltin Environmental Contamination and Toxicology*, Vol. 80, No. 4, 2008, pp. 378-384.

[13] H. H. Shi, X. R. Wang, Y. Luo and Y. Su, "Electron Paramagneticresonance Evidence of Hydroxyl Radical Generation and Oxidative Damage Induced by Tetrabromobisphenol A in Carassius Auratus," *Aquatic Toxicology*, Vol. 74, No.4, 2005, pp. 365–371.

[14] D. M. Miller and S. D. Aust, "Studies of Ascorbate-Dependent, Iron Catalyzed Lipid Peroxidation," *Archives Biochemistry and Biophysics*, Vol. 271, No.1, 1989, pp. 113–119.

[15] M. M. Bradford, "Rapid and Sensitive Method for Quantitation of Microgram Quantities of Protein Utilizing the Principle of Protein-dye Binding,"*Analytical Biochemistry*, Vol. 72, No. 1-2, 1976, pp. 248–254.

[16] Y. G. Xue, X. Y. Gu, X. R. Wang, C. Sun, X. H. Xu, J. Sun and B. G. Zhang, " The Hydroxyl Radical Generation and Oxidative Stress for the Earthworm Eisenia Fetida Exposed to Tetrabromobisphenol A," *Ecotoxicology*, Vol. 18, No. 6,2009, pp. 693–699.

[17] C. M. Howe, M. Berrill, B. D. Pauli, C. C. Helbing, K. Werry and N. Veldhoen, "Toxicity of Glyphosate-Based Pesti-Cides to Four North American Frog Species," *Environmental Toxicology and Chemistry*, Vol. 23, No. 8, 2004, pp. 1928-1938.

[18] Y. Yin, H. X. Jia, Y. Y. Sun, H. X. Yu, X. R. Wang, J. C. Wu and Y. Q. Xue, "Bioaccumulation and ROS Generation in Liver of Carassius Auratus, Exposed to Phenanthrene," *Comparative Biochemistry and Physiology C-Toxicology & Pharamacology*, Vol. 145, No. 2, 2007, pp. 288–293.

[19] Y. Luo, Y. Su, R. Z. Lin, H. H. Shi and X. R. Wang,"2-Chlorophenol Induced ROS Generation in Fish Carassius Auratus Based on the EPR Method," *Chemosphere*, Vol. 65, No. 6, 2006, pp. 1064–1073

[20] Y. Luo, Y. X. Sui, X. R. Wang and Y. Tian, "2-Chlorophenol Induced Hydroxyl Radical Production in Mitochondria in Carassius Auratus and Oxidative Stress–An Electron Para-magnetic Resonance Study," *Chemosphere*, Vol. 71, No. 7, 2008, pp. 1260-1268.

[21] Y.Y. Sun, Y. Yin, J. F. Zhang, H. X. Yu, X. R. Wang, J. C. Wu and Y. Q. Xue,"Hydroxyl Radical Generation and Oxidative Stress in Carassius Auratus Liver, Exposed to Pyrene," *Ecotoxicology and Environmental Safety*, Vol. 71, No. 2, 2008, pp. 446–453.

[22] G. M. Williams, R. Kroes and I. C. Munro,"Safety Evaluation and Risk Assessment of the Herbicide Roundup and Its Active Ingredient, Glyphosate, for Humans," *Regulatory Tox-icology and Pharmacology*, Vol. 31, No. 2, 2000, pp. 117-165.

[23] F. Y. Li, L. L. Ji, Y. Luo and K. Oh, "Hydroxyl Radical Generation and Oxidative Stress in Carassius Auratus Liver as Affected by2,4,6-Trichlorophenol," *Chemosphere,* Vol. 67, No.1, 2007, pp. 13-19.

[24] J. P. Giesy, S. Dobson and K. R. Solomon, "Ecotoxicological Risk Assessment for Roundup Herbicide," *Reviews of Environmental Contamination and Toxicology*, Vol. 167, 2000, pp. 35–120.

[25] E. Oruc, "Effects of Diazinon on Antioxidant Defense System and Lipid Peroxidation in the Liver of Cyprinus

Carpio (L.)," *Environmental Toxicology*, Vol. 26, No. 6, 2010, pp. 571-578.

[26] L. Glusczak, D. S. Miron, M. Crestani, M. M. Fonseca, F. A. Pedron, M. F. Duarte and V. L. P. Vieira, "Effect of Glyphosate Herbicide on Acetylcholinesterase Activity and Metabolic and Hematological Parameters in Piava (Leporinus Obtusidens)," *Ecotoxicology and Environmental Safety*, Vol. 65, No. 2, 2006, pp. 237–241.

[27] K. A. Modesto and C. B. R. Maritnea, "Effects of Roundup Transorb Onfish: Hematology, Antioxidant Defenses and Acetylcholinesterase Activity," *Chemosphere*, Vol. 81, No. 6, 2010, pp. 781-787.

[28] E. O. Oruc and D. Usta, "Evaluation of Oxidative Stress Responses and Neurotoxicity Potential of Diazinon in Different Tissues of Cyprinus Carpio," *Environmental Toxicology and Pharmacology*, Vol. 23, No. 1, 2007, pp. 48–55.

13

Protecting Water Quality and Public Health Using a Smart Grid

Ken Thompson[1], **Raja Kadiyala**[2]
[1]Intelligent Water Solutions, CH2M HILL, Englewood, Colorado, USA
[2]Intelligent Water Solutions, CH2M HILL, Oakland, California, USA

ABSTRACT

After the attacks on September 11, 2001 and the follow-up risk assessments by utilities across the United States, securing the water distribution system against malevolent attack became a strategic goal for the U.S. Environmental Protection Agency. Following 3 years of development work on a Contamination Warning System (CWS) at the Greater Cincinnati Water Works, four major cities across the United States were selected to enhance the CWS development conducted by the USEPA. One of the major efforts undertaken was to develop a process to seamlessly process "Big Data" sets in real time from different sources (online water quality monitoring, consumer complaints, enhanced security, public health surveillance, and sampling and analysis) and graphically display actionable information for operators to evaluate and respond to appropriately. The most significant finding that arose from the development and implementation of the "dashboard" were the dual benefits observed by all four utilities: the ability to enhance their operations and improve the regulatory compliance of their water distribution systems. **Challenge:** While most of the utilities had systems in place for SCADA, Work Order Management, Laboratory Management, 311 Call Center Management, Hydraulic Models, Public Health Monitoring, and GIS, these systems were not integrated, resulting in duplicate data entry, which made it difficult to trace back to a "single source of truth." Each one of these data sources can produce a wealth of raw data. For most utilities, very little of this data is being translated into actionable information as utilities cannot overwhelm their staffs with manually processing the mountains of data generated. Instead, utilities prefer to provide their staffs with actionable information that is easily understood and provides the basis for rapid decision-making. Smart grid systems were developed so utilities can essentially find the actionable needle in the haystack of data. Utilities can then focus on rapidly evaluating the new information, compare it known activities occurring in the system, and identify the correct level of response required. **Solution:** CH2M HILL was engaged to design, implement, integrate, and deploy a unified spatial dashboard/smart grid system. This system included the processes, technology, automation, and governance necessary to link together the disparate systems in real time and fuse these data streams to the GIS. The overall solution mapped the business process involved with the data collection, the information flow requirements, and the system and application requirements. With these fundamentals defined, system integration was implemented to ensure that the individual systems worked together, eliminating need for duplicate data entry and manual processing. The spatial dashboard was developed on top of the integration platform, allowing the underlying component data streams to be visualized in a spatial setting. **Result:** With the smart grid system in place, the utilities had a straightforward method to determine the true operating conditions of their systems in real time, quickly identify a potential non-compliance problem in the early stages, and improve system security. The smart grid system has freed staff to focus on improving water quality through the automation of many mundane daily tasks. The system also plays an integral role in monitoring and optimizing the utilities' daily operations and has been relied on during recovery operations, such as those in response to recent Superstorm Sandy. CH2M HILL is starting to identify the processes needed to expand the application of the smart grid system to include real-time water demands using AMI/AMR and real-time energy loads from pumping facilities. Once the smart grid system has been expanded to include Quality-Quantity-Energy, CH2M HILL can apply optimization engines to provide utility operations staffs with a true optimization tool for their water systems.

Keywords: Smart Grid; On-line Water Quality Monitoring; OWQM; Event Detection System; EDS; TEVA-SPOT

1. Introduction

Continuous monitoring of distribution system water quality was rarely conducted prior to the terrorist attacks of September 2001 on the United States. Following those events and the completion and review of risk assessments for all public water systems (PWSs) serving a population greater than 3,300, the distribution system was identified as the most vulnerable area of attack.

Homeland Security Presidential Directive 9 required the U.S. Environmental Protection Agency (EPA) to develop a process for utilities to improve the protection of their water distribution systems. In response, distribution system water quality monitoring pilot projects were conducted, which were funded by the EPA Water Security Division Water Security (WS) initiative. [1] As a result, continuous monitoring systems are in operation in Cincinnati, Dallas, New York, Philadelphia, and San Francisco. Independent from the WS initiative program, some PWSs and U.S. government agencies have been developing similar programs. Benefits of these systems include improvement of water treatment processes, increased efficiency of water utility operations, more assured quality of water delivered to consumers, and increased protection of public health.

2. Benefits of Distribution System Monitoring

Benefits from continuous distribution system on-line water quality monitoring (OWQM) may be categorized as operational enhancements, regulatory compliance, and contamination warning.

Operational enhancements include continuous indication of water quality in the distribution system beyond that which is possible through routine regulatory sampling. Early indication may be provided of unusually low residual chlorine levels, impending nitrification (elevated ammonia), turbidity excursions caused by main breaks, and other unusual water quality changes. This monitoring is achieved through measurement of several water quality parameters with which water utilities are already familiar (e.g., chlorine residual) and other parameters that are relatively new to this application (e.g., total organic carbon).

Regulatory compliance benefits of OWQM include improving the ability to maintain chlorine residual as part of the Total Coliform Rule (TCR) [2] and maintaining proper pH control to avoid potential violations of the Lead and Copper rule [3].

Warning of intentional or unintentional contamination in the distribution system is somewhat more complex. Specialized analyzers are available, including gas chromatographs that may detect specific contaminants and toxicity monitors of many types that can provide a general warning of contamination. Due to the large number of potential contaminants, rather than attempting to specifically identify a contaminant, it is more practical to monitor for an indication of contamination through changes in many of the same water quality parameters, or surrogates, that are used for operational benefit monitoring.

Beyond monitoring for operational and contamination purposes, utilities should consider OWQM as part of the Distribution System Optimization program of the Partnership for Safe Water [4]. The Partnership is a voluntary effort between EPA, the American Water Works Association (AWWA), several other drinking water organizations, and more than 200 water utilities throughout the United States. The goal of the Partnership is to provide a new measure of safety to millions of Americans by implementing prevention programs where legislation or regulation does not exist to do so. The preventative measures are focused on optimizing treatment plant performance and distribution system operation.

3. Selection of Water Quality Parameters

A water utility that is embarking on design of distribution monitoring must first decide which water quality parameters to monitor. Parameters that are typically included in OWQM systems include:

- Total organic carbon (TOC)
- Residual chlorine
- Conductivity
- pH
- Turbidity

These parameters are of interest to utilities from a distribution system operational and regulatory perspective and provide critical information including:

- TOC – Elevated turbidity excursions can be associated with a breakthrough at the water treatment plant (WTP) or scouring and release of biofilm within the distribution system.
- Residual chlorine – A sudden loss in residual could promote biofilm growth and potential violation of the TCR.
- Conductivity – This measurement provides an easy method for identifying mixing or different water sources, which can have a significant impact on many industrial operations.
- pH – pH is controlled for disinfection and corrosion control. The formation of some disinfection byproducts is pH dependent.
- Turbidity – This parameter provides warning of a system disruption created by a surge or reversal in flow that scours the pipeline. This could be caused by a pipeline break, hydrant knockover, or other problems that will impact chlorine residual and customer satisfaction.

Utilities that use chloramines for disinfection also measure for ammonia, nitrates, and dissolved organic carbon (DOC) to provide early warning of nitrification in the distribution system. The first water quality indicator of nitrification will be the increase of ammonia, which will occur before nitrites and nitrates begin to increase.

The myriad potential contaminants have been classified among twelve categories by EPA [5]. Laboratory testing has concluded that three of these water quality parameters—TOC, residual chlorine, and conductivity—respond to the presence of contaminants from ten of the twelve categories, so along with operational monitoring, broad contaminant coverage is also provided with a minimum of instrumentation. The relative change in quality of water with either chlorine or chloramines has been investigated by EPA and the pilot cities to serve as a general guidance for evaluating a water quality anomaly.

Most OWQM systems include monitoring for absorbance of ultra-violet (UV) light. UV analyzers, which operate by measuring absorption at the single 254-nanometer (nm) wavelength, are generically referred to as UV-254 analyzers. Instruments that operate through analysis of a broad spectrum from 200 to 720 nm are referred to as spectral analyzers.

UV absorbance has been shown to be strongly correlated to TOC content [6]. UV-254 analyzers measure absorbance at the 254-nm wavelength as this is the radiation emitted by a common mercury-based UV source lamp. The absorbance of UV light by TOC or other contaminants contained within the water sample is reported as a percent of the uninhibited lamp intensity, ranging from 0 to 100 percent.

Spectral analyzers utilize a xenon lamp to produce a light source across UV and visible light wavelengths from 200 to 720 nm. Measurement of absorbance at 254 discrete wavelengths across this range enables construction of an absorbance, or spectral, curve (**Figure 1**). Due to the substantially greater information provided by spectral analysis, the broadband spectrum enables measurement of TOC based on calculation of the numerous UV/visible light wavelengths that are associated with this parameter. Similarly, turbidity is also calculated based on analysis of numerous wavelengths. Subtraction of the turbidity component enables derived measurement of nitrate and DOC. Other parameters, not typically included in OWQM analysis, also may be derived from broad spectral absorbance of UV and visible light.

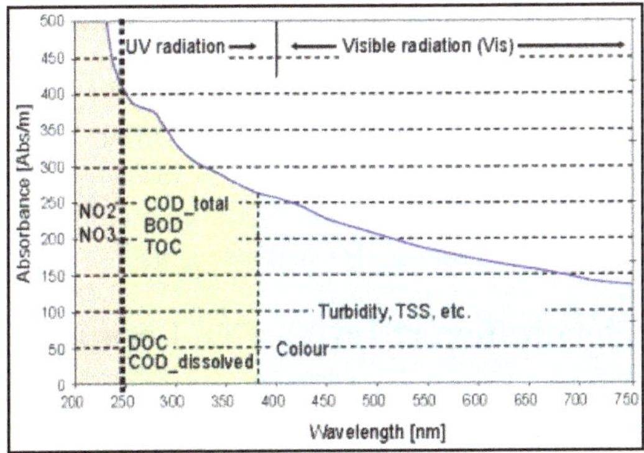

Figure 1. Ultraviolet/Visible Absorbance spectrum enables calculation and derivation of numerous water quality parameters.

Not all sources of TOC are revealed by UV/visible light absorbance, but a large enough percentage are detected that these technologies are generally accepted for the OWQM application. This limitation of correlated or spectral indication of TOC must be considered when selecting OWQM parameters.

4. Selection of Water Quality Analyzers

Once water quality parameters to be used for OWQM monitoring are selected, specific instruments to be used for this purpose must be selected. All utilities are under pressure to minimize capital and operating costs, so this consideration factors into analyzer selection. Addition of OWQM monitoring must be done without unnecessarily adding to the existing responsibilities of utility technicians, also impacting the selection of OWQM analyzers. One way to do this is to select sensors and analyzers that are reliable and inexpensive to operate, and require only infrequent direct attention or maintenance.

4.1. Chlorine Analyzers

The two most common methods for on-line chlorine analysis are amperometric and colorimetric detection. The colorimetric method requires use of chemical reagents to produce a reaction, which is measured and used to quantify chlorine content. Reagent reservoir levels must be regularly monitored and refilled by maintenance personnel. Additionally, reagents used for colorimetric tests have been found to degrade in environments that exceed 105 degrees, significantly affecting the quality of the analysis.

Amperometric sensors measure changes in electric current or potential and operate without use of chemical reagents. For monitoring residual chlorine by OWQM systems, these sensors reduce maintenance costs and activities, as well as operational risks from depletion of reagent reservoirs between service visits. Therefore amperometric technologies are most frequently used.

4.2. TOC Analyzers

Traditional TOC analyzers involve multiple electrochemical reactions for operation. Phosphoric acid is used for pH reduction and inorganic carbon removal, followed by oxidation of organic carbon to CO_2 by sodium or ammonium persulfate and heat or UV light. At least one manufacturer uses boron-doped electrodes to generate oxidation radicals in place of the persulfate solution.

CO_2 is directly detected by a non-dispersive infrared (NDIR) detector, or converted to carbonic acid and measured by conductance. Operation requires periodic replenishment of acid and, if used, persulfate reagents. The UV lamp, when used, also requires replacement.

Prefiltration of the analyzer sample stream is frequently required to prevent plugging of the micro-tubing that is included in the analyzer's internal construction. In some water that includes a substantial level of inorganic carbon, additional filters to remove this carbon component must be included at the inlet to the analyzer. These filtering systems are generally available from the analyzer manufacturer, but constitute a maintenance activity and additional cost that varies depending upon the particular nature of the sample stream.

Mechanically, TOC analyzers are highly complex and require substantial technical training and experience to ensure proper operation over extended periods [7]. While analyzers that are mechanically less intricate are now commercially available, they remain complex in operation.

4.3. UV Analyzers

When UV_{254} analysis is used for an OWQM system, the absorbance measurement alone may be used as a general indication of water quality. Due to the potential difficulties involved with operating and maintaining TOC analyzers, UV analyzers are sometimes used to provide a TOC measurement or indication, depending on the particular technology selected.

Several factors must be considered when selecting a UV absorbance analyzer for use in OWQM systems.

Some UV_{254} analyzers apply a correlation coefficient to the 254 nm absorbance reading to generate a TOC measurement. However, the accuracy of the correlation is dependent upon the stability of the correlation. For systems where the source water is subject to change, the correlation may change, and without adjustment to the programmed coefficient, the TOC measurement may be inaccurate.

TOC correlation to UV_{254} absorbance is also impacted by the turbidity of the sample stream. Some UV_{254} analyzers include automated turbidity compensation, while others do not.

Spectral analyzers calculate TOC and turbidity measurement from the UV and visible light absorption measurements that make up the spectral curve. Turbidity readings from these analyzers are applied to the TOC calculation to achieve a turbidity-compensated TOC measurement. These analyzers can also derive nitrate, DOC and other water quality parameters. Specific water contaminants (e.g., ricin [8]) can be calculated based on their specific absorbance spectrum, similar to the method used for TOC determination.

EPA studies have found that UV/visible light absorbance-based TOC readings will not detect all TOC compounds, but they do detect a large enough portion of potential TOC contaminants that these instruments are valid for contamination monitoring [9]. Similarly, TOC readings by these instruments are suitable for indicating distribution system water quality because drinking water TOC is typically made up of humic and fulvic acids, which are very accurately detected by UV absorption.

Absorbance-based UV and TOC measurements have the potential to be affected by deposition of mineral content on optical surfaces. Highly dependent on the water being tested, mineral deposits on lenses will impact absorbance readings. Optically based analyzers frequently include automated cleaning systems ranging from periodic flushing with various solutions to mechanical wipers or brushes, or continuous operation of ultrasonic wave generators. These have been found to be of varying effectiveness.

Some UV-based optical analyzers include automated compensation for variation of the UV lamp output over time. This function is essential as lamp output is known to decrease over time, requiring periodic lamp replacement.

4.4. Ammonia Analyzers

Ammonia sensors may be reagent or non-reagent based. For OWQM systems, reagentless technologies that use ion-selective electrodes are preferred to minimize maintenance activities and costs.

The ammonia measurement should include pH compensation, which may be integral to the ammonia sensor or a separate pH sensor with signal input to the ammonia analyzer. The pH signal can also be separately used for the OWQM pH measurement. Ammonia sensors may also include potassium compensation because elevated potassium levels will interfere with low-level ammonia measurements, such as those close to the sensor detection limit where OWQM systems typically operate. The potassium signal is not an OWQM parameter and is not separately reported by the OWQM system.

4.5. Conductivity, pH, Turbidity Analyzers

These sensors operate using standard, proven electrochemical and optical technologies that water utilities have deployed in WTPs and other facilities for many years. Specific analyzers to be used in an OWQM station should be those that the user has found to be reliable in service and to provide accurate readings with minimal maintenance requirements. For these parameters, use of a utility's standard sensors is usually acceptable.

5. Prioritization of Installation Locations

Selection of installation locations for OWQM stations involves important considerations to reduce cost and provide an environment conducive to long-term and successful operations.

Monitoring stations are typically installed at the discharge of each WTP or at wholesale connection interties to indicate baseline conditions entering the distribution system for comparison with downstream measurements. These stations also indicate results of changes to the treatment process and warn of conditions in the treatment process that may otherwise go undetected. Some OWQM process measurements may already be made at the WTP, and these existing measurements can be used for OWQM purposes. When adding other OWQM parameters that are not already monitored at the WTP, some utilities use installation of OWQM stations as an opportunity to upgrade older instruments or to convert to reagentless technologies to reduce labor and maintenance costs.

OWQM stations are frequently installed at the discharge of distribution system reservoirs and chlorine boosting pump stations. Measurement may be upstream of chlorine addition to provide a measure of the quality of the water in the reservoir and that further upstream. The OWQM station could alternately be installed downstream of chlorine addition at a booster station to provide a baseline measurement for comparison to OWQM stations further downstream.

OWQM stations can also be installed at critical nodes in the distribution system, and different approaches may be taken in selecting these locations. Distribution system managers generally have a good understanding of the operation of the piping network and can often identify the nodes of interest based on experience. A more scientific approach to OWQM siting is frequently conducted through the use of the Threat Ensemble Vulnerability Analysis and Sensor Placement Optimization Tool (TEVA-SPOT) software [10]. This analytical package, developed by EPA, Sandia Laboratories and others, analyzes a distribution system network and identifies critical nodes that will represent water quality impacting the largest number of consumers. TEVA-SPOT analysis results often validate the operational understanding expressed by distribution system managers, but also frequently identify nodes otherwise not understood to be critical or recommend subtle location changes compared to the managers' recommendations.

The analysis enables prioritization of a selected number of OWQM stations to meet the budget available for a project and identify stations to be added as a monitoring system expands over a period of years.

6. Requirements for Data Communication and Analysis

Several products are available for analysis of OWQM data and alarming of unusual conditions. The most common software event detection systems (EDSs) are commercially available through s::can, Whitewater, and Hach. Also available is the Sandia National Laboratories freeware system *Canary*, which was developed as part of the WS initiative. Each EDS has strengths and weaknesses associated with its performance under distribution system operations. Additional information can be obtained through the EPA Water Security Division.

OWQM alarms are generally based on more than simple alarm setpoints for parameter measurements. Typically the alarms or alerts are associated with *pattern alarms*, where multiple parameters change in a manner that is atypical of their normal relationship. As an example, if TOC increased, it would also be expected that DOC would increase in a proportional manner. When a utility implements enhanced coagulation, the TOC-to-DOC relationship changes, generating an alert at the water quality monitoring stations. Broadband UV/visible systems also produce a *spectral alarm* that is initiated if the normal spectral fingerprint displays an unusual shift.

OWQM data are collected more frequently than typical Supervisory Control and Data Acquisition (SCADA) monitoring data, and therefore OWQM data usually cannot be communicated over traditional low-bandwidth SCADA networks. Additionally, spectral data cannot be communicated over the typical SCADA data collection network, so separate communication pathways must be established. Many utilities, therefore, include all OWQM data on the alternate path and keep monitoring and maintenance of this information separate from operational SCADA data, although there is no requirement to maintain this separation. If T-1 or optical connections to the central monitoring facility are available, these pathways may be used, but typically the OWQM measurements are still transmitted as a separate data stream from SCADA parameters.

Water quality analysis is conducted locally at the OWQM station, and measured values and alarms are communicated to a central historian and display. Typically a long-term database is used for storage and retrieval of data and a short-term cache for short-term (30-day) trending.

For OWQM stations at water utility locations, the data may be communicated over a virtual private network (VPN) set up on the existing utility network, if available. For locations that do not have access to the utility's network, data are frequently communicated over commercial digital radio or cable service connections.

7. Fabricated On-line Water Quality Monitoring Stations

Installed OWQM stations take several forms, depending on the parameters and analyzers selected for use. Outdoor installations are generally fabricated in enclosed cabinets for protection and security. Because water flows inside the cabinet with an open drain, ventilation of the cabinet is required to dissipate moisture that may accumulate. In hot southern climates, temperatures inside the enclosed cabinets are also of concern, so ventilation is also necessary and shade from direct solar illumination is recommended. High internal temperatures can also impact the stability of chemical reagents, so reagentless analyzers are also preferred.

Many of these issues may be avoided by installing OWQM stations indoors. For secure indoor locations, such as those owned and controlled by the utility, the station may be configured as a wall-mount panel or open frame system. Heat and moisture are usually no longer an issue, but the potential for tampering or vandalism of equipment may be more of a concern. Indoor locations that are not owned by the utility, such as those at fire departments, police stations, hospitals or other host facilities, should be configured as enclosed cabinets that will still require ventilation to remove humidity. Indoor locations avoid the OWQM environmental impacts of excessive heat or cold, and also make it easier for utility technicians to conduct routine calibration and service. However, indoor installation at any facility requires that access be available on a short-notice basis to retrieve automatically collected water samples in the event potential contamination is detected. Continuous access is a priority for locating OWQM stations that hold contamination monitoring as a mission-critical consideration. For those focused only on operational benefits, quick access is less of a siting priority.

8. Operational Benefits

The most important aspect of OWQM systems are the benefits of assured and improved water quality provided to the consumer. Such monitoring may provide recommendations for adjustment of the water treatment process and feedback of the results of treatment process changes. OWQM stations can provide early warning of water main breaks, low or high chlorine conditions, nitrification, and other conditions and thereby not only improve operations, but can also save considerable costs to the utility.

When the EDS identifies an anomaly in a data stream, an alert is sent to the centralized distribution system monitoring dashboard, which allows the operator to spatially correlate information from other data streams (consumer complaints, enhanced security alarms, public health alerts) in real time. Operations staff can rapidly and independently query each alert to evaluate trends and relate the anomalies to explainable events that may be occurring in the water distribution system (e.g., a pipeline break). During this operational evaluation, it is possible that the cause of the alerts cannot be explained by known activities. When this occurs, the utility can proceed with a tiered response, i.e., a Consequence Management Plan, that becomes increasingly aggressive as more information related to a potential contamination event is received.

The ability of rapidly converting very large data streams into actionable information and providing the consolidated alert information on a user-friendly dashboard significantly decreases a utility's response time. The ability to identify trends and nuances in the data supports operational benefits not previously available. Examples of collected data associated with distribution system events are presented below.

Figure 2 is a plot of spectral absorbance that indicated a peak characteristic of iron oxide. Plots of the data in five-day increments showed the size of the peak was progressively increasing. This trend was ultimately identified as accumulating deposition of iron on the analyzer optics, caused by aggressive water unexpectedly leaching from ductile iron pipe. Identifying and addressing the problem early saved the utility an estimated $20 million in early pipe replacement costs.

In **Figure 3**, DOC and TOC plots were used to optimize granular activated carbon (GAC) filter performance. In this case the utility developed correlations between total trihalomethane (TTHM) production and effluent DOC for use in determining when to change the GAC and maintain system compliance. In this example the utility reduced the annual replacement costs by $100,000 at each WTP.

Figure 4 illustrates how OWQM data can be used to track water age. In this case, nitrate profiles were compared over time to determine travel time between sites. The data were validated through the utility's calibrated model and similar readings from other distribution system locations were used to verify the hydraulic model.

Figure 5 shows an example of how spectral absorbance changes indicated failure of treatment plant controls early enough for the problem to be resolved before major damage occurred. In this case, the failure allowed spent brine solution to flow into the distribution system reservoir. The immediate result was a spectral change associated with the highly colored brine solution blocking the

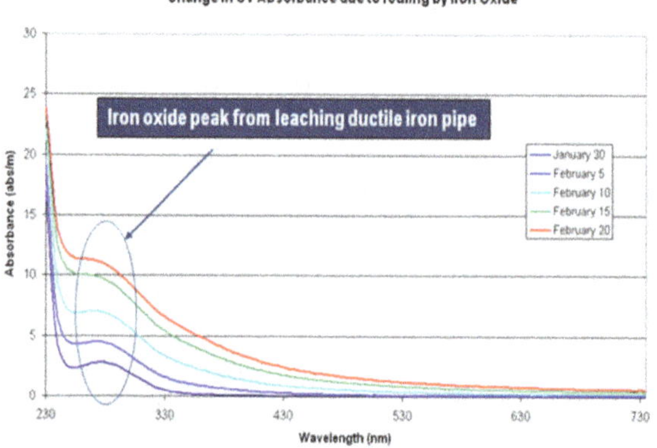

Figure 2. On-line water quality monitoring ultraviolet absorbance spectrum enables identification of distribution system accelerated corrosion.

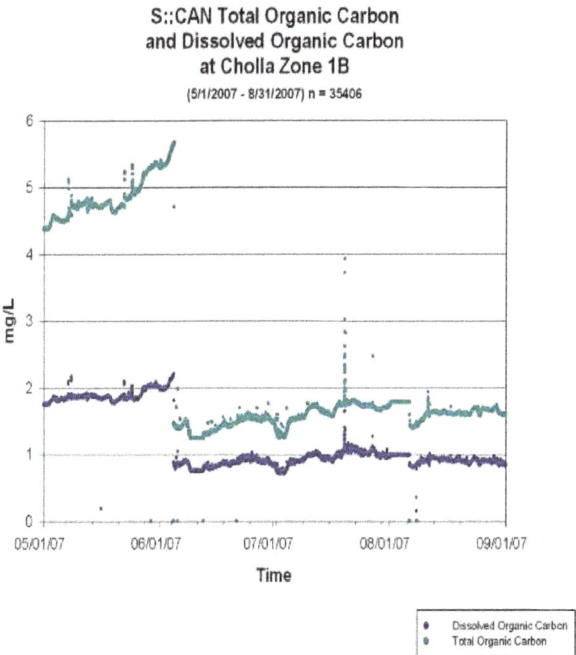

Figure 3. On-line water quality monitoring dissolved organic carbon and total organic carbon plots used to optimize treatment plant maintenance activities.

UV transmittance and creating a spectral alarm and notification. The graph on the left indicates normal absorbance (blue trace at bottom, behind and coincident with green trace). Over a period of several minutes, the absorbance increases, with maximum disruption indicated by the orange and purple marks indicating high absorbance of 30 to 50 percent for three minutes. Thirteen minutes after the start of the event, the condition was resolved, the brine solution had passed the monitoring station, and the absorbance spectrum returned to normal (green trace at bottom, coincident with the pre-event absorbance).

The right graph shows conditions the next day at monitoring station in the distribution system. The con-

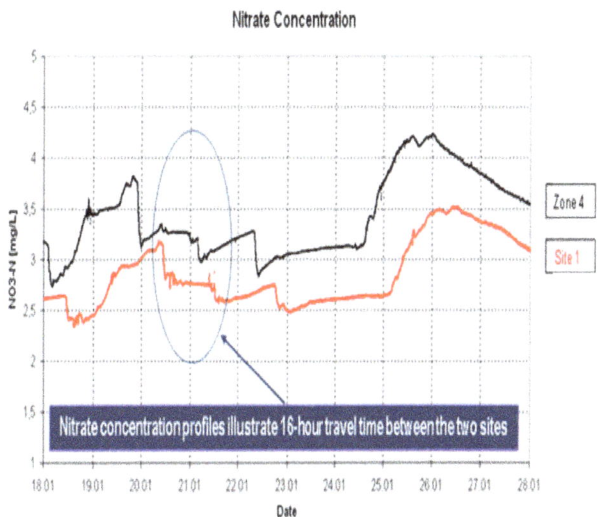

Figure 4. On-line water quality monitoring data enables comparison of nitrate profiles and hydraulic model verification.

taminant is shown to be considerably diluted, with the maximum absorption reduced to 30 to 40 percent and spread over seven minutes. The event is seen to take a total of four hours to pass the station from start of the event to return to normal water quality.

A military base that uses water from a local utility often had reports of water quality problems that could not be readily attributed to a known cause. Monitoring of the water inlet to the base identified changes in the water supply as shown in **Figure 6**. The water provider occasionally changed the source of the water and where it entered the distribution system. The result was heavy scouring in the pipe, directly impacting water quality on the base. Once the problem was understood, the base was able to work with the water provider to inform them of the impact associated with flow reversals in the distribution system. The water provider also implemented a flushing program to reduce the sediment that had accumulated in the distribution system pipelines.

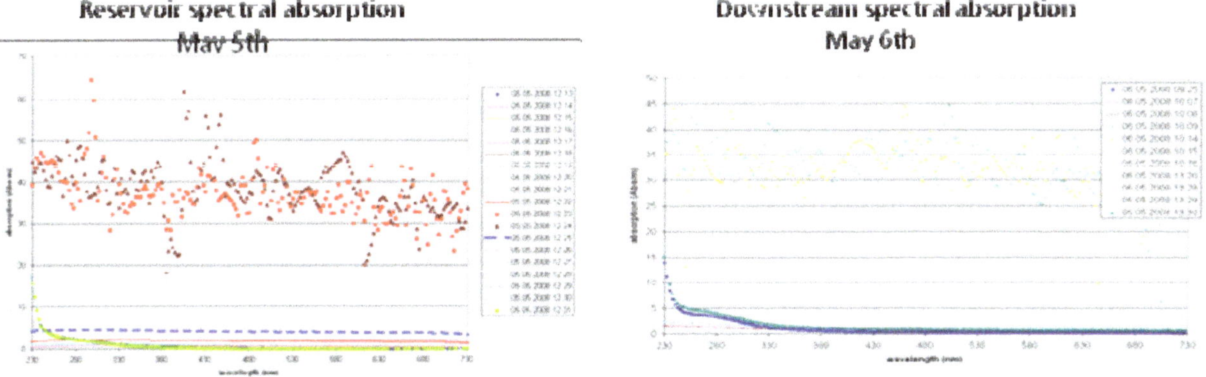

Figure 5. On-line water quality monitoring-identified changes in spectral characteristics indicate equipment failure.

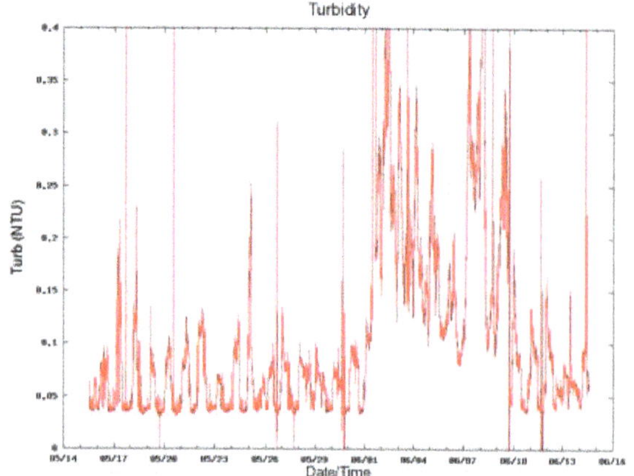

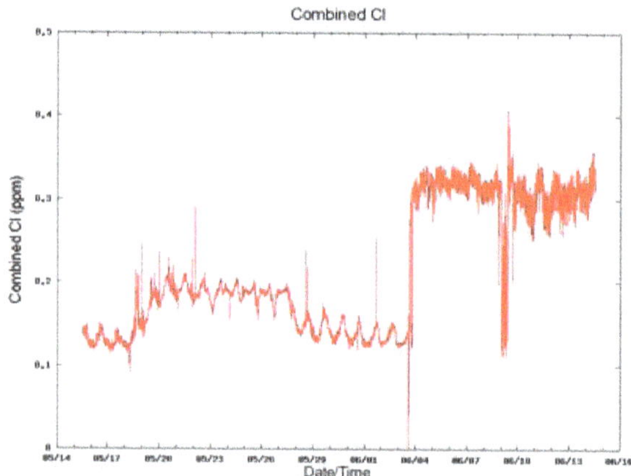

Figure 6. On-line water quality monitoring system identifies changes in source water as the cause of water quality problems.

9. Operational Benefits

Installation and operation of OWQM stations at WTP and reservoir outlets, and at strategic locations throughout the distribution system provides an understanding of water delivery conditions that may have been previously unknown or incompletely understood. These systems enable correction of problems and improvement in delivered water quality, provide added protection of public health, and produce a substantial savings in operational and maintenance/replacement costs. A mix of well known and new technology sensors and analyzers may be used to generate the substantial amount of data required for a complete picture of distribution system operation. The data can be displayed in an optimized manner on a central dashboard as well as through mobile technologies to support water utility operations.

REFERENCES

[1] CH2M HILL, "Contamination Warning System Demonstration Pilot Project: Implementation and Assessment," 2013.

[2] EPA. (n.d.) "The Effectiveness of Disinfectant Residuals in the Distribution System," Washington, DC: US EPA Office of Water, Office of Ground Water and Drinking Water. http://www.epa.gov/ogwdw/disinfection/tcr/pdfs/issuepaper_effectiveness.pdf

[3] EPA, "Revised Guidance Manual for Selecting Lead and Copper Control Strategies," Washington, DC: US EPA Office of Water, Office of Ground Water and Drinking Water, 2003.

[4] AWWA. (n.d.) Partnership for Safe Water Program Fact Sheet: Treatment Plant and Distribution System Optimization. http://www.awwa.org/Portals/0/files/resources/water%20utiity%20management/partnership%20safe%20water/files/FactSheetRevSept20102.pdf

[5] EPA, "Water Sentinel Online Water Quality Monitoring as an Indicator of Drinking Water Contamination," Washington, DC: US EPA Water Security Division, 2005.

[6] J. Dilling and K. Kaiser, "Estimation of the hydrophobic fraction of dissolved organic matter in water samples using UV photometry," *Water Resourse*, Vol. 36, No. 20, 2002, pp. 5037-5044.

[7] J. Hall and J. Szabo, "Distribution System Water Quality Monitoring: Sensor Technology Evaluation Methodology and Results," Cincinnati, OH: EPA, 2009.

[8] J. van den Broeke, "Spectral RICIN Measurement - Detection Limits," s: can Messtechnik, GmbH, 2009.

[9] EPA, "Water Security Initiative: Interim Guidance on Planning for Contamination Warning System Deployment," Cincinnati, OH: EPA Water Security Division, 2007.

[10] J. E. Berry, TEVA-SPOT Toolkit and User's Manual. Washington, DC: EPA, 2008.

14

Explosion of Sun

Alexander Bolonkin, Joseph Friedlander*
Strategic Solutions Technology Group, New York, USA

ABSTRACT

The Sun contains ~74% hydrogen by weight. The isotope hydrogen-1 (99.985% of hydrogen in nature) is a usable fuel for fusion thermonuclear reactions. This reaction runs slowly within the Sun because its temperature is low (relative to the needs of nuclear reactions). If we create higher temperature and density in a limited region of the solar interior, we may be able to produce self-supporting detonation thermonuclear reactions that spread to the full solar volume. This is analogous to the triggering mechanisms in a thermonuclear bomb. Conditions within the bomb can be optimized in a small area to initiate ignition, then spread to a larger area, allowing producing a hydrogen bomb of any power. In the case of the Sun certain targeting practices may greatly increase the chances of an artificial explosion of the Sun. This explosion would annihilate the Earth and the Solar System, as we know them today. The reader naturally asks: Why even contemplate such a horrible scenario? It is necessary because as thermonuclear and space technology spreads to even the least powerful nations in the centuries ahead, a dying dictator having thermonuclear missile weapons can proce (with some considerable mobilization of his military/industrial complex)—an artificial explosion of the Sun and take into his grave the whole of humanity. It might take tens of thousands of people to make and launch the hardware, but only a very few need know the final targeting data of what might be otherwise a weapon purely thought of (within the dictator's defense industry) as being built for peaceful, deterrent use. Those concerned about Man's future must know about this possibility and create some protective system—or ascertain on theoretical grounds that it is entirely impossie. Humanity has fears, justified to greater or lesser degrees, about asteroids, warming of Earthly climate, extinctions, etc. which have very small probability. But all these would leave survivors—nobody thinks that the terrible annihilation of the Solar System would leave a single person alive. That explosion appears possible at the present time. In this paper is derived the "AB-Criterion" which shows conditions wherein the artificial explosion of Sun is possible. The author urges detailed investigation and proving or disproving of this rather horrifying possibility, so that it may be dismissed from mind—or defended against.

Keywords: Artificial Explosion of Sun; Annihilation of Solar System; Criterion of Nuclear Detonation; Nuclear Detonation Wave; Detonate Sun; Artificial Supernova

1. Introduction

Information about Sun. The Sun is the star at the center of the Solar System. The Earth and other matter (including other planets, asteroids, meteoroids, comets and dust) orbit the Sun, which by itself accounts for about 99.8% of the solar system's mass. Energy from the Sun—in the form of sunlight—supports almost all life on Earth via photosynthesis, and drives the Earth's climate and weather.

The Sun is composed of hydrogen (about 74% of its mass, or 92% of its volume), helium (about 25% of mass, 7% of volume), and trace quantities of other elements. The Sun has a spectral class of G2V. G2 implies that it has a surface temperature of approximately 5500 K (or approximately 9600 degrees Fahrenheit/5315 Celsius), Sunlight is the main source of energy to the surface of Earth. The solar constant is the amount of power that the Sun deposits per unit area that is directly exposed to sunlight. The solar constant is equal to approximately 1370 watts per square meter of area at a distance of one AU from the Sun (that is, on or near Earth). Sunlight on the surface of Earth is attenuated by the Earth's atmosphere so that less power arrives at the surface—closer to 1000 watts per directly exposed square meter in clear conditions when the Sun is near the zenith.

*J. Friedlander corrected the author's English, wrote together with author Abstract, Sections 8, 10 ("Penetration into Sun" and "Results"), and wrote Section 11 "Discussion" as the solo author.

The Sun is about halfway through its main-sequence evolution, during which nuclear fusion reactions in its core fuse hydrogen into helium. Each second, more than 4 million tonnes of matter are converted into energy within the Sun's core, producing neutrinos and solar radiation; at this rate, the sun will have so far converted around 100 earth-masses of matter into energy. The Sun will spend a total of approximately 10 billion years as a main sequence star.

The core of the Sun is considered to extend from the center to about 0.2 solar radii. It has a density of up to 150,000 kg/m^3 (150 times the density of water on Earth) and a temperature of close to 13,600,000 kelvins (by contrast, the surface of the Sun is close to 5785 kelvins (1/2350th of the core)). Through most of the Sun's life, energy is produced by nuclear fusion through a series of steps called the *p-p* (proton-proton) chain; this process converts hydrogen into helium. The core is the only location in the Sun that produces an appreciable amount of heat via fusion: the rest of the star is heated by energy that is transferred outward from the core. All of the energy produced by fusion in the core must travel through many successive layers to the solar photosphere before it escapes into space as sunlight or kinetic energy of particles [1].

About 3.4 × 10^{38} protons (hydrogen nuclei) are converted into helium nuclei every second (out of about ~8.9 × 10^{56} total amount of free protons in Sun), releasing energy at the matter-energy conversion rate of 4.26 million tonnes per second, 383 yottawatts (383 × 10^{24} W) or 9.15 × 10^{10} megatons of TNT per second. This corresponds to extremely low rate of energy production in the Sun's core—about 0.3 μW/cm^3, or about 6 μW/kg. For comparison, an ordinary candle produces heat at the rate 1 W/cm^3, and human body—at the rate of 1.2 W/kg. Use of plasma with similar parameters as solar interior plasma for energy production on Earth is completely impractical—as even a modest 1 GW fusion power plant would require about 170 billion tonnes of plasma occupying almost one cubic mile. Thus all terrestrial fusion reactors require much higher plasma temperatures than those in Sun's interior to be viable.

The rate of nuclear fusion depends strongly on density (and particularly on temperature), so the fusion rate in the core is in a self-correcting equilibrium: a slightly higher rate of fusion would cause the core to heat up more and expand slightly against the weight of the outer layers, reducing the fusion rate and correcting the perturbation; and a slightly lower rate would cause the core to cool and shrink slightly, increasing the fusion rate and again reverting it to its present level.

The high-energy photons (gamma and X-rays) released in fusion reactions are absorbed in only few millimeters of solar plasma and then re-emitted again in random direction (and at slightly lower energy)—so it takes a long time for radiation to reach the Sun's surface. Estimates of the "photon travel time" range from as much as 50 million years to as little as 17,000 years. After a final trip through the convective outer layer to the transparent "surface" of the photosphere, the photons escape as visible light. Each gamma ray in the Sun's core is converted into several million visible light photons before escaping into space. Neutrinos are also released by the fusion reactions in the core, but unlike photons they very rarely interact with matter, so almost all are able to escape the Sun immediately.

This reaction is very slowly because the solar temperatute is very lower of Coulomb barrier.

The Sun's current age, determined using computer models of stellar evolution and nucleocosmochronology, is thought to be about 4.57 billion years.

Astronomers estimate that there are at least 70 sextillion (7 × 10^{22}) stars in the observable universe. That is 230 billion times as many as the 300 billion in the Milky Way [2].

Atmosphere of Sun. The parts of the Sun above the photosphere are referred to collectively as the solar atmosphere. They can be viewed with telescopes operating across the electromagnetic spectrum, from radio through visible light to gamma rays, and comprise five principal zones: the *temperature minimum*, the chromosphere, the transition region, the corona, and the heliosphere.

The chromosphere, transition region, and corona are much hotter than the surface of the Sun; the reason why is not yet known. But their density is low.

The coolest layer of the Sun is a temperature minimum region about 500 km above the photosphere, with a temperature of about 4000 K.

Above the temperature minimum layer is a thin layer about 2,000 km thick, dominated by a spectrum of emission and absorption lines. It is called the *chromosphere* from the Greek root *chroma*, meaning color, because the chromosphere is visible as a colored flash at the beginning and end of total eclipses of the Sun. The temperature in the chromosphere increases gradually with altitude, ranging up to around 100,000 K near the top.

Above the chromosphere is a transition region in which the temperature rises rapidly from around 100,000 K to coronal temperatures closer to one million K. The increase is because of a phase transition as helium within the region becomes fully ionized by the high temperatures. The transition region does not occur at a well-defined altitude. Rather, it forms a kind of nimbus around chromospheric features such as spicules and filaments, and is in constant, chaotic motion. The transition region is not easily visible from Earth's surface, but is readily observable from space by instruments sensitive to the far ultraviolet portion of the spectrum.

The corona is the extended outer atmosphere of the Sun, which is much larger in volume than the Sun itself. The corona merges smoothly with the solar wind that fills the solar system and heliosphere. The low corona, which is very near the surface of the Sun, has a particle density of 10^{14} m^{-3} - 10^{16} m^{-3}. (Earth's atmosphere near sea level has a particle density of about 2 × 10^{25} m^{-3}.) The temperature of the corona is several million kelvin. While no complete theory yet exists to account for the temperature of the corona, at least some of its heat is known to be from magnetic reconnection [3].

Physical characteristics of Sun: Mean diameter is 1.392 × 10^6 km (109 Earths). Volume is 1.41 × 10^{18} km³ (1,300,000 Earths). Mass is 1.988 435 × 10^{30} kg (332,946 Earths). Average density is 1408 kg/m³. Surface temperature is 5785 K (0.5 eV). Temperature of corona is 5 MK (0.43 keV). Core temperature is ~13.6 MK (1.18 keV). Sun radius is R = 696 × 10^3 km, solar gravity g_c = 274 m/s². Photospheric composition of Sun (by mass): Hydrogen 73.46%; Helium 24.85%; Oxygen 0.77%; Carbon 0.29%; Iron 0.16%; Sulphur 0.12%; Neon 0.12%; Nitrogen 0.09%; Silicon 0.07%; Magnesium 0.05%.

Sun photosphere has thickness about 7 × 10^{-4} R (490 km) of Sun radius R, average temperature 5.4 × 10^3 K, and average density 2 × 10^{-7} g/cm³ (n = 1.2 × 10^{23} m^{-3}). Sun convection zone has thickness about 0.15 R, average temperature 0.25 × 10^6 K, and average density 5 × 10^{-7} g/cm³. Sun intermediate (radiation) zone has thickness about 0.6 R, average temperature 4 × 10^6 K, and average density 10 g/cm³. Sun core has thickness about 0.25 R, average temperature 11 × 10^6 K, and average density 89 g/cm³.

Detonation is a process of combustion in which a supersonic shock wave is propagated through a fluid due to an energy release in a reaction zone. This self-sustained detonation wave is different from a deflagration, which propagates at a subsonic rate (*i.e.*, slower than the sound speed in the material itself).

Detonations can be produced by explosives, reactive gaseous mixtures, certain dusts and aerosols.

The simplest theory to predict the behavior of detonations in gases is known as Chapman-Jouguet (CJ) theory, developed around the turn of the 20th century. This theory, described by a relatively simple set of algebraic equations, models the detonation as a propagating shock wave accompanied by exothermic heat release. Such a theory confines the chemistry and diffusive transport processes to an infinitely thin zone.

A more complex theory was advanced during World War II independently by Zel'dovich, von Neumann, and Doering. This theory, now known as ZND theory, admits finite-rate chemical reactions and thus describes a detonation as an infinitely thin shock wave followed by a zone of exothermic chemical reaction. In the reference frame in which the shock is stationary, the flow following the shock is subsonic. Because of this, energy release behind the shock is able to be transported acoustically to the shock for its support. For a self-propagating detonation, the shock relaxes to a speed given by the Chapman-Jouguet condition, which induces the material at the end of the reaction zone to have a locally sonic speed in the reference frame in which the shock is stationary. In effect, all of the chemical energy is harnessed to propagate the shock wave forward.

Both CJ and ZND theories are one-dimensional and steady. However, in the 1960s experiments revealed that gas-phase detonations were most often characterized by unsteady, three-dimensional structures, which can only in an averaged sense be predicted by one-dimensional steady theories. Modern computations are presently making progress in predicting these complex flow fields. Many features can be qualitatively predicted, but the multi-scale nature of the problem makes detailed quantitative predictions very difficult [1-4].

2. Statement of Problem, Main Idea and Our Aim

The present solar temperature is far lower than needed for propagating a runaway thermonuclear reaction. In Sun core the temperature is only ~13.6 MK (0.0012 MeV). The Coulomb barrier for protons (hydrogen) is more then 0.4 MeV. Only very small proportions of core protons take part in the thermonuclear reaction (they use a tunnelling effect). Their energy is in balance with energy emitted by Sun for the Sun surface temperature 5785 K (0.5 eV).

We want to clarify: If we create a zone of limited size with a high temperature capable of overcoming the Coulomb barrier (for example by insertion of a thermonuclear warhead) into the solar photosphere (or lower), can this zone ignite the Sun's photosphere (ignite the Sun's full load of thermonuclear fuel)? Can this zone self-support progressive runaway reaction propagation for a significant proportion of the available thermonuclear fuel?

If it is possible, researchers can investigate the problems: What will be the new solar temperature? Will this be metastable, decay or runaway? How long will the transformed Sun live, if only a minor change? What the conditions will be on the Earth?

Why is this needed?

As thermonuclear and space technology spreads to even the least powerful nations in the decades and centuries ahead, a dying dictator having thermonuclear weapons and space launchers can produce (with some considerable mobilization of his military/industrial complex)—

the artificial explosion of the Sun and take into his grave the whole of humanity.

It might take tens of thousands of people to make and launch the hardware, but only a very few need know the final targeting data of what might be otherwise a weapon purely thought of (within the dictator's defense industry) as being built for peaceful, "business as usual" deterrent use. Given the hideous history of dictators in the twentieth century and their ability to kill technicians who had outlived their use (as well as major sections of entire populations also no longer deemed useful) we may assume that such ruthlessness is possible.

Given the spread of suicide warfare and self-immolation as a desired value in many states, (in several cultures—think Berlin or Tokyo 1945, New York 2001, Tamil regions of Sri Lanka 2006) what might obtain a century hence? All that is needed is a supportive, obedient defense complex, a "romantic" conception of mass death as an ideal—even a religious ideal—and the realization that his own days at power are at a likely end. It might even be launched as a trump card in some (to us) crazy internal power struggle, and plunged into the Sun and detonated in a mood of spite by the losing side. "*Burn baby burn*"!

A small increase of the average Earth's temperature over 0.4 K in the course of a century created a panic in humanity over the future temperature of the Earth, resulting in the Kyoto Protocol. Some stars with active thermonuclear reactions have temperatures of up to 30,000 K. If not an explosion but an enchanced burn results the Sun might radically increase in luminosity for say—a few hundred years. This would suffice for an average Earth temperature of hundreds of degrees over 0°C. The oceans would evaporate and Earth would bake in a Venus like greenhouse, or even lose its' atmosphere entirely.

Thus we must study this problem to find methods of defense from human induced Armageddon.

The interested reader may find needed information in [4-9].

3. Theory Estimations and Computation

1) Coulomb barrier (repulsion). Energy is needed for thermonuclear reaction may be computed by equations

$$E = \frac{kZ_1Z_2e^2}{r} = 2.3\times10^{-28}\frac{Z_1Z_2}{r}[\mathrm{J}]$$

$$\text{or} \quad E = \frac{kZ_1Z_2e}{r} = 1.44\times10^{-9}\frac{Z_1Z_2}{r}[\mathrm{eV}], \qquad (1)$$

$$\text{where} \quad r = r_1 + r_2, \quad r_i = (1.2 \div 1.5)\times10^{-15}\sqrt[3]{A_i}$$

where E is energy needed for forcing contact between two nuclei, J or eV; $k = 9 \times 10^9$ is electrostatic constant, $\mathrm{N{\cdot}m^2/C^2}$; Z is charge state; $e = 1.6 \times 10^{-19}$ is charge of proton, C; r is distance between nucleus centers, m; r_i is radius of nucleus, m; $A = Z + N$ is nuclei number, N is number neutrons into given ($i = 1, 2$) nucleus.

The computations of average temperature (energy) for some nucleus are presented in **Table 1** below. We assume that the first nucleus is moving; the second (target) nucleus is motionless.

In reality the temperature of plasma may be significantly lower than in table 1 because the nuclei have different velocity. Parts of them have higher velocity (see Maxwell distribution of nuclei speed in plasma), some of the nuclei do not (their energy are summarized), and there are tunnel effects. If the temperature is significantly lower, then only a small part of the nuclei took part in reaction and the fuel burns very slowly. This case we have—happily in the present day Sun where the temperature in core has only 0.0012 MeV and the Sun can burn at this rate for billions of years [5,6].

The ratio between temperatures in eV and in K is

$$T_K = 1.16\times10^4 T_e, \quad T_e = 0.86\times10^{-4} T_K. \qquad (2)$$

2) The energy of a nuclear reaction. The energy and momentum conservation laws define the energetic relationships for a nuclear reaction [1,2].

When a reaction $A(a,b)B$ occurs, the quantity

$$Q = \left[(M_A + M_a) - (M_B + M_b)\right]c^2, \qquad (3)$$

where M_i are the masses of the particles participating in the reaction and c is the speed of light, Q is the reaction energy.

Usually *mass defects* ΔM are used, instead of masses, for computing Q:

$$Q = (\Delta M_A + \Delta M_a) - (\Delta M_B + \Delta M_b). \qquad (4)$$

The mass defect is the quantity $\Delta M = M - A$ where M is the actual mass of the particle (atom), A is the so-called mass number, *i.e.* the total number of nucleons (protons and neutrons) in the atomic nucleus. If M is expressed in atomic mass units (a.m.u.) and A is assigned the same unit, then ΔM is also expressed in a.m.u. One a.m.u. represent 1/12 of the ^{12}C nuclide mass and equals $1.6605655 \times 10^{-27}$ kg. For calculations of reaction energies it is more convenient to express ΔM in kilo-electronvolts: a.m.u. = 931501.59 keV.

Employing the mass defects, one can handle numbers that are many times smaller than the nuclear masses or the binding energies.

Table 1. Columb barrier of some nuclei pairs.

Reaction	E, MeV	Reaction	E, MeV	Reaction	E, MeV	Reaction	E, MeV
p + p	0.53	T + p	0.44	^{6}L + p	1.13	^{13}C + p	1.9
D + p	0.47	D + d	0.42	^{7}Be + p	1.5	^{12}C + ^{4}He	3.24

Kinetic energy may be released during the course of a reaction (exothermic reaction) or kinetic energy may have to be supplied for the reaction to take place (endothermic reaction). This can be calculated by reference to a table of very accurate particle rest masses (see http://physics.nist.gov/PhysRefData/Compositions/index.html). The reaction energy (the "Q-value") is positive for exothermal reactions and negative for endothermal reactions.

The other method calculate of thermonuclear energy is in [1]. For a nucleus of atomic number Z, mass number A, and Atomic mass $M(Z,A)$, the binding energy is

$$Q=\left[ZM\left({}^{1}\mathrm{H}\right)+\left(A-Z\right)m_n-M\left(Z,A\right)\right]c^2, \quad (5)$$

where $M({}^1H)$ is mass of a hydrogen atom and m_n is mass of neutron. This equation neglects a small correction due to the binding energy of the atomic electrons.

The binding energy per nucleus Q/A, varies only slightly in the range of 7 - 9 MeV for nuclei with $A > 12$.

The binding energy can be approximately calculated from Weizsacker's semiempirical formula:

$$Q=a_v A-a_s A^{2/3}-a_c Z\left(Z-1\right)A^{-1/3} \\ -a_{sym}\left(A-2Z\right)^2/A+\delta \quad (6)$$

where δ accounts for pairing of like nucleons and has the value $+a_pA^{-3/4}$ for Z and N both even, $-a_pA^{-3/4}$ for Z and N both odd, and zero otherwise (A odd). The constants in this formula must be adjusted for the best agreement with data: typical values are $a_v = 15.5$ MeV, $a_s = 16.8$ MeV, $a_c = 0.72$ MeV, $a_{sym} = 23$ MeV, and $a_p = 34$ MeV.

The binding energy per nucleon of the helium-4 nucleus is unusually high, because the He-4 nucleus is doubly magic. (The He-4 nucleus is unusually stable and tightly-bound for the same reason that the helium atom is inert: each pair of protons and neutrons in He-4 occupies a filled **1s** nuclear orbital in the same way that the pair of electrons in the helium atom occupies a filled **1s** electron orbital). Consequently, alpha particles appear frequently on the right hand side of nuclear reactions [7,8].

The energy released in a nuclear reaction can appear mainly in one of three ways:

- kinetic energy of the product particles.
- emission of very high energy photons, called gamma rays.
- some energy may remain in the nucleus, as a metastable energy level.

When the product nucleus is metastable, this is indicated by placing an asterisk ("*") next to its atomic number. This energy is eventually released through nuclear decay.

If the reaction equation is balanced, that does not mean that the reaction really occurs. The rate at which reactions occur depends on the particle energy, the particle flux and the reaction cross section.

In the initial collision which begins the reaction, the particles must approach closely enough so that the short range strong force can affect them. As most common nuclear particles are positively charged, this means they must overcome considerable electrostatic repulsion before the reaction can begin. Even if the target nucleus is part of a neutral atom, the other particle must penetrate well beyond the electron cloud and closely approach the nucleus, which is positively charged. Thus, such particles must be first accelerated to high energy, for example by very high temperatures, on the order of millions of degrees, producing thermonuclear reactions

Also, since the force of repulsion is proportional to the product of the two charges, reactions between heavy nuclei are rarer, and require higher initiating energy, than those between a heavy and light nucleus; while reactions between two light nuclei are commoner still.

Neutrons, on the other hand, have no electric charge to cause repulsion, and are able to affect a nuclear reaction at very low energies. In fact at extremely low particle energies (corresponding, say, to thermal equilibrium at room temperature), the neutron's de Broglie wavelength is greatly increased, possibly greatly increasing its capture cross section, at energies close to resonances of the nuclei involved. Thus low energy neutrons *may* be even more reactive than high energy neutrons [9].

3) Distribution of thermonuclear energy between particles. In most cases, the result of thermonuclear reaction is more than one product. As you see in **Table 2** that may be "He" and neutron or proton. The thermonuclear energy distributes between them in the following manner:

$$\text{From}\quad E=E_1+E_2=\frac{m_1V_1^2}{2}+\frac{m_2V_2^2}{2},\quad m_1V_1=m_2V_2, \\ \text{we have}\quad \frac{E_1}{E}=\frac{m_2}{m_1+m_2}=\frac{\mu_2}{\mu_1+\mu_2},\quad E_2=E-E_1, \quad (7)$$

where m is particle mass, kg; V is particle speed, m/s; E is particle energy, J; $\mu = m_i/m_p$ is relative particle mass. Lower indexes "$_{1,\,2}$" are number of particles.

After some collisions the energy $E = kT$ (temperature) of different particles may be closed to equal.

4) The power density produced in thermonuclear reaction may be computed by the equation

$$P=En_1n_2\left\langle\sigma v\right\rangle, \quad (8)$$

where E is energy of single reaction, eV or J; n_1 is density (number particles in cm^3) the first component; n_2 is density (number particles in cm^3) the second component; $\langle\sigma v\rangle$ is reaction rate, in cm^3/s; σ is cross section of reaction, cm^2, 1 barn = 10^{-24} cm^2; v is speed of the first component, cm/s; P is power density, eV/cm^3 or J/cm^3. Cross section of reaction before σ_{max} very strongly

Table 2. Exothermic thermonuclear reactions.

№	Reaction	Energy of reaction MeV	σ_{max} barn $E \le 1$ MeV	E of σ_{max} MeV	№	Reaction MeV	Energy of Reaction MeV	σ_{max} barn $E \le 1$ MeV	E of σ_{max} MeV
1	$p + p \to d + e^+ + \nu$	2.2	10^{-23}	-	15	$d + {}^6Li \to {}^7Li + p$	5.0	0.01	1
2	$p + d \to {}^3He + \gamma$	5.5	10^{-6}	-	16	$d + {}^6Li \to 2{}^4He$	22.4	0.026	0.60
3	$p + t \to {}^4He + \gamma$	19.7	10^{-6}	-	17	$d + {}^7Li \to 2{}^4He + n$	15.0	10^{-3}	0.2
4	$d + d \to t + p$	4.0	0.16	2	18	$p + {}^9Be \to 2{}^4He + d$	0.56	0.46	0.33
5	$d + d \to {}^3He + n$	3.3	0.09	1	19	$p + {}^9Be \to {}^6Li + {}^4He$	2.1	0.34	0.33
6	$d + d \to {}^4He + \gamma$	24	-	-	20	$p + {}^{11}B \to 3{}^4He$	8.7	0.6	0.675
7	$d + t \to {}^4He + n$	17.6	5	0.13	21	$p + {}^{15}N \to {}^{12}C + {}^4He$	5.0	0.6	1.2
8	$t + d \to {}^4He + n$	17.6	5	0.195	22	$d + {}^6Li \to {}^7Be + n$	3.4	0.01	0.3
9	$t + t \to {}^4He + 2n$	11.3	0.1	1	23	${}^3He + t \to {}^4He + d$	14.31	0.7	≈1
10	$d + {}^3He \to {}^4He + p$	18.4	0.71	0.47	24	${}^3H + {}^4He \to {}^7Li + \gamma$	2.457	$7 \cdot 10^{-5}$	≈3
11	${}^3He + {}^3He \to {}^4He + 2p$	12.8	-	-	25	${}^3H + d \to {}^4He$	17.59	$5 \cdot 10^{-4}$	≈2
12	$n + {}^6Li \to {}^4He + t$	4,8	2.6	0.26	26	${}^{12}C + p \to {}^{13}N + \gamma$	1.944	10^{-6}	0.46
13	$p + {}^6Li \to {}^4He + {}^3He$	4,0	10^{-4}	0.3	27	${}^{13}C + p \to {}^{14}N + \gamma$	7.55	10^{-4}	0.555
14	$p + {}^7Li \to 2{}^4He + \gamma$	17.3	$6 \cdot 10^{-3}$	0.44	28	${}^3He + {}^4He \to {}^7Be + \gamma$	1.587	10^{-6}	≈8

Here are: p (or ${}^1H$)—proton, d (or D, or ${}^2H$)—deuterium, t (or T, or ${}^3H$)—tritium, n—neutron, He—helium, Li—lithium, Be—beryllium, B—barium, C—carbon, N—hydrogen, ν—neutrino, γ—gamma radiation.

depends from temperature and it is obtainable by experiment. They can have the maximum resonance. For very high temperatures the σ may be close to the nuclear diameter.

The terminal velocity of the reaction components (electron and ions) are

$$v_{Te} = \left(kT_e/m_e\right)^{1/2} = 4.19\times10^7 T_e^{1/2}.\ \text{cm/s}, \qquad (9)$$

$$v_{Ti} = \left(kT_i/m_i\right)^{1/2} = 9.79\times10^7 \left(T_i/\mu_i\right)^{1/2}.\ \text{cm/s}, \qquad (10)$$

where T is temperature in eV; $\mu_i = m_i/m_p$ is ratio of ion mass to proton mass.

The sound velocity of ions is

$$v = \left(\frac{\gamma z k T_k}{m_i}\right)^{1/2}, \qquad (11)$$

where $\gamma \approx (1.2 - 1.4)$ is adiabatic coefficient; z is number of charge ($z = 1$ for p), T_k is plasma temperature in K; m_i is mass of ion.

The deep of penetration of outer radiation into plasma is

$$d = 5.31\times10^5 n_e^{-1/3}, [\text{cm}]$$

where n_e is number of electrons in unit of volume.

In internal plasma detonation there is no loss in radiation because the plasma reflects the radiation.

4. Possible Thermonuclear Reactions to Power a Hypothetical Solar Explosion

The Sun mass is ~74% hydrogen and 25% helium.

Possibilities exist for the following self-supporting nuclear reactions in the hydrogen medium: proton chain reaction, CNO cycle, Triple-alpha process, Carbon burning process, Neon burning process, Oxygen burning process, Silicon burning process.

For our case of particular interest (a most probable candidate) the proton-proton chain reaction. It is more exactly the reaction $p + p$.

The proton-proton chain reaction is one of several fusion reactions by which stars convert hydrogen to helium, the primary alternative being the CNO cycle. The proton-proton chain dominates in stars the size of the Sun or less.

The first step involves the fusion of two hydrogen nuclei ${}^1H$ (protons) into deuterium ${}^2H$, releasing a positron and a neutrino as one proton changes into a neutron.

$${}^1H + {}^1H \to {}^2H + e^+ + \nu_e. \qquad (12)$$

with the neutrinos released in this step carrying energies up to 0.42 MeV.

The positron immediately annihilates with an electron, and their mass energy is carried off by two gamma ray photons.

$$e^+ + e^- \to 2\gamma + 1.02\ \text{MeV}. \qquad (13)$$

After this, the deuterium produced in the first stage can fuse with another hydrogen to produce a light isotope of helium, ${}^3He$:

$${}^2H + {}^1H \to {}^3He + \gamma + 5.49\ \text{MeV}. \qquad (14)$$

From here there are three possible paths to generate helium isotope ${}^4He$. In pp1 helium-4 comes from fusing two of the helium-3 nuclei produced; the pp2 and pp3 branches fuse ^{3}He with a pre-existing ^{4}He to make Beryllium-7. In the Sun, branch pp1 takes place with a frequency of 86%, pp2 with 14% and pp3 with 0.11%. There is also an extremely rare pp4 branch.

1) The pp I branch

$$^3\text{He} + {}^3\text{He} \to {}^4\text{He} + {}^1\text{H} + {}^1\text{H} + 12.86\ \text{MeV}$$

The complete pp I chain reaction releases a net energy of 26.7 MeV. The pp I branch is dominant at temperatures of 10 to 14 megakelvins (MK). Below 10 MK, the PP chain does not produce much ^{4}He.

2) The pp II branch

$^3\text{He} + {}^4\text{He}$	$\to$	$^7\text{Be} + \gamma$
$^7\text{Be} + \text{e}^-$	$\to$	$^7\text{Li} + \nu_e$
$^7\text{Li} + {}^1\text{H}$	$\to$	$^4\text{He} + {}^4\text{He}$

The pp II branch is dominant at temperatures of 14 to 23 MK. 90% of the neutrinos produced in the reaction $^7\text{Be}(\text{e}^-,\nu_e)^7\text{Li}^*$ carry an energy of 0.861 MeV, while the remaining 10% carry 0.383 MeV (depending on whether lithium-7 is in the ground state or an excited state, respectively).

3) The pp III branch

$^3\text{He} + {}^4\text{He}$	$\to$	$^7\text{Be} + \gamma$
$^7\text{Be} + {}^1\text{H}$	$\to$	$^8\text{B} + \gamma$
^8B	$\to$	$^8\text{Be} + \text{e}^+ + \nu_e$
^8Be	$\leftrightarrow$	$^4\text{He} + {}^4\text{He}$

The pp III chain is dominant if the temperature exceeds 23 MK.

The pp III chain is not a major source of energy in the Sun (only 0.11%), but was very important in the solar neutrino problem because it generates very high energy neutrinos (up to 14.06 MeV).

4) The pp IV or hep

This reaction is predicted but has never been observed due to its great rarity (about 0.3 parts per million in the Sun). In this reaction, Helium-3 reacts directly with a proton to give helium-4, with an even higher possible neutrino energy (up to 18.8 MeV).

$$^3\text{He} + {}^1\text{H} \to {}^4\text{He} + \nu_e + \text{e}^+$$

5) Energy release

Comparing the mass of the final helium-4 atom with the masses of the four protons reveals that 0.007 or 0.7% of the mass of the original protons has been lost. This mass has been converted into energy, in the form of gamma rays and neutrinos released during each of the individual reactions.

The total energy we get in one whole chain is

$$4\,{}^1\text{H} \to {}^4\text{He} + 26.73\ \text{MeV}\ .$$

Only energy released as gamma rays will interact with electrons and protons and heat the interior of the Sun. This heating supports the Sun and prevents it from collapsing under its own weight. Neutrinos do not interact significantly with matter and do not help support the Sun against gravitational collapse. The neutrinos in the ppI, ppII and ppIII chains carry away the 2.0%, 4.0% and 28.3% of the energy respectively.

This creates a situation in which stellar nucleosynthesis produces large amounts of carbon and oxygen but only a small fraction of these elements is converted into neon and heavier elements. Both oxygen and carbon make up the *ash* of helium burning. Those nuclear resonances sensitively are arranged to create large amounts of carbon and oxygen, has been controversially cited as evidence of the anthropic principle.

About 34% of this energy is carried away by neutrinos. That reaction is part of solar reaction, but if initial temperature is high, the reaction becomes an explosion.

The detonation wave works a short time. That supports the reactions (12)-(13). They produce energy up to 1.44 MeV. The reactions (12)-(14) produce energy up to 5.8 MeV. But after detonation wave and the full range of reactions the temperature of plasma is more than the temperature needed to pass the Coulomb barrier and the energy of explosion increases by 20 times [10-12].

5. Theory of Detonation

The one dimensional detonation wave may be computed by equations (see **Figure 1**):

1) Law of mass

$$\frac{D}{V_1} = \frac{v}{V_3}, \tag{15}$$

where D—speed of detonation, m/s; v—speed of ion sound, m/s about the front of detonation wave (Equation (11)); V_1, V_3 specific density of plasma in points 1, 3 respectively, kg/m^3.

2) Law of momentum

$$p_1 + \frac{D^2}{V_1} = p_3 + \frac{v^2}{V_3}, \tag{16}$$

where p_1, p_3 are pressures, N/m^2, in point 1, 3 respectively.

3) Law of energy

$$E_3 - E_1 = Q + 0.5(p_3 + p_1)(V_1 - V_3), \tag{17}$$

where E_3, E_1—internal energy, J/kg, of mass unit in point 3, 1 respectively, Q is nuclear energy, J/kg.

4) Speed of detonation is

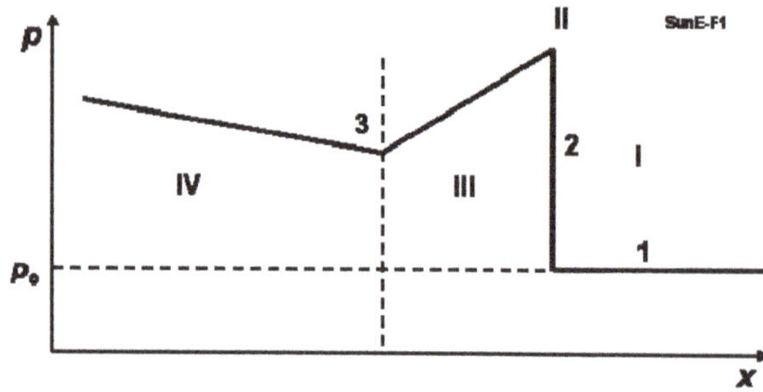

Figure 1. Pressure in detonation wave. I—plasma, II—front of detonation wave, III—zone of the initial thermonuclear fusion reaction, IV—products of reaction and next reaction, p_o—initial pressure, x—distance.

$$D = \sqrt{2Q\left(\gamma^2 - 1\right)}, \quad (18)$$

$\gamma \approx 1.2$ - 1.4 is adiabatic coefficient [13].

6. Model of Artificial Sun Explosion. Estimation of Ignition

Thermonuclear reactions proceeding in the Sun's core are under high temperature and pressure. However the core temperature is substantially lower than that needed to overcome the Columb barrier. That way the thermonuclear reaction is very slow and the Sun's life cycle is about 10 billion years. But that is enough output to keep the Sun a plasma ball, hot enough for life on Earth to exist. Now we are located in the middle of the Sun's life and have about 5 billions years until the Sun becomes a Red Giant.

However, this presumes that the Sun is stable against deliberate tampering. Supposing our postulations are correct, the danger exists that introducing a strong thermonuclear explosion into the Sun which is a container of fuel for thermonuclear reactions, the situation can be cardinally changed. For correct computations it is necessary to have a comprehensive set of full initial data (for example, all cross-section areas of all nuclear reactions) and supercomputer time. The author does not have access to such resources. That way he can only estimate probability of these reactions, their increasing or decreasing. Supportive investigations are welcome in order to restore confidence in humanity's long term future [14].

7. AB-Criterion for Solar Detonation

A self-supporting detonation wave is possible if the speed of detonation wave is greater or equals the ion sound speed:

$$D \geq v, \text{ where } D = \sqrt{2Q\left(\gamma^2 - 1\right)}, v = \left(\frac{\gamma z k T_k}{m_i}\right)^{1/2}. \quad (19)$$

Here Q is a nuclear specific heat [J/kg], $\gamma = 1.2$ - 1.4 is adiabatic coefficient (they are noted in (17)-(18)); z is number of the charge of particle after fusion reaction ($z = 1$ for ^{2}H), $k = 1.36 \times 10^{-23}$ is Boltzmann constant, J/K; T_k is temperature of plasma after fusion reaction in Kelvin degrees; $m_i = \mu m_p$ is mass of ion after fusion reaction, kg; $m_p = 1.67 \times 10^{-27}$ kg is mass of proton; μ is relative mass, $\mu = 2$ for ^{2}H.

When we have sign ">" the power of the detonation wave increases, when we have the sign "<" it decreases.

Substitute two last equations in the first equation in (19) we get

$$D^2 \geq v^2, 2Q\left(\gamma^2 - 1\right) \geq \frac{\gamma z k T_k}{m_i}.$$
$$\text{where } Q = \frac{f e E \tau}{n m_p} = \frac{1}{4} n^2 e E \langle \sigma v \rangle \frac{\tau}{n m_p} \quad (20)$$

where f is speed of nuclear reaction, s/m^3; $e = 1.6 \times 10^{-19}$ is coefficient for converting the energy from electron-volts to joules; E is energy of reaction in eV; n is number particles (p - protons) in m^3; $\langle \sigma v \rangle$ is reaction rate, m^3/s (**Figure 2**), $m_i = 2m_p$, τ is time, sec.

From (20) we get the AB-Criterion for artificial Sun explosion:

$$n\tau \geq \frac{\gamma z k T_k}{\left(\gamma^2 - 1\right) e E \langle \sigma v \rangle} = \frac{1.16 \times 10^4 \gamma z k T_e}{\left(\gamma^2 - 1\right) e E \langle \sigma v \rangle} = \frac{\gamma z T_e}{\left(\gamma^2 - 1\right) E \langle \sigma v \rangle} \quad (21)$$

where T_e is temperature of plasma after reaction in eV.

The offered AB-Criterion (21) is different from the well-known Lawson criterion

$$n_e \tau_e > \frac{12 k_B T_k}{E_{ch} \langle \sigma v \rangle},$$

where E_{ch} is energy of reaction in keV, k_B is Boltzmann

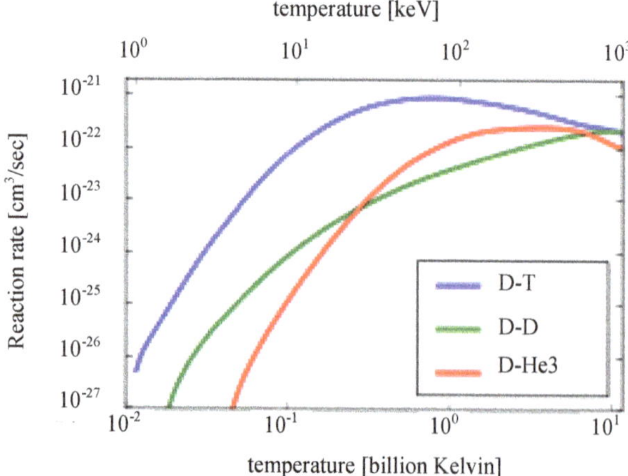

Figure 2. Reaction rate $\langle \sigma v \rangle$ via plasma temperature for D-T (top), D-D (middle) and D-^{3}He (bottom in left side).

constant.

The offered AB-Criterion contains the γ adiabatic coefficient and z—number of electric charge in the electron charges. It is not surprising because Lawson derived his criterion from the condition where the energy of the reaction must be greater than the loss of energy by plasma into the reactor walls, where

$$W_{\text{reaction}} > W_{\text{loss}} .$$

In our case no the reactor walls and plasma reflects the any radiation.

The offered AB-Criterion is received from the condition (19): Speed of self-supporting detonation wave must be greater than the speed of sound where

$$D > v .$$

For main reaction $p + p$ the AB-Criterion (21) has a form

$$n\tau \geq \frac{T_e}{E\langle \sigma v \rangle} . \tag{21a}$$

Estimation. Let us take the first step of the reaction $^1H + {}^1H$ (12)-(13) having in point 3 (**Figure 1**) $T_e = 10^5$ eV, $E \approx 1.44 \times 10^6$ eV, $<\sigma v> \approx \times 10^{-22}$. Substituting them in Equation (21) we receive

$$n\tau > 0.7 \times 10^{21} . \tag{22}$$

The Sun surface (photosphere) has density $n = 10^{23}$ 1/m^3, the encounter time of protons in the hypothetical detonation wave III (**Figure 1**) may be over 0.01 sec. The values in left and right sides of (22) have the same order. That means a thermonuclear bomb exploded within the Sun may conceivably be able to serve as a detonator which produces a self-supported nuclear reaction and initiates the artificial explosion of the Sun.

After the initial reaction the temperature of plasma is very high (>1 MeV) and time of next reaction may be very large (hundreds of seconds), the additional energy might in these conditions increase up to 26 MeV.

A more accurate computation is possible but will require cooperation of an interested supercomputer team with the author, or independent investigations with similar interests [15].

8. Penetration of Thermonuclear Bomb into Sun

The Sun is a ball of plasma (ionized gases), not a solid body. A properly shielded thermonuclear bomb can permeate deep into the Sun. The warhead may be protected on its' way down by a special high reflectivity mirror offered, among others, by author A.A. Bolonkin in 1983 [11] and described in [7] Chapters 12, 3A, [8] Ch.5 (see also [9-15]). This mirror allows to maintain a low temperature of the warhead up to the very boundary of the solar photosphere. At that point its' velocity is gigantic, about 617.6 km/s, assuring a rapid penetration for as far as it goes.

The top solar atmosphere is very rarefied; a milliard (US billion) times less than the Earth's atmosphere. The Sun's photosphere has a density approximately 200 times less than the Earth's atmosphere. Some references give a value of only 0.0000002 gm/cm^3 (0.1 millibar) at the photosphere surface. Since present day ICBM warheads can penetrate down (by definition) to the 1 bar level (Earth's surface) and that is by no means the boundary of the feasible, the 10 bar level may be speculated to be near-term achievable. The most difficult entry yet was that of the Galileo atmospheric probe on Dec. 7, 1995 [16]. The Galileo Probe was a 45° sphere-cone that entered Jupiter's atmosphere at 47.4 km/s (atmosphere relative speed at 450 km above the 1 bar reference altitude). The peak deceleration experienced was 230 g (2.3 km/s^2). Peak stagnation point pressure before aeroshell jettison was 9 bars (900 kPa). The peak shock layer temperature was approximately 16000 K (and remember this is into hydrogen (mostly) the solar photosphere is merely 5800 K). Approximately 26% of the Galileo Probe's original entry mass of 338.93 kg was vaporized during the 70 second heat pulse. Total blocked heat flux peaked at approximately 15000 W/cm² (hotter than the surface of the Sun).

If the entry vehicle was not optimized for slowdown as the Galileo Probe but for penetration like a modern ICBM warhead, with extra ablatives and a sharper cone half-angle, achievable penetration would be deeper and faster. If 70 seconds atmospheric penetration time could be achieved, (with minimal slowdown) perhaps up to 6% of the way to the center might be achieved by near term technology.

The outer penetration shield of the warhead may be made from carbon (which is an excellent ablative heat protector). The carbon is also an excellent nuclear catalyst of the nuclear reactions in the CNO solar thermonuclear cycle and may significantly increase the power of the initial explosion [17].

A century hence, what level of penetration of the solar interior is possible? This depth is unknown to the author, exceeding plausible engineering in the near term. Let us consider a hypothetical point (top of the radiation layer) 30 percent of the way from the surface to the core, at the density of 0.2 g/cm^3 with a temperature of 2,000,000°C. No material substance can withstand such heat—for extended periods.

We may imagine however hypothetical penetration aids, analogous to ICBM techniques of a half century ago. Shock waves bear the brunt of the encountered heat and force it aside, the opacity shielding the penetrator. A form of multiple disposable shock cones may be employed to give the last in line a chance to survive; indeed the destruction of the next to last may arm the trigger.

If the heat isolation shield and multiple penetration aids can protect the bomb at near entry velocity for a hellish 10 *minute interval*, (which to many may seem impossible but which cannot be excluded without definitive study—remember we are speaking now of centuries hence, not the near term case above—see reference 14) that means the bomb may reach the depth of 350 thousands kilometers or 0.5*R*, where R = 696 × 103 km is Sun's radius.

The Sun density via relative Sun depth may be estimated by the equation

$$n = n_s e^{20.4\bar{h}}, \quad \text{where } \bar{h} = h/R, \qquad (23)$$

where $n_s \approx 10^{23}$ 1/m^3 is the plasma density on the photosphere surface; h is deep, km; R = 696 × 10^3 is solar radius, km. At a solar interior depth of $h = 0.5R$ the relative density is greater by 27 *thousand times* than on the Sun's surface.

Here the density and temperature are significantly more than on the photosphere's surface. And conditions for the detonation wave and thermonuclear reaction are "better"—from the point of view of the attacker.

9. Estimation of Nuclear Bomb Needed for Sun Explosion

Sound speed into plasma headed up T = 100 °K million degrees is about

$$v \approx 10^2 T^{0.5} \text{ m/s} = 10^6 \text{ m/s}. \qquad (24)$$

Time of nuclear explosion (a full nuclear reaction of bomb) is less $t = 10^{-4}$ sec. Therefore the radius of heated Sun photosphere is about $R = vt$ = 100 m, volume V is about

$$V = \frac{4}{3}\pi R^3 \approx 4 \times 10^6 \text{ m}^3. \qquad (25)$$

Density of Sun photosphere is $p = 2 \times 10^{-4}$ kg/m^3. Consequently the mass of the heated photosphere is about $m = pV$ = 1000 kg.

The requested power of the nuclear bomb for heating this mass for temperature $T = 10^4$ eV (100 K million degrees) is approximately

$$E = 10^3 \times 10^4 / 1.67 \times 10^{-27} \text{ eV} \approx 0.6 \times 10^{34} \text{ eV} \\ \approx 2 \times 10^{15} \text{ J} \approx 0.5 \text{ Mt} \qquad (26)$$

The requested power of nuclear bomb is about 0.5 Megatons. The average power of the current thermonuclear bomb is 5 - 10 Mt. That means the current thermonuclear bomb may be used as a fuse of Sun explosion. That estimation needs in a more complex computation by a power computer.

10. Results of Research

The Sun contains 73.46% hydrogen by weight. The isotope hydrogen-1 (99.985% of hydrogen in nature) is usable fuel for a fusion thermonuclear reaction.

The p-p reaction runs slowly within the Sun because its temperature is low (relative to the temperatures of nuclear reactions). If we create higher temperature and density in a limited region of the solar interior, we may be able to produce self-supporting, more rapid detonation thermonuclear reactions that may spread to the full solar volume. This is analogous to the triggering mechanisms in a thermonuclear bomb. Conditions within the bomb can be optimized in a small area to initiate ignition, build a spreading reaction and then feed it into a larger area, allowing producing a "solar hydrogen bomb" of any power—but not necessarily one whose power can be limited. In the case of the Sun certain targeting practices may greatly increase the chances of an artificial explosion of the entire Sun. This explosion would annihilate the Earth and the Solar System, as we know them today.

Author A.A. Bolonkin has researched this problem and shown that an artificial explosion of Sun cannot be precluded. In the Sun's case this lacks only an initial fuse, which induces the self-supporting detonation wave. This research has shown that a thermonuclear bomb exploded within the solar photosphere surface may be the fuse for an accelerated series of hydrogen fusion reactions.

The temperature and pressure in this solar plasma may achieve a temperature that rises to billions of degrees in which all thermonuclear reactions are accelerated by many thousands of times. This power output would further heat the solar plasma. Further increasing of the plasma temperature would, in the worst case, climax in a solar explosion.

The possibility of initial ignition of the Sun significantly increases if the thermonuclear bomb is exploded under the solar photosphere surface. The incoming bomb has a diving speed near the Sun of about 617 km/sec. Warhead protection to various depths may be feasible-ablative cooling which evaporates and protects the warhead some minutes from the solar temperatures. The deeper the penetration before detonation the temperature and density achieved greatly increase the probability of beginning thermonuclear reactions which can achieve explosive breakout from the current stable solar condition.

Compared to actually penetrating the solar interior, the flight of the bomb to the Sun, (with current technology requiring a gravity assist flyby of Jupiter to cancel the solar orbit velocity) will be easy to shield from both radiation and heating and melting. Numerous authors, including A. A. Bolonkin in works [7-12] offered and showed the high reflectivity mirrors which can protect the flight article within the orbit of Mercury down to the solar surface.

The author A. A. Bolonkin originated the AB Criterion, which allows estimating the condition required for the artificial explosion of the Sun.

11. Discussion

If we (humanity—unfortunately in this context, an insane dictator representing humanity for us) create a zone of limited size with a high temperature capable of overcoming the Coulomb barrier (for example by insertion of a specialized thermonuclear warhead) into the solar photosphere (or lower), can this zone ignite the Sun's photosphere (ignite the Sun's full load of thermonuclear fuel)? Can this zone self-support progressive runaway reaction propagation for a significant proportion of the available thermonuclear fuel?

If it is possible, researchers can investigate the problems: What will be the new solar temperature? Will this be metastable, decay or runaway? How long will the transformed Sun live, if only a minor change? What the conditions will be on the Earth during the interval, if only temporary? If not an explosion but an enhanced burn results the Sun might radically increase in luminosity for-say—a few hundred years. This would suffice for an average Earth temperature of hundreds of degrees over 0°C. The oceans would evaporate and Earth would bake in a Venus like greenhouse, or even lose its' atmosphere entirely.

It would not take a full scale solar explosion, to annihilate the Earth as a planet for Man. (For a classic report on what makes a planet habitable, co-authored by Issac Asimov, see http://www.rand.org/pubs/commercial_books/2007/RAND_CB179-1.pdf).

Converting the sun even temporarily into a "superflare" star, (which may hugely vary its output by many percent, even many times) over very short intervals, not merely in heat but in powerful bursts of shorter wavelengths) could kill by many ways, notably ozone depletion—thermal stress and atmospheric changes and hundreds of others of possible scenarios—in many of them, human civilization would be annihilated. And in many more, humanity as a species would come to an end.

The reader naturally asks: Why even contemplate such a horrible scenario? It is necessary because as thermonuclear and space technology spreads to even the least powerful nations in the centuries ahead, a dying dictator having thermonuclear missile weapons can produce (with some considerable mobilization of his military/industrial complex)—the artificial explosion of the Sun and take into his grave the whole of humanity. It might take tens of thousands of people to make and launch the hardware, but only a very few need know the final targeting data of what might be otherwise a weapon purely thought of (within the dictator's defense industry) as being built for peaceful, deterrent use.

Those concerned about Man's future must know about this possibility and create some protective system—or ascertain on theoretical grounds that it is entirely impossible, which would be comforting.

Suppose, however that some variation of the following is possible, as determined by other researchers with access to good supercomputer simulation teams. What, then is to be done?

The action proposed depends on what is shown to be possible.

Suppose that no such reaction is possible—it dampens out unnoticeably in the solar background, just as no fission bomb triggered fusion of the deuterium in the oceans proved to be possible in the Bikini test of 1946. This would be the happiest outcome.

Suppose that an irruption of the Sun's upper layers enough to cause something operationally similar to a targeted "coronal mass ejection"—CME—of huge size targeted at Earth or another planet? Such a CME like weapon could have the effect of a huge electromagnetic pulse. Those interested should look up data on the 1859 solar superstorm, the Carrington event, and the Stewart Super Flare. Such a CME/EMP weapon might target one hemisphere while leaving the other intact as the world turns. Such a disaster could be surpassed by another step up the escalation ladder—by a huge hemisphere killing thermal event of ~12 hours duration such as postulated by science fiction writer Larry Niven in his 1971 story "Inconstant Moon"—apparently based on the Thomas Gold theory (ca. 1969-70) of rare solar superflares of 100 times normal luminosity. Subsequent research[18] (Wdowczyk and Wolfendale, 1977) postulated horrific levels of solar activity, ozone depletion and other such consequences might cause mass extinctions. Such an improbable event might not occur naturally, but could it be triggered by an interested party? A triplet of satellites monitoring at all times both the sun from Earth orbit and the "far side" of the Sun from Earth would be a good investment both scientifically and for purposes of making sure no "creative" souls were conducting trial CME eruption tests!

Might there be peaceful uses for such a capability? In the extremely hypothetical case that a yet greater super-scale CME could be triggered towards a given target in space, such a pulse of denser than naturally possible gas might be captured by a giant braking array designed for such a purpose to provide huge stocks of hydrogen and helium at an asteroid or moon lacking these materials for purposes of future colonization.

A worse weapon on the scale we postulate might be an asymmetric eruption (a form of directed thermonuclear blast using solar hydrogen as thermonuclear fuel), which shoots out a coherent (in the sense of remaining together) burst of plasma at a given target without going runaway and consuming the outer layers of the Sun. If this quite unlikely capability were possible at all (dispersion issues argue against it—but before CMEs were discovered, they too would have seemed unlikely), such an apocalyptic "demo" would certainly be sufficient emphasis on a threat, or a means of warfare against a colonized solar system. With a sufficient thermonuclear burn—and if the condition of nondispersion is fulfilled—might it be possible to literally strip a planet—Venus, say—of its' atmosphere? (It might require a mass of fusion fuel—and a hugely greater non-fused expelled mass comparable in total to the mass to be stripped away on the target planet.)

It is not beyond the limit of extreme speculation to imagine an expulsion of this order sufficient to strip Jupiter's gas layers off the "Super-Earth" within.—To strip away 90% or more of Jupiter's mass (which otherwise would take perhaps ~400 Earth years of total solar output to disassemble with perfect efficiency and neglecting waste heat issues). It would probably waste a couple Jupiter masses of material (dispersed hydrogen and helium). It would be an amazing engineering capability for long term space colonization, enabling substantial uses of materials otherwise unobtainable in nearly all scenarios of long term space civilization.

Moving up on the energy scale—"boosting" or "damping" a star, pushing it into a new metastable state of greater or lesser energy output for times not short compared with the history of civilization, might be a very welcome capability to colonize another star system—and a terrifying reason to have to make the trip.

And of course, in the uncontrollable case of an induced star explosion, in a barren star system it could provide a nebula for massive mining of materials to some future super-civilization. It is worth noting in this connection that the Sun constitutes 99.86 percent of the material in the Solar System, and Jupiter another.1 percent. Literally a thousand Earth masses of solid (iron, carbon) building materials might be possible, as well as thousands of oceans of water to put inside space colonies in some as yet barren star system.

But here in the short-term future, in our home solar system, such a capability would present a terrible threat to the survival of humanity, which could make our own solar system completely barren.

The list of possible countermeasures does not inspire confidence. A way to interfere with the reaction (dampen it once it starts)? It depends on the spread time, but seems most improbable. We cannot even stop nuclear reactions once they take hold on Earth—the time scales are too short.

Is defense of the Sun possible? Unlikely—such a task makes missile defense of the Earth look easy. Once a gravity assist Jupiter flyby nearly stills the velocity with which a flight article orbits the Sun, it will hang relatively motionless in space and then begin the long fall to fiery doom. A rough estimate yields only one or two weeks to intercept it within the orbit of Mercury, and the farther it falls the faster it goes, to science fiction-like velocities sufficient to reach Pluto in under six weeks before it hits.

A perimeter defense around the Sun? The idea seems impractical with near term technology.

The Sun is a hundred times bigger sphere than Earth in every dimension. If we have 10,000 ready to go interceptor satellites with extreme sunshields that function a few solar radii out each one must be able to intercept with 99% probability the brightening light heading toward its' sector of the Sun over a circle the size of Earth, an incoming warhead at around 600 km/sec.

If practical radar range from a small set is considered (4th power decline of echo and return) as 40,000 km then only 66 seconds would be available to plot a firing solution and arm for a destruct attempt. More time would be available by a telescope looking up for brightening, infalling objects—but there are many natural incoming objects such as meteors, comets, etc. A radar might be needed just to confirm the artificial nature of the in-falling object (given the short actuation time and the limitations of rapid storable rocket delta-v some form of directed nuclear charge might be the only feasible countermeasure) and any leader would be reluctant to authorize dozens of nuclear explosions per year automatically (there would be no time to consult with Earth, eight light-minutes away—and eight more back, plus decision time). But the cost of such a system, the reliability required to function endlessly in an area in which there can presumably be no human visits and the price of its' failure, staggers the mind. And such a "thin" system would be not difficult to defeat by a competent aggressor...

A satellite system near Earth for destroying the rockets moving to the Sun may be a better solution, but with more complications, especially since it would by definition also constitute an effective missile defense and space blockade. Its' very presence may help spark a war. Or if only partially complete but under construction, it may invite preemption, perhaps on the insane scale that we here discuss…

Astronomers see the explosion of stars. They name these stars novae and supernovae—"New Stars" and try to explain (correctly, we are sure, in nearly all cases) their explosion by natural causes. But some few of them, from unlikely spectral classifications, may be result of war between civilizations or fanatic dictators inflicting their final indignity upon those living on planets of the given star. We have enough disturbed people, some in positions of influence in their respective nations and organizations and suicide oriented violent people on Earth. But a nuclear bomb can destroy only one city. A dictator having possibility to destroy the Solar System as well as Earth can blackmail all countries—even those of a future Kardashev scale 2 star-system wide civilization—and dictate his will/demands on any civilized country and government. It would be the reign of the crazy over the sane.

Author A.A. Bolonkin already warned about this possibility in 2007 (see his interview http://www.pravda.ru/science/planet/space/05-01-2007/208894-sun_detonation-0 [15] (in Russian) (A translation of this is appended at the end of this article) and called upon scientists and governments to research and develop defenses against this possibility. But some people think the artificial explosion of Sun impossible. This led to this current research to give the conditions where such detonations are indeed possible. That shows that is conceivably possible even at the present time using current rockets and nuclear bombs—and only more so as the centuries pass. Let us take heed, and know the risks we face—or disprove them.

The first information about this work was published in [15]. This work produced the active Internet discussion in [18]. Among the raised questions were the following:

1) It is very difficult to deliver a warhead to the Sun. The Earth moves relative to the Sun with a orbital velocity of 30 km/s, and this speed should be cancelled to fall to the Sun. Current rockets do not suffice, and it is necessary to use gravitational maneuvers around planets. For this reason (high delta-V (velocity changes required) for close solar encounters, the planet Mercury is so badly investigated (probes there are expensive to send).

Answer: The Earth has a speed of 29 km/s around the Sun and an escape velocity of only 11 km/s. But Jupiter has an orbital velocity of only 13 km/sec and an escape velocity of 59.2 km/s. Thus, the gravity assist Jupiter can provide is more than the Earth can provide, and the required delta-v at that distance from the Sun far less—enough to entirely cancel the sun-orbiting velocity around the Sun, and let it begin the long plunge to the Solar orb at terminal velocity achieving Sun escape speed 617.6 km/s. Notice that for many space exploration maneuvers, we require a flyby of Jupiter, exactly to achieve such a gravity assist, so simply guarding against direct launches to the Sun from Earth would be futile!

2) Solar radiation will destroy any a probe on approach to the Sun or in the upper layers of its photosphere.

Answer: It is easily shown, the high efficiency AB-reflector can full protection the apparatus. See [7] Chapters 12, 3A, [8] Ch.5, (see also [9-12].

3) The hydrogen density in the upper layers of the photosphere of the Sun is insignificant, and it would be much easier to ignite hydrogen at Earth oceans if it in general is possible.

Answer: The hydrogen density is enough known. The Sun has gigantic advantage—that is PLASMA. Plasma of sufficient density reflects or blocks radiation—it has opacity. That means: **no radiation losses in detonation**. It is very important for heating. The AB Criterion in this paper is received for PLASMA. Other planets of Solar system have MOLECULAR atmospheres which passes radiation. No sufficient heating—no detonation! The water has higher density, but water passes the high radiation (for example γ-radiation) and contains a lot of oxygen (89%), which may be bad for the thermonuclear reaction. This problem needs more research.

12. Summary

This is only an initial investigation. Detailed supercomputer modeling which allows more accuracy would greatly aid prediction of the end results of a thermonuclear explosion on the solar photosphere.

Author invites the attention of scientific society to detailed research of this problem and devising of protection systems if it proves a feasible danger that must be taken seriously. The other related ideas author Bolonkin offers in [5-15].

13. Acknowledgements

The author wishes to acknowledge Alexei Turchin (Russia) for discussing the problems in this article [18].

REFERENCES

[1] AIP Physics Desk Reference, 3rd Edition, Spring, Berlin, 888 Pages.

[2] I. Grigoriev, "Handbook of Physical Quantities," CRC Press, Boca Raton, 1997.

[3] I. K. Kikoin, "Tables of Physical Values," Atomizdat, Moscow Ctiy, 1975, 1006 Pages (in Russian).

[4] K. Nishikawa and M. Wakatani, "Plasma Physics," Spring, Berlin, 2000.

[5] A. A. Bolonkin, "New AB-Thermonuclear Reactor for Aerospace," *AIAA*-2006-7225 *to Space*-2006 *Conference*, San Jose, 19-21 September 2006.
http://arxiv.org/ftp/arxiv/papers/0706/0706.2182.pdf
http://arxiv.org/ftp/arxiv/papers/0803/0803.3776.pdf

[6] A. A. Bolonkin, "Simplest AB-Thermonuclear Space Propulsion and Electric Generator".
http://arxiv.org/ftp/physics/papers/0701/0701226.pdf

[7] A. A. Bolonkin, "Non-Rocket Space Launch and Flight," Elsevier, Amsterdam, 2006, 488 Pages (The Book Contains Theories of the More than 20 New Revolutionary Author Ideas in Space and Technology).
http://Bolonkin.narod.ru/p65.htm
http://www.scribd.com/doc/24056182

[8] A. A. Bolonkin, "New Concepts, Ideas and Innovations in Aerospace and Technology, Nova, 2007," The Book Contains Theories of the More than 20 New Revolutionary Author Ideas in Space and Technology.
http://Bolonkin.narod.ru/p65.htm
http://www.scribd.com/doc/24057071

[9] A. A. Bolonkin and R. B. Cathcart, "Macro-Projects: Environment and Technology," NOVA, 2009, 536 Pages.
http://Bolonkin.narod.ru/p65.htm
Book contains many new revolutionary ideas and projects.
http://www.scribd.com/doc/24057930.

[10] A. A. Bolonkin, "High Speed AB Solar Sail," 42 *JOINT Propulsion Conferences*, Sacramento, 9-12 July 2006.
http://arxiv.org/ftp/physics/papers/0701/0701073.pdf

[11] A. A. Bolonkin, "Light Multi-Reflex Engine," *Journal of British Interplanetary Society*, Vol. 57, No. 9-10, 2004, pp. 353-359.

[12] A. A. Bolonkin, "Light Pressure Engine, Patent (Author Certificate) # 1183421, 1985, USSR (Priority on 5 January 1983).

[13] A. A. Bolonkin, "Converting of Matter to Nuclear Energy by AB-Generator," *American Journal of Engineering and Applied Science*, Vol. 2, No. 2, 2009, pp. 683-693.
http://www.scipub.org/fulltext/ajeas/ajeas24683-693.pdf
http://www.scribd.com/doc/24048466/

[14] A. A. Bolonkin, "Femtotechnology. Nuclear AB-Matter with Fantastic Properties," *American Journal of Engineering and Applied Sciences*, Vol. 2, No. 2, 2009, pp. 501-514.
http://www.scipub.org/fulltext/ajeas/ajeas22501-514.pdf
http://www.scribd.com/doc/24046679/

[15] A. A. Bolonkin, "Artificial Explosion of Sun," 2007 (in Russian).
http://www.pravda.ru/science/planet/space/05-01-2007/208894-sun_detonation-0

[16] Solar Physics Group at NASA's Marshall Space Flight Center Website for Solar Facts.
http://solarscience.msfc.nasa.gov/

[17] J. Wdowczyk and A. W. Wolfendale, "Cosmic Rays and Ancient Catastrophes," *Nature*, Vol. 268, 1977, Article ID: 510.
http://www.nature.com/nature/journal/v268/n5620/abs/268510a0.html

[18] A. V. Turchin, "The Possibility of Artificial Fusion Explosion of Giant Planets and Other Objects of Solar System," 2009.
http://www.scribd.com/doc/8299748/Giant-planets-ignition

15

New Approach to the Diagnosis of Control Hydraulic Systems

Noura Rezika Bellahsene Hatem[1], Mohamed Mostefai[2], Oum El-Kheir Aktouf[3]
[1]LTII Laboratory, A. Mira Béjaia University, Béjaia, Algeria
[2]Automatic Control Laboratory, University of Setif, Setif, Algeria
[3]LCIS Laboratory, INPG Valence, Valence, France

ABSTRACT

A dynamical hybrid system is described by a set of continuous variables and a set of discrete events interacting mutually. The reality imposes to take into account the failures of the components or the uncertainties on the knowledge of the system. Therefore, the systems can work in several modes. Some of these modes correspond to a normal functioning and the others represent failing modes. In this article, we are interested in the evaluation of the probability of occurrence of the failing events in a hybrid dynamic system. Furthermore, we propose an approach to detect on line the failed states and try to find the components responsible for this fault. This approach is based on the knowledge in priori of the system, at least from the point of view of the state of its components in normal functioning. This approach is here described through a simple case study taken from literature.

Keywords: Diagnosis; Dynamic System; Hybrid System; Reliability; System Safety

1. Introduction

The objective of this study is precisely to detect failed states in an embedded system and find the faulty components. Otherwise, we assess the probability of occurrence of these events that lead to the failed state (off line). The hardware and software is a component of a more complex system that must operate independently of human intervention. An embedded system may include some kind of operating systems but often it will be simple enough to be written as a single program. It is a hybrid system which exhibits both continuous and discrete dynamic behavior—a system that can be both flow described by a differential equation and jump described by control graph. Two basic hybrid system modeling approaches can be classified as an implicit and an explicit one. The explicit approach is often represented by a hybrid automata or a hybrid Petri net [1-3]. Once the model has been defined, our approach permits identifying the faulty states and tries to determine the components that are responsible. Furthermore, reliability of the system is calculated. A comparison with previous results of calculation of reliability is provided [4,5].

2. Description of the Method

We represent a system, in a diagnostic point of view as A life cycle, which is interrupted by the occurrence of an event that switches the operating mode of the system from a normal state to a state of failing functioning, is shown schematically in **Figure 1**. It is considered to be a life cycle which is interrupted by the occurrence of an event that switches the operating mode of the system

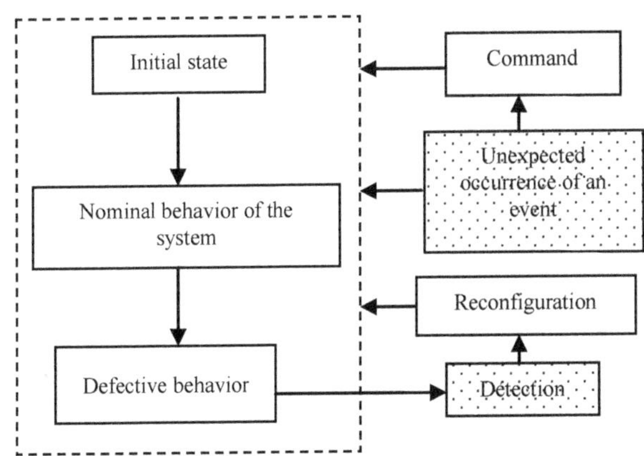

Figure 1. Global state of a dynamic system.

from a normal state to a failing state. By adopting this representation, the designer can easily notice that his contribution may not occur at the block level detection and then reconfigure the system. The occurrence of the event could not be more random. A system is a set of components and the component, in our case, is considered as a unit represented by a binary number (bit). The system operation is divided into time steps. Each step is considered as such with all its parameters. We consider a step function as a sub-state of the functioning that can be a normal state as it may be a failed state. n the simplest case, the system can be considered as a single step. The state parameters are the values of bits representing the system components (1 or 0 eg closed, open, off, on ...). The subsystem with the corresponding time step in question is then a combination of bits representative of the components. A step involving N components will be 2^N combinations representing 2^N deputy possible states (normal and faulty). Each combination indicates the status of each component and indirectly, it also informs us about the components that are the cause of the failed state. Once the sub-states are simulated, the results are analyzed and failed modes are identified. After that, a probability calculation is performed.

3. Algorithm

3.1. Previous Study

- Decompose the total functioning of the system in steps of operation where the components involved have a single state.
- Identify components involved in each step.
- Represent each component by a bit B {1.0} a: 0 state and asked.
- Represent step by a binary number where the number of bits is equal to the number of components involved in the stage in question.
- Expressing representative combinations of step (sub-states of the functioning step)

3.2. Off-Line Study

- Detect the failing states by analyzing the indicator of each failing event.
- Calculate the probability of occurrence of each failed state

3.3. On-Line Study

- Identify the failing states by analyzing the indicator of each failing events.
- List the components that are responsible for this state.
- Reconfiguration.

4. The Case Study

The test case consists of a tank containing a liquid whose level h must be maintained with a main pump P1, a backup pump and a drain valve V [5,6]. Each of the three components is controlled by a control loop containing a level detector (**Figure 2**).

The task ahead is to maintain the liquid level constant in the interval ([6,8]). If the level is below 6, the command closes the valve and opens the backup pump and if the level is above 8, the pumps are stopped and the valve is opened. Two extreme situations can occur: the drying up and overflow. These two cases occur when the order can no longer act on the system components that become failing (1 or more components). We then say that the system is in a failed state even dangerous. The three components of pump P1, P2 pump and valve V are mutually independent and non-repairable.

The continuous variable for the system is the liquid level h which depends on the status of components. Thus, the differential equation for the system is given by:

$$y(u) = \mathrm{d}h(t)/\mathrm{d}t \quad (1)$$

where $u = (u_{p1}, u_{p2}, u_V)$

And $u_c = 0$ if c is OFF and $u = 1$ if $c = ON$ (2)

There $y(u) = (u_{p1} + u_{p2} - u_V)D$

D is the flow rate of the elements P1, P2 and valve.

The generalized Equation (1) reflects the various possible operating modes of the process. It shows the influence of discrete phenomena on the evolution of the process through u_{p1}, u_{p2}, u_V. These can take, in the case of this example, the value 1 if the component is active or if a fault blocks it in the open or active position, and 0 otherwise, as expressed by the Equation (2). In nominal conditions, the flow of P1 is equal to the flow of P2 and V.

Then, D = 1.5 $m^3 \cdot h^{-1}$ for P_1, P_2 et V.

At time t = 0, the liquid level h = 7 m, the pump P1 works, P2 is stopped and the valve V is opened. The system control laws which define the state of the pumps and the valve as a function of liquid level are given in **Table 1** below:

The *state* of an embedded system is defined by the values of the *continuous variables* and a discrete *control mode*. The state changes either continuously, according to a flow, discretely according to a *control graph*. Continuous flow is permitted as long as so-called *invariants* hold, while discrete transitions can occur as soon as given *jump conditions* are satisfied. The system thus defined may be considered as a hybrid controller which takes into account the different modes of continuous operation of the system and the transition from one another on the occurrence of deterministic events which are produced by crossing the thresholds of continuous variables. The elementary automata are given in **Figure 3** and the automaton representing the command is given in **Figure 4**. Events associated with P1, P2, P3 are:

- a_P1 and a_P2: stopping the pumps P1 and P2

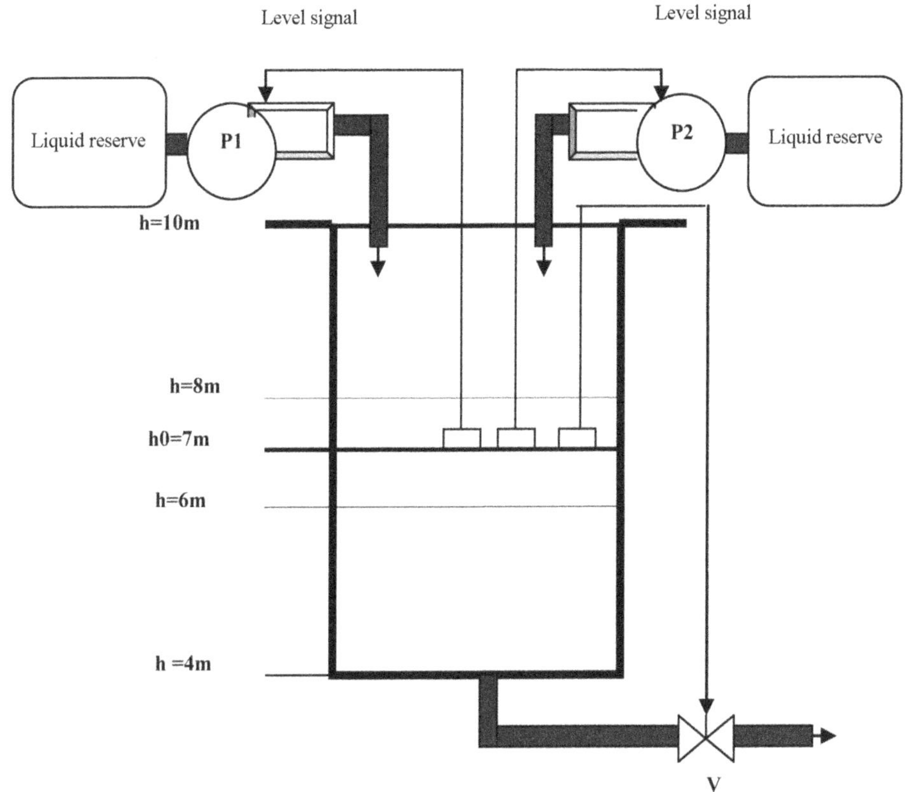

Figure 2. Level control system.

Table 1. Control laws of the system.

Level h (m)	Pump P1	Pump P2	Pump P3
h < 6	Opened	Opened	Closed
6 < h > 8	Opened	Closed	Opened
h > 8	Closed	Closed	opened

- d_P1 et d_P2: start of pumps P1 and P2
- o_V: opening of the valve and
- f_V: closing of the valve

We distinguish for the components P1, P2 and V the following states:

- ON-P1, ON-P2 and ON-V: active pumps and valve opened,
- OFF-P1, OFF-P2 and OFF-V: pumps and valve closed.

For the reservoir we have the states:

- N_n: Normal level of the reservoir ($6 \leq h \leq 8$)
- N_ass: Level of drying ($h \leq 4$) et
- N_deb: Overflow level ($h \geq 10$).

Control laws of the command:

- Initial: C0 – P1 active, P2 stopped et V open,
- If $h < 6$: C2 → C3, P1 active; C3 → C4, P2 active et C4 → C5, V stopped,
- If $h > 8$: C6 → C7, P1 stopped; C7 → C8, P2 stopped et C8 → C1, V open.

5. Main Results

5.1. Online Monitoring: Detection of Failed States

The total operating system can be described by a single step or a single state that is maintain the liquid level h in the interval [6,8]. The simulation of the system without the occurrence of unwanted events gives the normal state, knowing that the initial level is h = 7 m. The system seen like that, keeps the liquid level constant. The simulation of a failure in the 9th iteration (blocked valve closing) the liquid level increases indefinitely and the overflow indicator equals 0 (**Figure 5**). The simulation of another fault (pump P1 is locked in opening and the pump P2 is locked in the closure, the valve opened), the level drops to the lowest level and the indicator of drying has become 0 (**Figure 6**). It should be noted that indicators of the two states have been initialized to 1.

5.2. Offline Monitoring: Calculation of Probability

The system starts with the normal state (101: Pump open P1, P2 pump closed and valve V open). At t = 9th iteration, there is the occurrence of the failed state (one, two or all three components are not in the expected state). To avoid ambiguity with the role of the command, it is in-

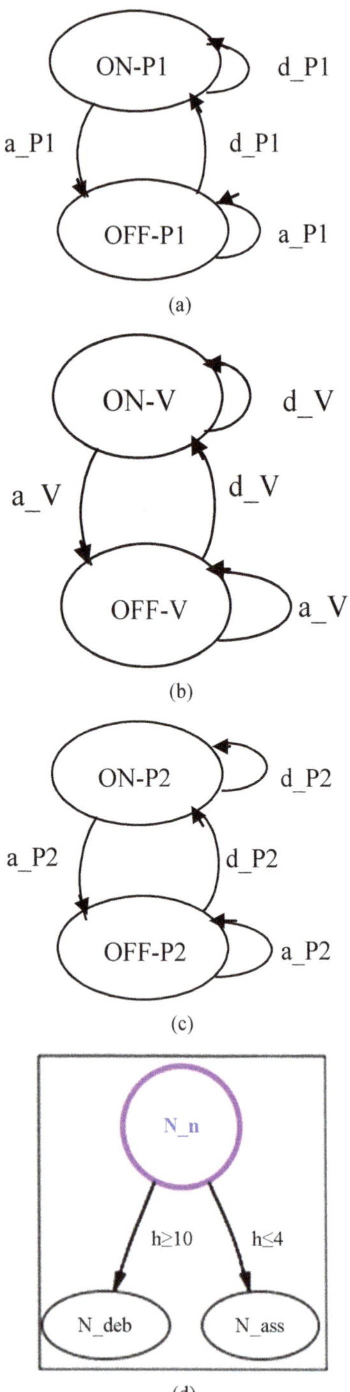

Figure 3. Finite state automata of the three components. (a) Automataof the pump P1; (b) Automataof the pump P2; (c) Automata of the valve; (d) Automataofthe reservoir.

habited to the occurrence of the feared event. Moreover, we are not interested here by the physical cause of the failure. This may be due, in fact, by an external action (closing or opening accidentally), the blocking component's state (open or closed) or simply, the failure was the consequence of the life of the component (wear).

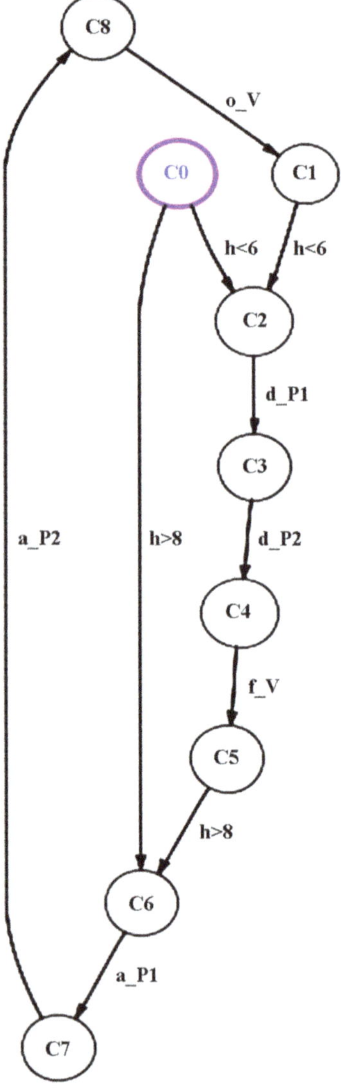

Figure 4. Automata of the command.

From the normal state, we look at all possibilities of failure. These modes are summarized in the following, noted that only the state 101 is the nominal operating condition.

000: Pump P1 is closed, the pump P2 is closed and the valve is closed. **Steady state in the initial state.**

001 Pump P1 closed, the pump P2 closed and the valve opened: **Drying**.

010 Pump P1 is closed, the pump P2 is opened and the valve is closed: **Overflow**.

011 Pump P1 is closed, pump P2 is opened and the valve is opened. **Normal state (rescue)**.

The backup pump takes over the pump P.

100 Pump P1 is opened, pump P2 is closed and the valve closed: **Overflow**.

110 Pump P1 is opened, pump P2 is opened and the valve closed: **Overflow**.

111 Pump P1 is opened, pump P2 is opened and the valve opened: **Overflow**.

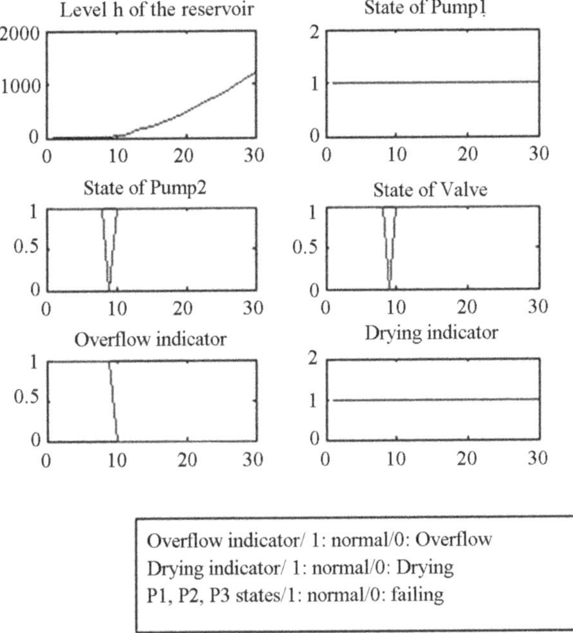

Figure 5. Overflow state of the system.

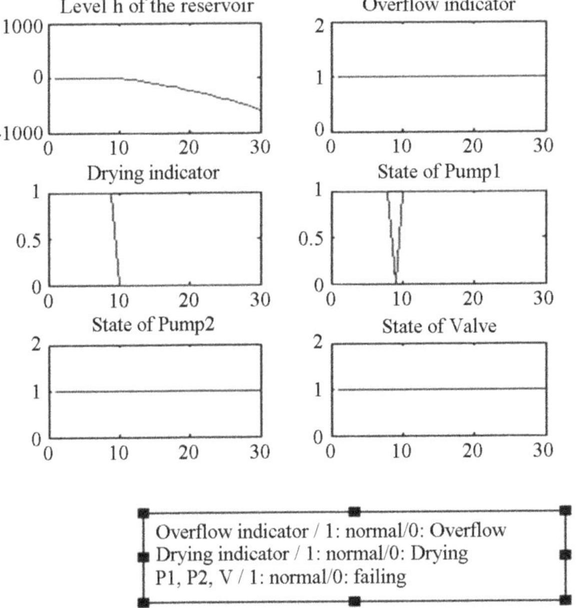

Figure 6. Drying state of the system.

The simulation results have shown that all faulty modes were detected. Thus, overflow mode appears in four cases, the drying mode appears in one case and the normal state in three cases (**Table 2**).

Mode corresponding to the combination 000 where all components are closed, the liquid level does not vary from the initial state. For mode (011), the pump P1 is closed and the pump P2 takes over, this state corresponds to the normal state and the liquid level is constant.

The calculation of probability of the occurrence of-

Table 2. Summary table of states.

Normal state	Drying	Overflow
000	011	010
011		100
101		110
		111

Table 3. Probability values of each state.

State	*Normal*	*Drying*	*Overflow*
Probability	0.375	0.125	0.5

Table 4. Probabilities of access to feared states.

Temps (h)	*Débordement*			*Assèchement*		
	PMDPM	RdP	ASH	PMDPM	RdP	ASH
1000	0.486	0.486	0.475	0.118	0.118	0.117

faulty modes gives explicitly the values shown in **Table 3** and the **Table 4** gives the probabilities of access to feared states in previous studies [4,8]

6. Discussion

The faulty modes are detected and the components responsible of these modes are listed on line. The system can be reconfigured. Furthermore, the results show that the probability of the state drying is 0.125 and the probability of occurrence of the overflow condition is 0.5. The values obtained in previous studies are not accurate and the number of iterations is very important. At 1000 h, the value is not exact. It may take more than this duration to obtain the final value.

7. Conclusion

Our approach allowed us to assess the probability of occurrence of undesirable events. Thus, we have introduced a method leading to identify and locate faulty modes. This approach can be applied to any hybrid system described by its dynamic equations since these equations are considered as the discrete parts of the system.

REFERENCES

[1] N. R. Bellahsene Hatem, O. Aktouf and M. Mostefai, "Aide au Diagnostic des Systèmes Dynamiques Hybrides," *Colloque d'Informatique, Automatique et Electronique CIAE* 2011, Casablanca, Moroccoc 24-25 Mars 2011.

[2] N. R. Bellahsene Hatem, O. Aktouf and M. Mostefai, "Contribution au Diagnostic des Evénements Défaillants Dans un Système Dynamique Hybride," *International congress for Global Science and Technology* (*ICGST*), Istanbul, 19-21 Décembre 2011, pp. 137-142.

[3] F. Thedjani, K. Laabidi and S. Zidi. "Classification non Supervisee Pour la Surveillance d'un Système Automatique 8e," *Conférence Internationale de MO- délisation et SIMulation—MOSIM'*10—10 au 12 mai Hammamet, 2010, pp. 125-131.

[4] J. Clarhaut, "Prise en Compte des Séquences de Défaillances Pour la Conception de Systèmes d'Automatisationque," *Thèse de Doctorat, University of Lille*, Lille, 2009.

[5] E. Borgonovo and M. Marseguera. "A Monte Carlo Methodological Approach to Plant Availability Modeling with Maintenance, Aging and Obsolescence," *Reliability Engineering and System Safety*, Vol. 67, 2000, pp. 61-73.

16

Silver Nanoparticle Adsorption to Soil and Water Treatment Residuals and Impact on Zebrafish in a Lab-scale Constructed Wetland

Angela Ebeling, Victoria Hartmann, Aubrey Rockman, Andrew Armstrong, Robert Balza, Jarrod Erbe, Daniel Ebeling
Biology, Chemistry, Biochemistry Departments, Wisconsin Lutheran College, Milwaukee, WI, USA

ABSTRACT

Nanoparticles (< 100 nm) are becoming more prevalent in residential and industrial uses and may enter the environment through wastewater. Although lab studies have shown that nanoparticles can be toxic to various organisms, limited research has been done on the effects of nanoparticles in the environment. Environmental conditions such as pH and ionic strength are known to alter the biotoxicity of nanoparticles, but these effects are not well understood. The objectives of this research were to determine the impacts of silver nanoparticles (AgNP) on zebrafish in the pseudo-natural environment of a lab-scale constructed wetland, and to investigate wastewater remediation through soil and water treatment residual (WTR) adsorption of AgNPs. Concurrently, the effect of particle size on AgNP sorption was examined. Researchers exposed adult zebrafish in a lab-scale constructed wetland to concentrations of AgNP ranging from 0 - 50 mg AgNP/L and compared them to negative controls with no silver exposure and to positive controls with exposure to silver nitrate. The results suggest that aggregated AgNP do not impact zebrafish. Separately, sorption experiments were carried out examining three media - a wetland soil, a silt loam soil, and a WTR - in their capacity to remove AgNPs from water. The silt loam retained less AgNPs from solution than did the wetland soil or the WTR. In the WTR AgNPs were associated with sand size particles (2 mm - 0.05 mm), but in the wetland soil and silt loam, approximately half of the AgNPs were associated with the sand-sized particles, while the rest were associated with silt sized (~0.05 mm) or smaller particles. The larger sorption capacity of the wetland soil and WTR was attributed to their higher carbon content. The sorption data indicate that AgNPs adsorbed to soil and WTRs and support the idea that natural and constructed wetlands can remove AgNPs from wastewater.

Keywords: Silver Nanoparticles; Soil; Water Treatment Residuals; Constructed Wetland; Zebrafish; Remediation

1. Introduction

Sufficient, clean, safe drinking water is increasingly scarce in many parts of the world [1-4]. Pollutants such as excess nutrients, particulates, and pathogens are known to cause environmental as well as human health problems [5-12]. Conventional and natural methods can effectively remediate wastewater of these types of pollutants [13-15]. Other pollutants such as pharmaceuticals (hormones, anti-depressants) and nanomaterials (materials less than 100 nm in at least one dimension [16]) in wastewater are becoming more common [17-21]. For example, nanoparticles are becoming more prevalent in residential use (e.g. sunscreen, textiles), medical applications (e.g. wound treatment), industry (e.g. sensors and solar cells), and environmental remediation [22-25]. As a result, nanoparticles are able to enter the environment via wastewater or improper disposal [16,19,25]. Unfortunately, little is known about the removal of these new types of pollutants using conventional or natural wastewater treatment methods.

Numerous studies have shown that many nanomaterials, including silver nanoparticles (AgNPs), have toxic effects on organisms (zebrafish, medaka, rainbow trout, crucian carp, flathead minnow [26]; rainbow trout [27]; bacteria [28]; zebrafish embryos [29]; plants [30]; zebrafish, microalga, water flea [31]; bacteria [32]). Both engineered and natural nanoparticles play important roles in the health of terrestrial and aquatic ecosystems [33-35]. Toxicity of metal nanoparticles is well known in laboratory settings as indicated in the previously cited studies, but the fate of nanoparticles in the environment and the potential for bioaccumulation is unclear and complicated [16,22,36].

Natural wetlands are known to have important roles in

wastewater remediation [37] and constructed wetlands have been increasingly been used to remove pollutants [13,15]. The use of natural media (e.g. soil) as well as engineered or recycled media (e.g. water treatment residuals, WTRs) could improve the effectiveness of the natural wastewater treatment and these media could have the potential to remove nanoparticles as well. The objectives of this research were to determine the effectiveness of several media (two soils and a water treatment residual) on AgNP removal from solution and to investigate AgNP effects on zebrafish in a (simulated) natural setting (lab-scale constructed wetland).

2. Materials and Methods

2.1. General

In general, two main experiments were conducted. The first experiment used three media to investigate the sorption of AgNPs out of wastewater: wetland soil, silt loam soil, and WTRs (waste product of drinking water treatment process). The second experiment investigated the effect of AgNPs on zebrafish (*Danio rerio*) living in the "natural" environment of a lab-scale constructed wetland.

2.2. Sorption Experiments

The sorption study was modeled after sorption experiments used to investigate phosphorus sorption to soil or water treatment residuals [38-40]. Three media were shaken with various concentrations of AgNPs in solution and the amount of AgNPs remaining in the solution after an equilibration time was measured.

All three media were sent to the Soil and Plant Analysis Lab in Madison, WI for elemental analysis (total minerals by Inductively Coupled Plasma Optical Emission Spectrometry, total N by Total/Kjeldahl, and total C by LECO CNS-2000 analyzer) and particle size analysis (hydrometer method). A wetland soil was used because this sorption experiment was designed to be compared with the constructed wetland study explained in the following section. The wetland soil was used in both the sorption study and in the constructed wetland and was obtained from Certified Products, New Berlin, WI (Black Topsoil: http://www.certifiedproductswi.com/). The second medium, a silt loam soil (Plano silt loam, fine-silty, mixed, superactive, mesic, typic arguidoll) was chosen because it represents a typical soil in Wisconsin. The third medium, WTR, was chosen because it has previously been shown to have the ability to remove phosphorus from wastewater and could be used in constructed wetlands for the removal of that nutrient [41]. If this material could also be used to remove AgNPs, it would provide an even greater incentive to beneficially reuse the material as an amendment to help clean wastewater. Little research has been done on the effects of WTRs on organisms in the environment [38], so to establish that these WTRs would not induce mortality, preliminary laboratory experiments using *Escherichia coli* and native soil bacteria were conducted and gave no evidence that moderate levels of WTRs (0.05 g WTR/ml) negatively impact bacterial growth (data not shown). The nanoparticles used in the sorption experiment were 10 nm monodispersed silver nanoparticles in 2 mM sodium citrate buffer obtained from Nano Composix (10 nm citrate NanoXact Silver, JMW1148).

To conduct the sorption experiment, 0.5 g (dry weight) of each medium was weighed into 15 ml conical tubes. Then 10 ml of AgNP solution was added to each tube. Five concentrations of silver nanoparticles were used: 0, 6, 15, 30, and 60 mg AgNP/L. The conical tubes were capped and shaken for 18 hours at 60 rpm on a Fotodyne Orbit shaker (Lab-Line Instruments, Inc. Model Number 3520) at 23℃ (room temperature). Each trial was replicated three times. After the 18 hour equilibration time, the supernatant in each tube was sampled three times –2 ml was removed after 30 s of particle settling, after 2 hours of particles settling, and after centrifuging at 2500 rpm for 10 min (IEC Centra-7R Refrigerated Centrifuge, S.N. 23601916). The supernatant was sampled at these times, because one objective in this research was to investigate AgNPs affinity for adsorbing to different size particles. According to Stoke's Law

$$v_s = \frac{(\rho_p - \rho_f) g^2 d_p}{18u} \quad (1)$$

where v_s is the particle settling velocity, ρ_p is the density of the particles, ρ_f the density of the fluid, g the acceleration due to gravity, d_p the diameter of the particle, and u the fluid viscosity, this equation predicts that sand size particles will settle after about 40 s (thus any AgNP measured in the supernatant would be either soluble, or adsorbed to silt or clay sized particles), silt size particles will settle after approximately 2 hours (thus any AgNP measured in the supernatant after 2 hours would be soluble or adsorbed to clay size particles), and after centrifuging all particles would be removed from the supernatant (and any AgNPs measured would be in solution) [42,43]. Using this method to determine particle size is based on an empirical method and cannot be used to accurately define the particle size [43], but for this research these sampling times give a rough estimation of the size of the particles with which AgNPs were associated.

A modified digestion method was used to quantify the amount of silver in the supernatants [44,45:EPA SW 846Method 3050B]. To each 2 ml supernatant sample, 1 ml of 6 M nitric acid was added using a repeat pipetter (Eppendorf, Repeater Plus, 2849689). The samples were placed into a water bath at 90℃ for one hour to digest before measurement of elemental silver on a Perkin Elmer AAnalyst 200 Atomic Absorption Spectrometer (S/N 20054062503). Silver nitrate ($AgNO_3$, Fischer Chemicals, Lot number 041796) was used to make standards. Samples and $AgNO_3$ standards were analyzed using a silver detection lamp (PerkinElmer Lumina Hollow Cathode Lamp, P/N N305-0120, S/N 030211-020140).

2.3. Lab-Scale Constructed Wetland Experiment

In this experiment, AgNP effects on zebrafish were investigated in a lab-scale constructed wetland monitored in a climate controlled greenhouse. Because most previous studies of AgNP impacts on organisms have taken place in petri dishes or other aseptic environments, this experiment was designed to investigate AgNP impacts in a more natural setting. Each AgNP treatment was applied in a separate constructed wetland that consisted of a five gallon bucket in which wetland media was placed in a polyvinyl chloride (PVC) column (to keep the media separate from the zebrafish).

The five gallon bucket contained wetland media in a 11.4 cm (4.5 in) (diameter) by 38.1 cm (15 in) (height) capped PVC column (**Figure 1**). To make the wetland media, each PVC column was filled with 20 cm of 1.3 cm (0.5 inch) washed gravel and 8 cm of wetland soil (~800 g). The gravel and wetland soil were obtained from Certified Products, New Berlin, WI. Each bucket also contained a circulating pump (Mini-Jet 404, Marineland) and a small amount of aquarium gravel to support the column from tipping over. A 1.3 cm (0.5 inch) diameter tube was directed from the pump to a split which distributed the water pressure; the split was controlled with a pinch clamp with one half dispersing oxygenated water back to the fish and the other half entering the top of the column. The column had multiple 1.3 cm (0.5 inch) holes covered with mesh to allow water to be circulated from the bucket through the soil media (**Figure 1(b)**). This allowed AgNPs to have contact with the soil media (which is more similar to a natural environment than having zebrafish and nanoparticles in isolation).

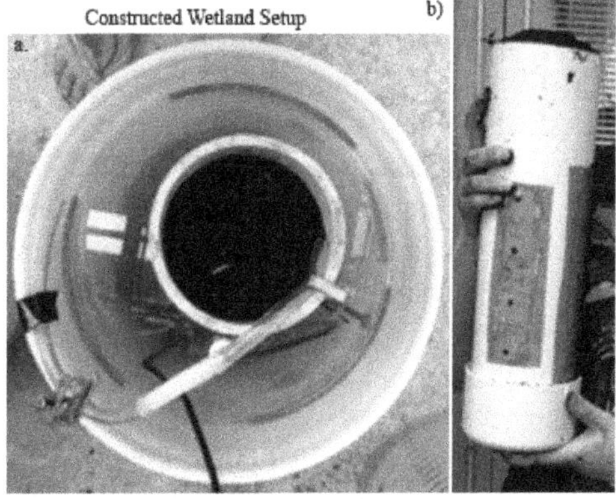

Figure 1. Photos of (a) the top view of the lab-scale constructed wetland: five gallon bucket with soil column in the middle and tubing along the side connected to the circulating pump and (b) the side view of the soil column: 20 cm of gravel on the bottom and 8 cm of wetland soil at the top; mesh covering 1.3 cm holes.

Each bucket contained 12 L of deionized (DI) water with 3.8 g of ocean salt and 0.5 g of pH 7 buffer. There were six treatments: negative control (ocean salt and no AgNP), negative control with dispersing agent (ocean salt with 2 mg/L Tide and no AgNP), and 15 mg AgNP/L (< 90 nm powder, mKnano), 25 mg AgNP/L, 50 mg AgNP/L, and a positive control (15 mg $AgNO_3$/L). Each AgNP treatment also had ocean salt and 2mg/L Tide. The liquid laundry detergent Tide (Tide® Active with Febreze) was used as the dispersing agent for the AgNP powder, both because it was shown to be an effective dispersing agent in preliminary trials and because it mimicked one of the ways nanoparticles might enter wastewater (*i.e.* through residential laundry). Tide is a commercial laundry detergent commonly used in American households. After the constructed wetland was prepared (PVC column of wetland media, circulating pump, and treatment addition), five zebrafish, three female and two male, were placed into each bucket. The zebrafish were sexually mature fish of at least 2 months of age and supplied from Aquatics Unlimited (Greenfield, WI). The constructed wetlands were placed in a temperature regulated greenhouse room kept at 27 ± 5℃. The temperature, pH, silver content of the water, and the health of the fish were monitored daily. The fish were kept in the constructed wetland exposed to AgNP for one week. The zebrafish were fed daily with Zeigler adult zebrafish diet (Zeigler product #AH271). Institutional approval from the on-campus Animal Care and Use Committee was received before carrying out this research.

At the end of the week remaining live zebrafish were euthanized and a soil sample taken from the wetland media column at 3 depths (surface, center, bottom). The water and soil samples were digested with 6 M nitric acid and analyzed with atomic absorption spectroscopy using the method described above in the AgNP sorption experiment section. For the water samples, 2 ml of sample were digested with 1 ml of 6 M nitric acid. The soil samples were air dried after which 0.5 g of soil was digested with 2 ml of 6 M nitric acid and 10 ml of 2 mM citrate solution before elemental silver analysis on the AA. This experiment was replicated three times in consecutive weeks.

3. Results and Discussion

3.1. Physical and Chemical Characteristics of the Media

Chemical and physical analysis of the media used in both

the sorption experiment and lab-scale constructed wetland experiment is reported in **Table 1**. The texture of the wetland soil, silt loam soil, and WTR was sandy loam, silt loam, and loamy sand, respectively. The wetland soil and WTR had very similar sand content (71% and 75%, respectively), the same silt content (23%), and correspondingly different clay content (6% and 2%, respectively). The wetland soil had the highest carbon content (332,600 mg/kg) as would be expected. The WTR had lower carbon content (76,800 mg/kg), and the silt loam had the lowest (19,850 mg/kg). These media had varying nutrient contents, e.g. the wetland soil had the highest nitrogen and phosphorus content (24,320 mg N/kg and 967 mg P/kg) but the lowest magnesium and aluminum content (2782 mg Mg/kg and 4035 mg Al/kg). The WTR had aluminum content an order of magnitude higher than the silt loam (110,532 vs. 15,480 mg Al/kg, respectively) and two orders of magnitude higher than the wetland soil (4035 mg Al/kg). This was not surprising because aluminum salts are used as coagulants in water treatment.

3.2. Sorption Experiments

The purpose of the sorption experiments was to determine if the media had the ability to remove AgNPs from water as well as to investigate which soil particle size (sand, silt, or clay) AgNPs adsorb too preferentially. The horizontal axes in **Figure 2** are the initial AgNP concentrations in the solution shaken with each medium (mg AgNP/L solution). The vertical axes are the mass of AgNP adsorbed per mass of medium (mg AgNP/kg medium). Straight lines indicate that the medium still has the ability adsorb more AgNP; a curve bending to the right (becoming horizontal) indicates that the medium had a diminished capacity to adsorb more AgNP, thus leaving more in solution. This occurs because the sorption sites gradually become filled. None of the media in this experiment show much curve (**Figures 2(a), (b), (c)**), indicating that they all have the potential to adsorb more AgNPs from solutions with concentrations of AgNPs higher than the highest used in this study (60 mg AgNP/L). However, both the wetland soil and the WTR adsorbed a much greater total mass of AgNPs per mass of media than did the silt loam soil (1250 and 1000 mg AgNP/kg vs. 220 mg AgNP/kg, respectively). The average percentage of AgNP adsorbed to all size particles at the highest initial solution concentration (60 mg AgNP/L) in the wetland soil, WTR, and silt loam soil (determined by subtracting the concentration of AgNP in solution after shaking from the initial concentration of AgNP in the shaking solution and dividing by the initial concentration) was 100%, 81%, and 18%, respectively (**Table 2**). This data also shows that the silt loam soil adsorbed approximately half of the total AgNP from the lowest initial concentration (49%) and progressively adsorbed a smaller percentage as the initial AgNP concentration increased. The WTR showed a similar phenomenon adsorbing 93% of the total AgNP available at lowest initial concentration decreasing to adsorbing 81% of the total AgNP in solution at the highest initial concentration. These results suggest that the wetland soil and the WTR are able to remove substantial amounts of AgNP from water.

Looking more closely at the impact of the size of the media particles in removing nanoparticles, the data show that all of AgNPs were associated with sand size particles in the WTR (**Figure 2(c)**). The data points for each sampling time (30 s, 2 hrs, and after centrifuging) of the WTR are very similar, indicating that little more AgNPs were removed with the smaller sized particles. However, data from the silt loam and wetland soils show that approximately half of the AgNPs were removed from the 60 mg AgNP solution after 30 s of settling (100 out of 220 mg AgNP/kg and 680 out of 1250 mg AgNP/kg, respectively) (**Figures 2(a)** and **(b)**). A similar effect was seen at the lower initial AgNP solution concentrations. This indicates that about half of the AgNP in solution were adsorbed to sand sized particles and about half of the AgNP were adsorbed to silt and clay size particles.

The wetland soil and WTR differ from the silt loam soil in that they both have very similar sand content (> 70% sand) and both have a higher total carbon value compared to the silt loam soil (**Table 1**). The sand content cannot be responsible for the higher amount of AgNP sorption in the wetland soil and WTR since only half of the total AgNPs in solution were removed with sand particles in the wetland soil (**Figure 2(b)**). However, the trend correlates well with the increase in carbon con-

Table 1. Selected properties of the three media used in the sorption and constructed wetland experiments.

Parameter[a]	Silt Loam Soil	Wetland Soil	WTR
C (mg/kg)	19850	332600	76800
N (mg/kg)	1818	24320	4068
P (mg/kg)	389	967	788
Ca (mg/kg)	3459	23954	31010
Mg (mg/kg)	4263	2728	12901
S (mg/kg)	199	11705	1655
Fe (mg/kg)	18830	14504	6614
Al (mg/kg)	15480	4035	110532
Sand (%)	19	71	75
Silt (%)	63	23	23
Clay (%)	18	6	2
Soil Texture	Silt Loam	Sandy Loam	Loamy Sand

[a]All minerals are total elemental; C by LECO; N by Kjeldahl; P, Ca, Mg, S, Fe, and Al by ICP-OES. Sand, silt, and clay percentages were determined by the hydrometer method. All analyses were completed at the Soil and Plant Analysis Lab, Madison, WI.

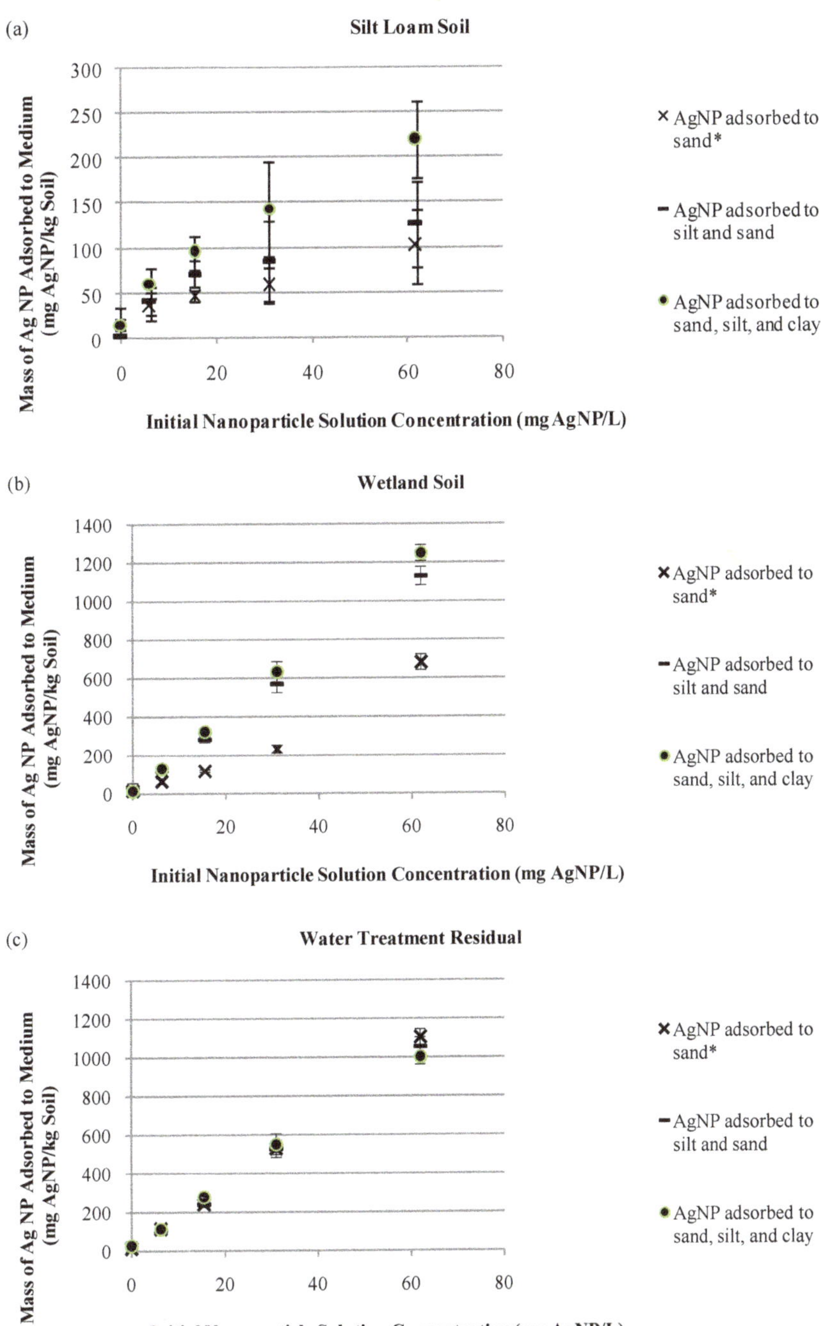

Figure 2. Silver nanoparticle (AgNP, 10 nm monodispersed) adsorption to three media (a) silt loam soil, (b) wetland soil, and (c) water treatment residual (WTR). Each data point is the average of three trials, with vertical error bars indicating standard deviation. The three curves on each graph indicate nanoparticles adsorbed to different size classes of soil. "AgNP adsorbed to sand" was measured from the solution concentration after 30 s (approximate time for sand to settle), "AgNP adsorbed to silt and clay" was measured from the solution concentration after 2 hours (approximate time for silt to settle, and "AgNP adsorbed to all media" was measured from the solution concentration after centrifugation (only dissolved nanoparticles in solution).

tent. The wetland soil has the highest carbon content and showed no reduction in ability to remove AgNP. The WTR showed a slight reduction and had lower carbon content, while the silt loam soil had the lowest amount of carbon and was the least capable of removing AgNPs (**Tables 1** and **2**, **Figure 2(a)**). Recently, researchers found that dispersion and toxicity of AgNP were dependent on the amount of humic acid present [46]. At high humic acid concentrations (> 20 mg total organic carbon/L), significant aggregation of AgNPs was ob-

Table 2. Percentage of the total silver nanoparticles (AgNP) adsorbed by each medium at each initial solution concentration. This was calculated from the average difference between the initial solution concentration and the final (equilibrium) solution concentration.

Initial AgNP Solution Concentration	Percentage of AgNP Adsorbed		
	Silt Loam Soil	Wetland Soil	WTR
	%	%	%
6 mg AgNP/L	49	100	93
15 mg AgNP/L	31	100	90
30 mg AgNP/L	16	100	89
60 mg AgNP/L	18	100	81

served. When studying the relative risk ratios for different metallic nanomaterials other researchers found AgNPs pose a greater environmental risk than either TiO_2 or ZnO nanoparticles, which highlights the importance of studying their fate in the environment [47]. However, they also note that their data overestimates the risk to the terrestrial environment, because in ecotoxicity studies there is an assumption that metallic silver is present, when in fact, other studies have shown that often AgNPs are converted to Ag_2S during wastewater treatment [48-50] and as such are much less soluble and therefore less toxic. Aggregation size was not measured in the data reported in this research, nor was the form of Ag after equilibration. In **Figure 2**, all of the removal of AgNPs from the solution is represented as adsorption to the media. Aggregation of AgNPs and settling from solution is not distinguished from adsorption, so adsorption amounts may be inflated. Future research that images the particles and characterizes the adsorption to the media would help determine this.

3.3. Lab-Scale Constructed Wetland Experiment

The purpose of the lab-scale constructed wetlands was to examine the toxicity of AgNPs to zebrafish living in a pseudo-natural environment rather than an aseptic, unnatural environment of a petri dish. As the results of the sorption experiment indicate, other environmental factors may impact the fate of AgNPs that enter an ecosystem, potentially rendering them less toxic or simply removed from the environment.

The results of this experiment are inconclusive but do shed light on the impact that environmental factors have on the fate of AgNPs. Fish mortality (**Figure 3**) only occurred in the positive control treatment (15 mg/L $AgNO_3$) (where in two out of the three weeks, two of the five fish did not survive to the end of the seven day experiment) and in the negative control (where one fish during one of the three weeks did not survive to the end of the experiment). There was no mortality in any of the

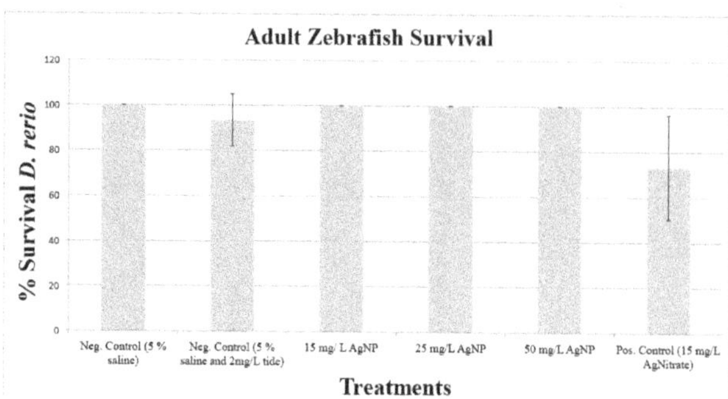

Figure 3. Adult zebrafish survival expressed as a % of fish surviving after 7 day exposure. Each bar is the average of three trials of 5 fish (2 male, 3 female) per lab-scale constructed wetland. Error bars indicate standard deviation; Kruskal-Wallis test gave a p-value of 0.134 between treatments.

AgNP treatments. At first, this data made sense in light of the sorption experiments explained above. The wetland soil had been shown to be able to remove AgNPs from water, so when analyzing the soil of the constructed wetland after the seven-day exposure, it was expected that it would contain AgNPs. However, after digesting the soil at three depth levels, there were no AgNPs measured at any depth (data not shown). Additionally, the daily water samples taken from the middle of the bucket also did not contain AgNPs (data not shown). If the AgNPs were not in the circulating water or in the soil, the conclusion was drawn that the dispersant (Tide detergent) may not have been strong enough to keep the AgNPs dispersed, and the AgNPs may have aggregated and sunk to the bottom of the bucket and stayed in the layer of pebbles. Preliminary trials indicated that the 2 ml Tide/L could keep AgNP in solution, although not as well as at higher concentrations. However, this level was also considered the detergent level where fish may absorb twice the amount of chemicals than they would normally absorb [51]. Tide was chosen as the dispersing agent in this study because it not only acted as a dispersant, but also is a likely way for AgNP to enter the environment through residential laundry.

Other research has shown that AgNPs aggregate in saline solutions [52], which support this conclusion. There was no feasible way found to measure the AgNPs in that layer of pebbles and confirm the conclusion. Recent research is beginning to focus on the fate of nanoparticles in a more natural setting, such as the study reported here. Reference [53] found that plants in a system could help decrease the toxicity of AgNPs because plants release dissolved organic matter which can bind with Ag ions. In a freshwater system, researchers found that sediments accumulated most of the ceria nanoparticles used in their aquatic system [54]. Additionally other researchers suggest that the chemistry of the nanoparticle capping agent plays an important role in the fate and transport of

AgNPs and environmental factors such as pH, ionic strength, and electrolyte composition can help predict the fate and transport of AgNPs [55]. Their data showed that positively charged branched polyethyleneimine stabilized AgNPs would most likely have limited mobility in soils, groundwater, and other environments, but sterically stabilized polyvinylpyrrolidone AgNPs may have the greatest potential for mobility and transport. Another recent study indicated that in a sandy loam soil AgNP concentrations eight times greater than $AgNO_3$ concentrations were needed to induce significant reproductive toxicity in earthworms [56]. A study such as the one reported here, although it investigates only a small fraction of the questions still remaining regarding the fate of nanoparticles in the environment provides valuable new information to help guide future studies. Some researchers [47] do not lament the idea that each nanomaterial may react differently depending on the material and the environmental properties and conditions, but instead they reinforce the importance of continuing to study these materials under many different conditions.

Although the lab-scale constructed wetland experiment did not lead to conclusive results regarding the fate of nanoparticles in this environment, it did indicate that the toxicity levels shown in laboratory conditions are not the same as in a more natural environment. Thus the need for similar experiments simulating natural environments is underscored. Simple, inexpensive experimental designs such as this can be implemented to investigate parameters that impact AgNP fate and biotoxicity. Simulated natural environments may prove useful in determining the mechanism of AgNP toxicity to fish. Is direct exposure to suspended AgNPs more or less toxic than dietary exposure to algae or zooplankton that has previously internalized nanoparticles? This question is especially relevant in light of the recent observation that a wide variety of living organisms do take AgNPs out of the water column [57]. Additionally, desorption experiments investigating how tightly bound AgNPs are to environmental media will help elucidate the effectiveness and lifetime of various media in a constructed wetland setting. Preliminary trials have indicated that AgNPs are bound most tightly to the wetland soil, followed by the silt loam soil, and least tightly to the WTR (data not shown), but further research is necessary to investigate this further. As is suggested by multiple groups of researchers [16,58, and others], a multidisciplinary approach is crucial to understanding nanoparticle risks in the environment and will involve collaborations between chemists, biologists, toxicologists, ecologists, engineers, and environmental scientists.

4. Conclusions

The results of the constructed wetland study indicate that silver nanoparticles appear to aggregate in a salt solution rendering them less toxic to zebrafish than would be expected from previous studies using silver nanoparticles and zebrafish in a pure media. This research did not measure the size of the nanoparticles after they were added to the lab-scale constructed wetland so the aggregation of the nanoparticles cannot be known for sure, but the lack of mortality in the zebrafish and the absence of silver nanoparticles in the water media and wetland soil after nanoparticle addition indicates that the particles most likely sank to the bottom of the buckets. More importantly, the sorption studies provide evidence that soil and water treatment residuals have the ability to remove silver nanoparticles from wastewater. This means that natural and constructed wetlands could be sinks for silver nanoparticles, removing them from wastewater. Using water treatment residuals in a constructed wetland would be a beneficial reuse of a waste product that would otherwise need to be disposed of. It is important to remember that after the silver nanoparticles are removed from the water by sorption to soil or other media, they still remain in the environment adsorbed to the media. Further studies are needed to investigate how tightly and for how long the silver nanoparticles are retained by soil and water treatment residuals. These desorption studies would shed light on the long-term sustainability of a wetland designed to remove nanoparticles and the kind of engineering that would be needed to best manage constructed wetlands.

5. Acknowledgements

The authors thank the Wisconsin Lutheran College Faculty Development committee for support of this research. They also wish to thank Michael Reep, Amanda Wagner, Krystal Weishaar, and Benjamin Tellier for their contributions to the preliminary experimental designs and test trials.

REFERENCES

[1] WHO: World Health Organization, "Small-scale Water Supplies in the pan-European Region," United Nations Economic Commission for Europe, 2010.

[2] T. Brick, B. Primrose, R. Chandrasekhar, S. Roy, J. Muliyil, G. Kang, "Water Contamination in Urban South India: Household Storage Practices and Their Implications for Water Safety and Enteric Infections," *Journal of Hygiene and Environmental Health*, Vol. 207, No. 5, 2004, pp. 473-480.

[3] R. B. Levin, P. R. Epstein, T. E. Ford, W. H. Harrington, E. Olson, E. G. Reichard, "U.S. Drinking Water Challenges in the Twenty-First Century," *Environmental Health Perspectives*, Vol. 100, 2002, pp. 43-52.

[4] A. J. Jowet, "China's Water Crisis," *The Geographical*

Journal, Vol. 152, No. 1, 1986, pp. 9-18.

[5] P. Vonlanthen, D. Bittner, A. G. Hudson, K. A. Young, R. Muller, B. Lundsgaard-Hansen, D. Roy, S. Di Piazza, C. R. Largiader and O. Seehausen, "Eutrophication Causes Speciation Reversal in Whitefish Adaptive Radiations", *Nature*, Vol. 482, 2012, pp. 357-362.

[6] S. S. Kaushal, W. M. Lewis Jr., and J. H. McCutehan Jr., "Land Use Change and Nitrogen Enrichment of a Rocky Mountain Watershed," *Ecological Applications*, Vol. 16, No. 1, 2006, pp. 299-312.

[7] J. Fawell and M. J. Nieuwenhuijsen, "Contaminants in Drinking Water," *British Medical Bulletin*, Vol. 68, No. 1, 2003, pp. 199-208.

[8] S. N. Levine and D. W. Schindler, "Influence of Nitrogen to Phosphorus Supply Ratios and Physicochemical Conditions on Cyanobacteria and Phytoplankton Species Composition in the Experimental Lakes Area, Canada," *Canadian Journal of Fisheries and Aquatic Sciences*, Vol. 56, No. 3, 1999, pp. 451-466.

[9] D. L. Correll, "The Role of Phosphorus in the Eutrophication for Receiving Waters: A Review," *Journal of Environmental Quality*, Vol. 27, No. 2, 1998, pp. 261-266.

[10] T. C. Daniel, A. N. Sharpley, and J. L. Lumunyon, "Agricultural Phosphorus and Eutrophication: A Symposium Overview," *Journal of Environmental Quality*, Vol. 27, No. 2, 1998, pp. 251-257.

[11] E. G. Srinath and S. C. Pillai, "Phosphorus in Sewage, Polluted Waters, Sludges, and Effluents," *The Quarterly Review of Biology*, Vol. 41. No. 4, 1966, pp. 384-407.

[12] C. N. Sawyer, H. B. Gotaas, and J. B. Lackey, "Factors Involved in Disposal of Sewage Effluents to Lakes," *Sewage and Industrial Wastes*, Vol. 26, No. 3, 1954, pp. 317-328.

[13] S. K. Liehr, "Natural Treatment and Onsite Processes," *Water Environment Research*, Vol. 77, No. 6, 2005, pp. 1389-1424.

[14] UN, "Waste-Water Treatment Technologies: A General Review," Economic and Social Commission for Western Asia, New York, 2003.

[15] U. Mander and P. D. Jenssen Eds., "Constructed Wetlands for Waste Water Treatment in Cold Climates," WIT Press, Southampton, 2002.

[16] E. S. Bernhardt, B. P. Colman, M. F. Hochella, Jr., B. J. Cardinale, R. M. Nisbet, C. J. Richardson, and L. Yin, "An Ecological Perspective on Nanomaterial Impacts in the Environment," *Journal of Environmental Quality*, Vol. 39, No. 6, 2010, pp. 1-12.

[17] A. B. A. Boxall, M. A. Rudd, B. W. Brooks, D. J. Caldwell, K. Choi, S. Hickmann, E. Innes, K. Ostapyk, J. P. Staveley, T. Verslycke, G. T. Ankley, K. F. Beazley, S. E. Benlanger, J. P. berninger, P. Carriquiriborde, A. Coors, P. DeLeo, S. D. Dyer, J. F. Ericson, J. Gagne, J. P. Biesy, T. Gouin, L. Hallstrom, M. V. Karlsson and D. G. J. Larsson, "Pharmaceuticals and Personal Care Products in the Environment: What Are the Big Questions?" *Environmental Health Perspectives*, Vol. 120, No. 9, 2012, pp. 1221-1229.

[18] S. Rodrigues-Mozaz and H. S. Weinberg, "Meeting Report: Pharmaceuticals in Water–An Interdisciplinary Approach to a Public Health Challenge," *Environmental Health Perspectives*, Vol. 118, No. 7, 2010, pp. 1016-1020.

[19] J. Fabrega, S. N. Luoma, C. R. Tyler, T. S. Galloway and J. R. Lead, "Silver Nanoparticles: Behaviour and Effects in the Aquatic Environment," *Environmental International*, Vol. 37, No. 2, 2011, pp.517-531.

[20] A. M. Comerton, R. C. Andrews and D. M. Bagley, "Practical Overview of Analytical Methods for Endocrine-Disrupting Compounds, Pharmaceuticals, and Personal Care Products in Water and Wastewater," *Philisophical Transacations: Mathematical, Physical, and Engineering Sciences*, Vol. 367, No. 1904, 2009, pp. 3923-3939.

[21] C. G. Daughton and T. A. Ternes, "Pharmaceuticals and Personal Care Products in the Environment: Agents of Subtle Change?" *Environmental Health Perspectives*, Vol. 107, 1999, pp. 907-938.

[22] N. B. Golovina and L. M. Kustov, "Toxicity of Metal Nanoparticles with a Focus on Silver," *Mendeleev Communications*, Vol. 23, No. 2, 2013, pp. 59-65.

[23] C. You, C. Han, X. Wang, Y. Zheng, Q. Li, X. Hu and H. Sun, "The Progress of Silver Nanoparticles in the Antibacterial Mechanism, Clinical Application, and Cytotoxicity," *Molecular Biology Reports*, Vol. 39, No. 9, 2012, pp. 9093-9201.

[24] K. Kulthong, S. Srising, K. Boonpavanitchak, W. Kangwansupamonkon and R. Maniratanachote, "Determination of Silver Nanoparticle Release from Antibacterial Fabrics into Artificial Sweat," *Particle Fibre Toxicology*, Vol. 7, 2010, pp. 8.

[25] B. Karn, T. Kuiken and M. Otto, "Nanotechnology and in Situ Remediation: A Review of the Benefits and Potential Risks," *Environmental Health Perspectives*, Vol. 117, No. 12, 2009, pp. 1823-1831.

[26] M. Yousefian and B. Payam, "Effects of Nanochemical Particles on Some Histological Parameters of Fish," *Advances in Environmental Biology*, Vol. 6, No. 3, 2012, pp. 1209-1215.

[27] F. Gagne, C. Andre, R. Skirrow, M. Gelinas, J. Auclair, G. van Aggelen, P. Turcotte and C. Gagnon, "Toxicity of Silvernanparticles to Rainbow Trout: a Toxicogenomic Approach," *Chemosphere*, Vol. 89, No. 5, 2012, pp. 615-622.

[28] S. W. Kim and Y. J. An, "Effect of ZnO and TiO2 Nanoparticles Preilluminated with UVA and UVB light on

Escherichia coli and Bacillus subtilis," *Applied Microbiology and Biotechnology*, Vol. 95, No. 1, 2012, pp. 243-253.

[29] D. A. Cowart, S. M. Guida, S. Ismat and A. G. Marsh, "Effects of Ag Nanoparticles on Survival and Oxygen Consumption of Zebrafish Embryos, Danio rerio," *Journal of Environmental Science and Health, Part A: Toxic/Hazardous Substances and Environmental Engineering*, Vol. 46, No. 10, 2011, pp. 1122-1128.

[30] D. Stampoulis, S. K. Sinha and J. White, "Assay-Dependent Phytotoxicity of Nanoparticles to Plants," *Environmental Science and Technology*, Vol. 43, No. 24, 2009, pp. 9473-9479.

[31] R. J. Griffit, J. Luo, J. Gao, J. C. Bonzongo and D. S. Barber, "Effects of Particle Composition and Species on Toxicity of Metallic Nanomaterials in Aquatic Organisms," *Nanomaterials in the Environment*, Vol. 27, No. 9, 2008, pp. 1972-1978.

[32] K. Y. Yoon, J. H. Byeon, J. H. Park and J. H. wang, "Susceptibility Constants of *Escherichia coli* and *Bacillus subtilis* to Silver and Copper Nanoparticles," *Science of the Total Environment*, Vol. 373, No. 2-3, 2007, pp. 572-575.

[33] A. S. Barnard and H. Guo, "Nature's Nanostructures," Pan Stanford Publishing, Singapore, 2012.

[34] N. J. Kagengi and A. Thompson, "The Emerging Emphasis on Nanometer-Scale Processes in Soil Environments," *Soil Science Society of America Journal*, Vol. 75, No. 2, 2011, pp. 333-334.

[35] B. K. G. Theng and G. Yuan, "Nanoparticles in the Soil Environment," *Elements*, Vol. 4, No. 6, 2008, pp. 395-399.

[36] J. M. Zook, M. D. Halter, D. Cleveland and S. E. Long, "Disentangling the Effects of Polymer Coatings on Silver Nanoparticle Agglomeration, Dissolution, and Toxicity to Determine Mechanisms of Nanotoxicity," *Journal of Nanoparticle Research*, Vol. 14, 2012, pp. 1165-1572

[37] K. R. Reddy, E. M. D'Angelo and W. G. Harris, "Biochemistry of Wetlands," In *Handbook of Soil Science*, M.E. Sumner (Ed.), CRC Press, 1999.

[38] J. A. Ippolito, K. A. Barbarick and H. A. Elliott, "Drinking Water Treatment Residuals: A Review of Recent Uses," *Journal of Environmental Quality*, Vol. 40, No.1, 2011, pp. 1-12.

[39] E. A. Dayton and N. T. Basta, "A Method for Determining the Phosphorus Sorption Capacity and Amorphous Aluminum of Aluminum-Based Drinking Water Treatment Residuals," *Journal of Environmental Quality*, Vol. 34, No.3, 2005, pp. 1112-1118.

[40] P. S. Nair, T. J. Logan, A. N. Sharpley, L. E. Sommers, M. A. Tabatabai and T. L. Yuan, "Interlaboratory Comparison of a Standardized Phosphorus Adsorption Procedure," *Journal of Environmental Quality*, Vol. 13, No. 4, 1984, pp. 591-595.

[41] H. A. Elliot, G. A. O'Connor, P. Lu and S. Brinton, "Influence of Water Treatment Residuals on Phosphorus Solubility and Leaching," *Journal of Environmental Quality*, Vol. 31, 2002, pp. 1362-0.69.

[42] G. J. Bouyoucos, "Hydrometer Method Improved for Making Particle Size Analysis of Soils," *Agronomy Journal*, Vol. 54, No. 5, 1962, pp. 464-465.

[43] G. W. Gee and J. W. Bauder, "Particle-size Analysis," In A. Klute, Ed., *Methods of soil analysis. Part 1. 2nd ed. Agronomy Monograph 9*. ASA and SSSA, Madison, WI, 1986, pp. 383-411.

[44] T. M. Benn and P. Westerhoff, "Nanoparticle Silver Released into Water from Commercially Available Sock Fabrics," *Environmental Science and Technology*, Vol. 42, No. 11, 2008, pp. 4133-4139.

[45] EPA. "Method 3050B (SW-846): Acid Digestion of Sediments, Sludges, and Soils," Revision 2, 1996. http://www.epa.gov/sam/pdfs/EPA-3050b.pdf

[46] J. Gao, K. Powers, Y. Wang, H. Zhou, S.M. Roberts, B. M. Moudgil, B. Kooopman and D. S. Barber, "Influence of Suwannee River Humic Acid on Particle Properties and Toxicity of Silver Nanoparticles," *Chemosphere*, Vol. 89, No. 1, 2012 pp. 96-101.

[47] F. Gottschalk, E. Kost and B. Nowack, "Engineered Nanomaterials in Waters and Soils: A Risk Quantification Based on Probabilistic Exposure and Effect Modeling," *Environmental Toxicology and Chemistry*, 2013, (In Press).

[48] B. C. Reinsch, C. Levard, Z. Li, R. Ma, A. Wise, K. B. Gregory, G. E. J. Brown and G. V. Lowry, "Sulfidation of Silver Nanoparticles Decreases Escherichia coli Growth Inhibition," *Environmental Science and Technology*, Vol. 46, No. 13, 2012, pp. 6992-7000.

[49] R. Kaegi, A. Boegelin, B. Sinnet, S. Zuleeg, H. Hagendorfer, M. Berkhardt and H. Siegrist, "Behavior of Metallic Silver Nanoparticles in a Pilot Wastewater Treatment Plant," *Environmental Science and Technology*, Vol. 45, No. 9, 2011, pp. 3902-3908.

[50] H. T. Ratte, "Bioaccumulation and Toxicity of Silver Compounds: A Review," *Environmental Toxicology and Chemistry*, Vol. 18, No. 2, 1999, pp. 89-108.

[51] Lenntech BV, "Detergents Occurring in Freshwater," 2012

[52] G. E. Batley, J. K. Kirby, M. J. McLaughlin, "Fate and Risks of Nanomaterials in Aquatic and Terrestrial Environments," *Accounts of Chemical Research*, Vol. 46, No. 3, 2013, pp. 854-862.

[53] J. M. Unrine, B .P. Colman, A. J. Bone, A. P. Gondikas and C. W. Matson, "Biotic and Abiotic Interactions in Aquatic Microcosms Determine Fate and Toxicity of Ag Nanoparticles. Part 1. Aggregation and Dissolution," *Environmental Science and Technology*, Vol. 46, No.13,

2012, pp. 6915-6924.

[54] P. Zhang, X. He, Y. Ma, K. Lu, Y. Zhao and Z. Zhang, "Distribution and Bioavailability of Ceria Nanoparticles in an Aquatic Ecosystem Model," *Chemosphere*, Vol. 89, No. 5, 2012, pp. 530-535.

[55] A. M. Badawy, T. P. Luxton, R. G. Silva, K. G. Scheckel, M. T. Suidan and T. M. Tolaymat, "Impact of Environmental Conditions (pH, Ionic Strength, and Electronlyte Type) on the Surface Charge and Aggregation of Silver Nanoparticles Suspensions," *Environmental Science and Technology*, Vol. 44, No. 4, 2010, pp. 1260-1266.

[56] W. A. Shoults-Wilson, B. C. Reinsch, O. V. Tysusko, P. M. Bertsch, G. V. Lowry and J. M. Unrine, "Role of Particle Size and Soil Type in Toxicity of Silver Nanoparticles to Earthworms," *Soil Science Society of America Journal*, Vol. 75, No. 2, 2011, pp. 365-377.

[57] D. Cleveland, S. E. Long, P. L. Pennington, E. Cooper, M. Fulton, G. I. Scott, T. Brewer, J. Davis, E. J. Petersen and L. Wood, "Pilot Estuarine Mesocosm Study on the Environmental Fate of Silver Nanomaterials Leached from Consumer Products," *Science of the Total Environment*, Vol. 421-422, No. 5, 2012, pp. 267-272.

[58] R. D. Handy and B. J. Shaw, "Ecotoxicity of Nanomaterials to Fish: Challenges for Ecotoxicity Testing," *Integrated Environmental Assessment and Management*, Vol. 3, No. 3, 2007, pp. 458-460.

17

Effect of Ambient Temperature on PUF Passive Samplers and PAHs Distribution in Puerto Rico

Nedim Vardar [1], Ziad Chemseddine[1], Juan Santos[2]
[1]School of Engineering, Inter American University, Bayamon, PR, 00957
[2]Department of Natural Science, Inter American University, Bayamon, PR, 00957

ABSTRACT

Passive sampling for the monitoring of organic pollutants (PAHs, PCBs, PBDEs) in ambient air has received increased attention in the last two decades. However, the accuracy of the concentration of organics obtained with passive samplers under varying environmental conditions is a subject of controversy. In this study, effect of ambient temperature on passive samplers was evaluated by using three different sampler configurations. Additionally, passive samplers with polyurethane disks (PUF) were applied throughout the Island for the determination of the airborne concentration of polycyclic aromatic hydrocarbons (PAHs). The passive samplers were deployed in seven municipalities for three-month periods in two different sampling campaigns, representing hurricane and non-hurricane seasons. Here we present preliminary results obtained from those sampling campaigns. The total concentrations of 15 PAHs varied from 3.1 to 19.6 and from 5.5 to 38.5 ng/m^3 for hurricane and non-hurricane seasons, respectively. Hurricane and non-hurricane season concentrations of PAH were significantly different for the samples taken in the northern municipalities of the Island. However, there was no significant difference in PAH concentrations between the hurricane and non-hurricane seasons for the southern sites. Increased rainfall and high-relative humidity during the hurricane season had an influence on the concentrations of PAHs derived by the passive PUF sampler.

Keywords: Passive Sampler; PAH; Puerto Rico; Hurricane; PUF; Ambient Temperature

1. Introduction

Passive samplers with polyurethane disks (PUF) have been widely used for the last decade in the measurement of semi volatile organics (SOCs), in part because they are easy to handle, operate independently for several months, and are inexpensive. Knowledge of the sampling rate of the device is necessary for an accurate conversion of the sampled mass to an ambient concentration. However, environmental conditions around the sampler influence the sampling rate and the performance with which the PUF sampler can be used. The reliability of passive sampling techniques under varying environmental conditions is therefore a subject of controversy. Environmental conditions that may affect the PUF sampler are wind speed, temperature (T), atmospheric pressure, sunlight/UV-light and humidity. T affects the compound specific molecular diffusion coefficients (D) of the pollutants by increasing D with T.

Polycyclic aromatic hydrocarbons (PAHs) are a natural component of most fossil fuels, and formed during the incomplete combustion of these fuels, or other organic substances. Some PAHs are manufactured as individual compounds for research purposes. The chemical structure of PAHs is comprised of two or more fused aromatic rings made entirely from carbon and hydrogen. PAHs are ubiquitous environmental pollutants and are formed from both natural and anthropogenic sources. Although natural sources such as forest fires and volcanic eruptions contribute to PAH formation, most PAHs in ambient air are the result of man-made processes. PAHs pose severe risks to human health and the environment due to their toxicity, persistence, ability to travel long distances on air and water currents [1-3]. Human exposure to these compounds may result in a variety of adverse health effects including damage to the central nervous and reproductive systems, development of lung cancer and genetic alterations. There are hundreds of PAH compounds in the environment, but only 16 of them are included in the priority pollutants list of the U.S. environment protection agency (USEPA) [4].

A number of studies have been conducted on the fate of PAHs in atmosphere during the past three decades [1-3, 5-9]. In most of these studies, a high-volume sampling technique using filter and adsorbent has been applied. However, the use of passive sampling methods to moni- tor airborne contaminants has greatly increased over the past few years. Passive samplers can be deployed at loca- tions where it is difficult or impractical to install and maintain Hi-Vol samplers. These devices are simple and inexpensive and do not need field calibration, electricity nor technical personnel at the sampling site. In addition, they can be deployed in many locations concurrently due to their low cost.

Puerto Rico is located in the Caribbean with mainly north-easterly trade winds. Due to its location, the Island enjoys a tropical climate and also experiences the Atlantic hurricane season. In this study, effect of ambient temperature on passive samplers and seasonal distribution of PAHs were evaluated Characterization of the PAHs in the ambient air has been studied for a long time in other parts of the world. A vast number of publications are available in the literature for both urban and rural areas from developed and developing countries [1-3,5-7]. How-ever, only limited data on atmospheric PAHs had been acquired for Puerto Rico, and to our knowledge, this is the first study to compare PAH concentrations spatially distributed across the island.

2. Materials and Method

2.1. Sampling Locations

Puerto Rico is a tropical island located in the Caribbean Sea with a population of around 4 million. It has 9100 km^2 total area and is one of the most densely populated islands in the world. PUF passive air samplers consisting of polyurethane foam disks of 14 cm diameter and 1.2 cm thickness housed in two stainless steel bowls were deployed in Bayamon, Cayey, Carolina, Guayama, Manati, San Sebastian, and Naguabo (**Figure 1**).

Two sampling campaigns were achieved in 2010 for three-month periods except in Naguabo where sampling took place for six motnhs. Sampling parameters for the samples are summarized in **Table 1**. Sampling sites of Cayey and Manati are considered as urban, medium density, and residential. Bayamon and Carolina is described as urban, high-density residential/industrial area. Naguabo site is characterized as low-density coastal/residential. San Sebastian is located in west part of the Island and considered as urban, medium density, and residential. Last, Guayama sampling site is situated in a coastal/rural region.

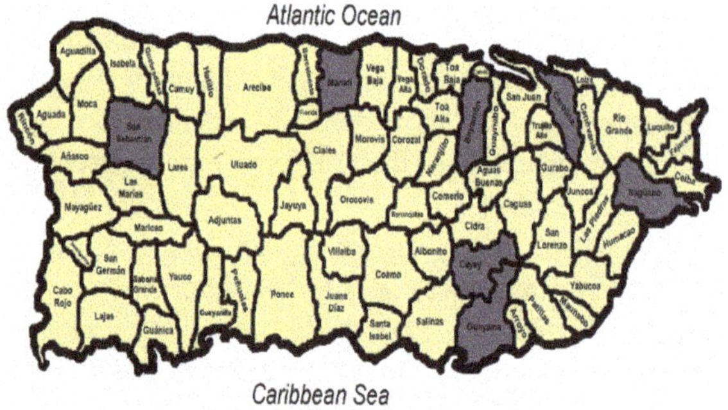

Figure 1. Map of Puerto Rico showing sampling locations.

Table 1. Passive Sampling Campaign Parameters.

Sampling Campaign	Start Date	End Date	Temp. (°C)	Rain (cm)
Non-Hurricane Season	March 2010	June 2010	27.7	17.9
Hurricane Season	June 2010	October 2010	28.6	21.8

The temperature in the south of the Island is usually a few degrees higher than the north. Between winter and summer, there is only a temperature swing of around 3˚C Rainfall tends to be evenly distributed throughout the year, but doubles during the months from May to October, as falls from November to April, with a driest period from January to April. The wind patterns across the island are basically zonal, from east to west.

2.2. Analytical Procedures

Information for sample preparation, extraction and analysis is given in detail elsewhere [8]. Briefly, each PUF was Soxhlet extracted with a 20:80 dichloromethane (DCM): petroleum ether (PE) solution for 24 h. All samples were spiked with PAHs surrogate standards prior to extraction. Four deuterated PAHs were used as surrogate standard and Pyrene-d_{10} for volumetric corrections. The recoveries of the following surrogate standards were used to correct the amounts of specific PAHs found in the samples: Acenaphthene-d_{10} for acenaphthene (ACE), acenaphthylene (ACT) and fluorine (FLN), phenanthrene-d_{10} for phenanthrene (PHE), anthracene (ANT), and fluoranthene (FL), chrysene-d_{12} for pyrene (PY), benz(a) anthracene (BaA) and chrysene (CHR), and perylene-d_{12} for benzo (b) fluoranthe (BbF), benzo(k)fluoranthe (BkF), benzo (a) pyrene (BaP), indeno (1,2cd) pyrene (IcdP), dibenz (a,h) anthracene (DahA), and benzo(ghi)perylene (BghiP). The average recoveries for surrogates in field samples were 79% ± 31% for acenaphthene-d_{10}, 86% ± 13% for phenanthrene-d_{10}, 84% ± 28% for chrysene-d_{12}, 80% ± 15% for perylene-d_{12}.

The analysis of the samples was performed using a Varian 450-GC coupled to an ion trap mass spectrometer Varian 240 MS. A Varian factor 4 capillary column (30 m, 0.25 mm, 0.25 μm) was used. The GC oven temperature was programmed from 60℃(held one minute) to 130℃ at 7℃ min^{-1}, then raised to 200℃ at 5℃ min^{-1}, and finally increased from 260℃ to 320℃ at 6℃ min^{-1}. The injector temperature was maintained at 295℃. The linearity in the response of the GC/MS system was evaluated with calibration standards, at five different levels of concentration (0.012, 0.06, 0.3, 0.6 and 1.2 ng/mL). The instrument was Auto tuned at the start of runs with perfluorotributylamine (PFTBA). Individual PAHs were identified based on the retention times of target ion peaks. Internal standard calibration procedure was used for quantifying the identified compounds.

Field blanks, which accompanied samples to the sampling site, were used to determine any contamination during sample handling and preparation. Newly cleaned PUFs were used as laboratory blanks. There was no statistically recognizable difference between field and laboratory blanks.

3. Results and Discussions

Passive sampling devices have been widely used for more than a decade for the measurement of semi volatile organics (SOCs) in air. PUF passive air sampler consists of two stainless steel bowls and all parts of the sampler are made from the stainless steel. In this study, twenty four hour temperature changes inside the passive sampler were characterized using HOBO pendant temperature sensor. In addition to stainless steel sampler, two more samplers, one consists of plastic bowls and another with cupper bowls, were used to characterize the effect material on the diurnal temperature changes inside the samplers. **Figure 2** shows the diurnal temperature changes inside a stainless steel sampler. Temperature variation inside and outside the sampler is shown with blue and green lines, respectively. Temperatures inside the sampler were significantly higher than outside temperatures during the midday due to intense solar radiation. During the early mornings and late afternoon hours no significant temperature differences were observed between inside and outside of the sampler.

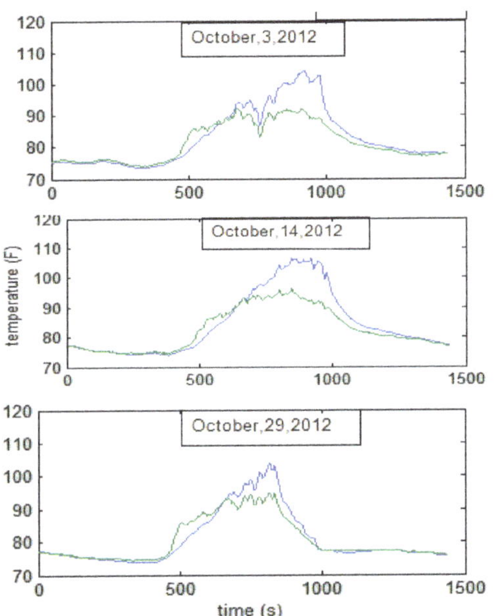

Figure 2. Temperature variation inside (blue line) and outside (green line) of a stainless steel sampler.

In order to minimize midday temperature differences between inside and outside of the sampler, choices of other feasible materials were investigated. Cupper and plastic sampling bowls were used in constructing the passive sampler. **Figure 3** compares the diurnal temperature changes among different samplers. It was expected that the lowest temperature would be observed inside the plastic passive sampler due to the lower conductivity of plastic. However, as it can be seen from the **Figure 3**, the highest temperature inside the samplers belongs to the one made of plastic bowls. A further study is needed to explain this finding. Out of these three samplers, stainless steel sampler gave the lowest internal temperatures.

A network of passive sampling spatially distributed throughout Puerto Rico including both urban and rural areas were conducted. The present study presents the first ambient air data for PAHs in Puerto Rico. Due to its location, Puerto Rico enjoys a tropical climate and also experiences the Atlantic hurricane season. Hurricane season extends from June 1st to November 30th. Therefore, samples taken in the first and second sampling campaign were classified as non-hurricane and hurricane samples, respectively. Higher PAH concentrations were obtained in all sampling sites for the non-hurricane samples except Cayey site. The total concentrations of 15 PAHs varied from 3.1 to 19.6 and from 5.5 to 38.5 ng/m^3 for hurricane and non-hurricane seasons, respectively (**Figure 4**).

Increased rainfall and high water vapor content during the hurricane season makes wet deposition very efficient, leading to a decreased atmospheric lifetime of PAHs for the Island. Carolina site had the highest PAH concentrations for both hurricane and non-hurricane season. Seasonal concentration differences of PAH were more significant for the samples taken in the northern municipalities (Bayamon, Manati, Carolina) than the southern municipalities (Cayey and Guayama) of the Island. The north

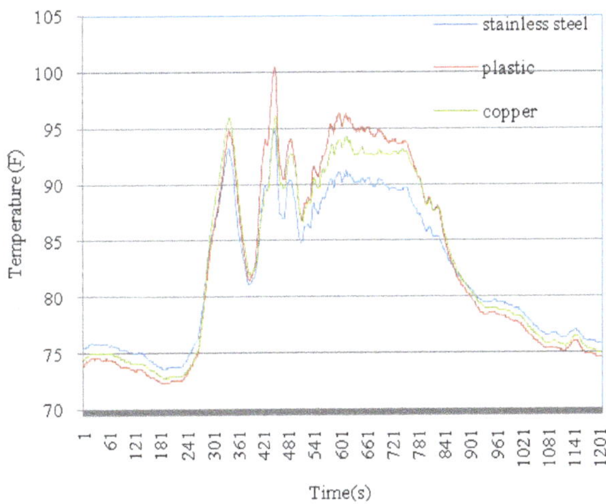

Figure 3. Diurnal temperature changes among different samplers.

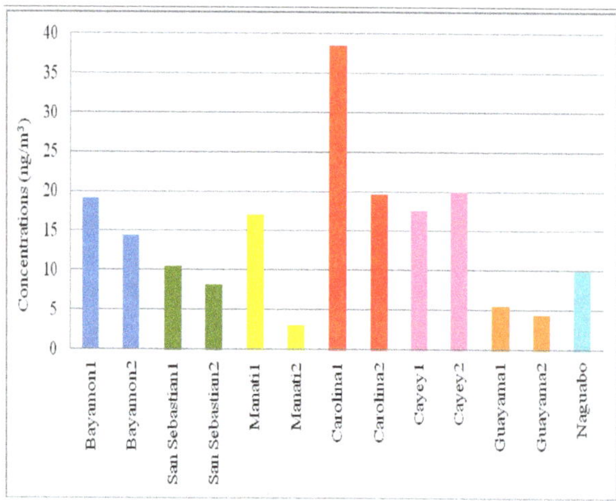

Figure 4. Total PAH concentrations for each sampling locations.

Table 2. Atmospheric PAH Concentrations Derived by Passive Samplers in Different Studies.

Location	Range of Total PAHs (ng/m^3)	Reference
Various Municipalities, PR		
Non-Hurricane Season	5.5 - 38.5	This study
Hurricane Season	3.1 - 19.6	
Toronto, Canada		
Summer	16.5 - 61.4	[1]
Winter	10.1 - 18.5	
Spring	3.53 - 18.8	
Harbin, China		
Spring	25 - 120	
Summer	13 - 50	[3]
Autumn	22 - 74	
Winter	81 - 240	
Taichung, Taiwan	387.70	[5]
Mexico City, Mexico	32 - 92	[6]
Athens, Philippines	28.44	[7]

coast of the Island gets twice as much rain as the south coast. Thus, increased precipitation as well as high- relative humidity during might have an influence on the concentrations of PAHs derived by the passive PUF sampler.

Table 2 provides a comparison of the ambient PAH concentrations measured by passive samplers in different part of the world. The PAH concentrations in the Island are almost similar to those reported by [4] for winter and spring seasons, but compared to summer season the Island concentrations are lower.

The PAH levels reported for Taichung, Taiwan are an order of magnitude higher than those measured in this study [14,15]. In Mexico, and Harpin, China, PAH concentrations significantly higher than the one measured in this study [16]. In a study conducted in Athens, Philippines, PAH concentrations were approximately twice higher than those measured in this study [17].

4. Conclusions

This study has provided some baseline data on the atmospheric PAH levels in Puerto Rico which could be useful for establishing long-term atmospheric monitoring program in the Island. In this study, the spatial characteristics of PAHs derived from PUF passive sampler were studied for two different sampling campaigns at the seven different sampling sites of the Island. It was found that hurricane and non-hurricane season concentrations of PAH were significantly different for the samples taken in the northern municipalities of the Island. However, there was no significant difference in PAH concentrations between the hurricane and non-hurricane seasons for the southern sites. Increased precipitation and high-relative humidity during the hurricane season might have an influence on the concentrations of PAHs derived by the passive PUF sampler.

Diurnal temperature variations inside the passive samplers constructed from stainless steel, cupper and plastic sampling bowls were investigated. Sampler made of stainless steel gave the lowest internal temperatures.

5. Acknowledgements

The authors would like to tank to the NSF-MRI program for funding the purchase of the GC/MS/MS and to Puerto Rico Louis Stokes Alliance for Minority Participation (PR-LSAMP) for supporting this study.

REFERENCES

[1] A. Motelay-Massei, T. Harner, M. Shoeib, M. Diamond, G. Stern and B. Rosenberg, "Using Passive Air Samplers to Assess Urban-Rural Trends for Persistent Organic Pollut- ants and Polycyclic Aromatic Hydrocarbons. 2. Seasonal Trends for PAHs, PCBs, and Organochlorine Pesticides," *Environmental Science and Technology*, Vol. 39, 2005, pp. 5763–5773.

[2] F. M. Jaward, N. J Farrar, T. Harner, A. J. Sweetmanand and K.C.Jones, *Environment Sciencean Technoogy*, Vol. 38, No. 1, 2004, pp. 34-41.

[3] P. Bohlin, K. Jones, H. Tovalin and B. Strandberg, "Observations on Persistent Organic Pollutants in Indoor and Outdoor Air Using Passive Polyurethane Foam Samplers. *Atmospheric Environment*, Vol.42, No. 31, 2008, pp. 7234-7241.

[4] USEPA, Provisional Guidance for Quantitative Risk Assessment of Polycyclic Aromatic Hydrocarbons, EPA/600/ R-93/089, US Environmental Protection Agency, Office of Research and Development, Washington, DC, 1993.

[5] G. C. Fang, Y. S. Wu, M. H. Chen, T. T. Ho, S. H. Huang

and J. Y. Rau, "Polycyclic Aromatic Hydrocarbons Study in Taichung, Taiwan, during 2002-2003," *Atmospheric Environment*, Vol. 38, 2004, pp. 3385-3391.

[6] M. E. Bartkow, K. Booij, K. E. Kennedy, J. F. Muller and D. W. Hawker, "Passive Air Sampling Theory for Semi-volatile Organic Compounds," *Chemosphere,* Vol. 60, 2005, pp. 170–176.

[7] E. Santiago and M. Cayetano, "Polycyclic Aromatic Hydrocarbons in Ambient Air in the Philippines Derived from Passive Sampler with Polyurethane Foam Disk," *Atmospheric Environment*, Vol. 41, No. 19, 2007, pp. 4138-4147.

[8] N. Vardar, F. Esen and Y. Tasdemir, "Seasonal Concentrations and Partitioning of PAHs in a Suburban Site of Bursa, Turkey," *Environmental Pollution,* Vol. 155, 2008, pp. 298-307.

[9] T. Gouin, L. Jantunen, T. Harner, P. Blanchard and T. Bidleman, "Spatial and Temporal Trends of Chiral Organochlorine Signatures in Great Lakes Air Using Passive Air Samplers," *Environmental Science and Technology,* Vol. 41, 2007, pp. 3877–3883.

Atmospheric Dispersion and Deposition of Radionuclides (^{137}Cs and ^{131}I) Released from the Fukushima Dai-ichi Nuclear Power Plant

Soon-Ung Park, Anna Choe, Moon-Soo Park
Center for Atmospheric and Environmental Modeling, Seoul, Korea

ABSTRACT

The Lagrangian Particle Dispersion Model (LPDM) in the 594 km × 594 km model domain with the horizontal grid scale of 3 km × 3 km centered at a power plant and the Eulerian Transport Model (ETM) modified from the Asian Dust Aerosol Model 2 (ADAM2) in the domain of 70° LAT × 140° LON with the horizontal grid scale of 27 km × 27 km have been developed. These models have been implemented to simulate the concentration and deposition of radionuclides (137Cs and 131I) released from the accident of the Fukushima Dai-ichi nuclear power plant. It is found that both models are able to simulate quite reasonably the observed concentrations of 137Cs and 131I near the power plant. However, the LPDM model is more useful for the estimation of concentration near the power plant site in details whereas the ETM model is good for the long-range transport processes of the radionuclide plume. The estimated maximum mean surface concentration, column integrated mean concentration and the total deposition (wet+dry) by LPDM for the period from 12 March to 30 April 2011 are, respectively found to be 2.975 × 102 Bq m^{-3}, 3.7 × 107 Bq m^{-2}, and 1.78 × 1014 Bq m^{-2} for 137Cs and 1.96 × 104 Bq m^{-3}, 2.24 × 109 Bq m^{-2} and 5.96 × 1014 Bq m^{-2} for 131I. The radionuclide plumes released from the accident power plant are found to spread wide regions not only the whole model domain of downwind regions but the upwind regions of Russia, Mongolia, Korea, eastern China, Philippines and Vietnam within the analysis period.

Keywords: Eulerian Transport Model; Fukushima Nuclear Power Plant; Lagrangian Particle Dispersion Model; Radionuclides of ^{137}Cs and ^{131}I

1. Introduction

On 11 March 2011, an extraordinary magnitude 9.0 earthquake occurred off the Sanriku about 180 km off the Pacific coast of Japan's main island Honshu, at 38.3 °N, 142.4 °E and followed by a large tsunami [1]. These events caused a station blackout at the Fukushima Dai-ichi nuclear power plant. As a consequence, four of the six Fukushima Dai-ichi nuclear power plants units heavily damaged, and causing a massive discharge of radionuclides into the air and into the ocean.

Fukushima Dai-ichi nuclear power plant consisted of six boiling water reactors lined up directly along the shore. The earthquake triggered the automatic shutdown of the chain reaction in the units 1 to 3 at 05:46 UTC (14:46 JST) on 11 March 2011. Outside power supply was lost and the emergency diesel generators started up. However, the tsunami arrived 50 minutes later and inundated the reactor sites and their auxiliary buildings and caused the total loss of AC power. Cooling of the reactor cores was lost, water levels in the reactor pressure vessels could not be maintained and the cores in all three units that had been under operation, were degraded and partially (or even completely) melted. The hydrogen produced in this process caused major explosions that massively damaged the upper parts of the reactor buildings of units 1 and 3. Damage to the upper parts of the reactor building could be prevented in unit 2, however, a hydrogen explosion were presumably damaged the suppression chamber [2].

During the Fukushima accident period, massive radioactive materials were released to the environment. Several studies have been devoted to estimate released radioactive materials from this accident. Stohl et al. (2012) have made a first guess of released rates based on fuel inventories and then subsequently improved by inverse modeling using the atmospheric transport model and measurement data. The release duration and radioactivity ratios of ^{131}I/^{137}Cs for the period between 05:00 JST on March 12 to 00:00 JST May 1 in 2011 have been reported [1,3-5]. However, the estimated radionuclides emission fluxes have been reported to be highly sensitive to the first guess of released rates. The estimated total ^{137}Cs emission flux ranged from 9.1 PBq (JAEA, 2011) for the period of 10-31 March to 36.6 PBq for the period of 10 March to 20 April [2].

To assess air and water contamination levels resulting from the Fukushima accident, as this will be an important consideration when evaluating the various applications to the mitigation measures, the estimates of the dispersion and deposition of radionuclides are required over the site and on the regional scale for a given emission flux. Since the topography in the vicinity of the Fukushima Dai-ichi plant is complex, the Eulerian transport model may not be useful to simulate the detailed concentration and deposition of radionuclides even though it may be reasonable to be used for the simulation of them in a regional scale.

For this purpose the Lagrangian Particle Dispersion Model (LPDM) developed by Park (1998) [6] in the 594 km × 594 km model domain centered at a power plant and the Eulerian transport model modified from the Asian Dust Aerosol Model 2 (ADAM2) [7] in the model domain of 140 degree longitudinal distance and 70 degree latitudinal distance co-centered with LPDM have been developed to estimate the concentration and deposition of radionuclides released from the nuclear power plant. The LPDM model and the Eulerian Transport Model (ETM) have, respectively, 3 km × 3 km and 27 km × 27 km grid spacing of a mesoscale meteorological model with 25 vertical layers. The ETM uses the estimated concentration by LPDM in the Lagrangian model domain as the source for the long-range transport.

The purpose of this study is to estimate concentrations and depositions of radionuclides of ^{131}I in the gas phase and ^{137}Cs in the aerosol phase released from the Fukushima accident for the period from 05:00 JST 12 March to 24:00 JST 30 April 2011 using both LPDM and ETM not only in the eastern Japan but in the regional domain including the Pacific Ocean and Asia.

2. Model Description

2.1. Meteorological Model

The meteorological model used in this study is the fifth-generation mesoscale model of non-hydrostatic version (MM5, PSU/NCAR) in the x, y, and coordinates [8, 9].

The model domains include the LPDM domain (**Figure 1(a)**) and the ETM domain (**Figure 1(b)**) centered by the Fukushima Dai-ichi nclear power plant (**Figure 1(b)**). The horizontal resolution of ETM is 27 km while that of the LPDM is 3 km with both 25 vertical layers. The simulations with both models have been conducted for the period of 05:00 JST 12 March to 24:00 JST 30 April 2011. The 6 hourly reanalyzed National Center for Environmental Program (NCEP) data are used for the initial and boundary conditions for the MM5 model. The results of the MM5 model are used for ETM. The nested MM5 model in the horizontal resolution of 3 km is used for the LPDM model.

2.2. Eulerian Transport Model (ETM)

The Eulerian transport model (ETM) has been obtained from the modification of the Asian Dust Aerosol Model 2 (ADAM2) [7]. The radionuclide concentrations estimated by LPDM in its domain are used for the source of ETM for the long-range transport of contaminants. The ADAM2 used 11 particle-size bins with near the same logarithmic intervals for the particles of 0.1 - 37 μm in radius [10,11]. This has been changed to the logarithmic size distribution with an aerodynamic mean diameter of 0.4 □m and a logarithmic standard deviation of 0.3 for ^{137}Cs [2], but for ^{131}I, the ADAM2 model has been changed to handle the gas phase contaminants.

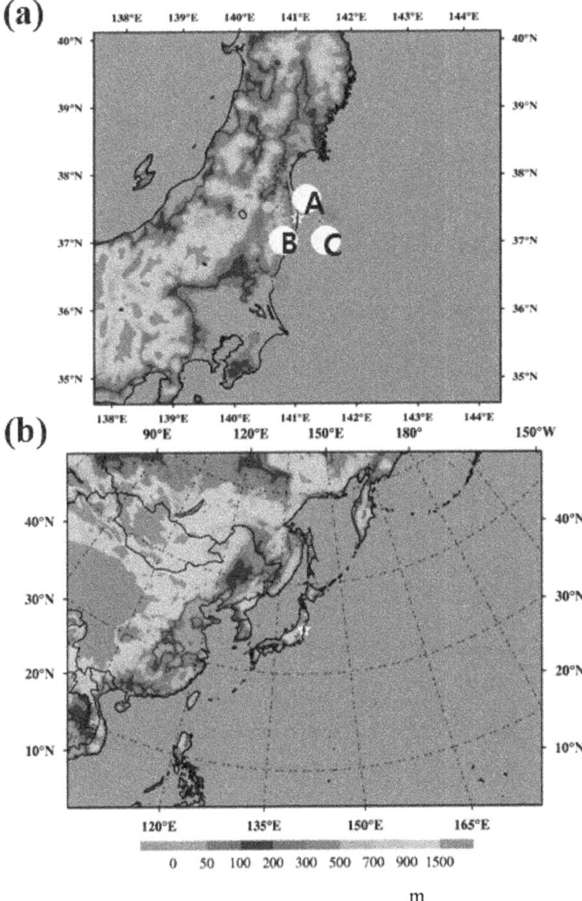

Figure 1. Domain for (a) the Langangian particle dispersion model and (b) the Eulerian transport model with the topography. Monitoring sites (A-C) and the Fukushima nuclear power plant site (white star) are indicated.

2.3. Lagrangian Particle Dispersion Model (LPDM)

The LPDM model [6,12-16] is based on conditioned particle concepts in which the conditioned particle is moving with the mean field velocity and the Lagrangian turbulent velocity. The released particle is continuously traced to find its position with time; that is

$$X_i(t+\Delta t) = X_i(t) + (\overline{u_i}(t) + u_i'(t))\Delta t\,,\quad i=1,2,3 \tag{1}$$

where X_i is the position of the Lagrangian particle, $\overline{u_i}$, and u_i' are the grid scale velocity component that are resolved by the meteorological model and the subgrid scale velocity component respectively, and Δt is the integral time step.

The subgrid scale velocity component is

$$\begin{aligned} u_i'(t+\Delta t) &= R_{L,i}(\Delta t)u_i'(t) + (1-R_{L,i}(\Delta t))^{1/2}u_i''(t) \\ &+ \delta_{i,3}(1-R_{L,i}(\Delta t))w_d \end{aligned} \tag{2}$$

where $R_{L,i}$ is the Lagrangian auto correlation coefficient, $\delta_{i,3}$ is the Kronecker delta, u_i'' is the random turbulent velocity component and W_d is the drift correction in vertically in homogeneous turbulence [17]. All necessary parameters and parameterizations are described in details in Park (1998) [6].

The LPDM model takes into account deposition processes; the dry deposition and wet deposition processes. Dry deposition of radionuclides is estimated with the dry deposition velocity, V_d multiplied by concentration of radionulides near the surface (h_s). The dry deposition velocity, V_d is parameterized with the use of the inferential method [18,19] with taking into account a gravitational settling velocity, V_t for the case of a particle in such a way that

$$V_d = \frac{1}{R_a + R_b + R_c} + V_t \tag{3}$$

where R_a is the aerodynamic resistance, R_b the quasi-laminar sublayer resistance, R_c the surface or canopy resistance and V_t the terminal velocity of a particle. The terminal velocity is given by

$$V_t = \frac{\rho_p g D_p^2}{18\mu} \tag{4}$$

where ρ_p is the density of a particle, g the gravity, D_p the diameter of a particle and μ the dynamic viscosity of air.

In this study, ^{137}Cs is assumed to be in a particle phase with the density of 1,900 kg m^{-3} and a logarithmic size distribution having an aerodynamic mean diameter of 0.4 μm and a logarithmic standard deviation of 0.3, while ^{131}I is assumed to be in a noble gas phase.

A Lagrangian particle with a hypothetical mass of Q_k positioned at (x_k, y_k, z_k) produces the near surface concentration (height of h_s) of

$$\begin{aligned} C_k(x,y,h_s) &= \frac{Q_k e^{-0.693t/\tau}}{(\sqrt{2\pi})^3 \sigma_{kx}\sigma_{ky}\sigma_{kz}} \cdot \\ &\exp[-\frac{(x_k-x)^2}{2\sigma_{kx}^2} - \frac{(y_k-y)^2}{2\sigma_{ky}^2} - \frac{(z_k-h_s)^2}{2\sigma_{kz}^2}] \end{aligned} \tag{5}$$

where σ_{kx}, σ_{ky}, and σ_{kz} are, respectively the standard deviation of the diffusion distance in the x, y, and z directions associated with the k's Largrangian particle, and τ is the half life time of the radionuclide. h_s is assumed to be 5 m above the ground.

The deposition flux due to the k's Lagrangian particle is

$$\begin{aligned} F_k &= \frac{Q_k V_d e^{-0.693t/\tau}}{(\sqrt{2\pi})^3 \sigma_{kx}\sigma_{ky}\sigma_{kz}} \cdot \\ &\exp[-\frac{(x_k-x)^2}{2\sigma_{kx}^2} - \frac{(y_k-y)^2}{2\sigma_{ky}^2} - \frac{(z_k-h_s)^2}{2\sigma_{kz}^2}] \end{aligned} \tag{6}$$

Therefore, the total mass deposition of the k's Lagrangian particle, Q_d is

$$\begin{aligned} Q_d &= \frac{Q_k \Delta t \exp[-0.693\frac{t+\Delta t}{\tau}]\exp[-\frac{(z_k-h_s)^2}{2\sigma_{kz}^2}]}{(\sqrt{2\pi})^3 \sigma_{kx}\sigma_{ky}\sigma_{kz}} \cdot \\ &\int_{-\infty}^{\infty}\int_{-\infty}^{\infty} V_d \exp[-\frac{(x_k-x)^2}{2\sigma_{kx}^2} - \frac{(y_k-y)^2}{2\sigma_{ky}^2}]dxdy \end{aligned} \tag{7}$$

Due to deposition the k's Lagrangian particle will lose the mass of Q_d after Δt time and results in a reduced mass of k particle after time interval Δt is $Q_k' = Q_k - Q_d$.

The k's Lagrangian particle is assumed to be totally deposited on the ground if z_k is negative and $|z_k| \geq 3\sigma_{kz}$, otherwise it will be reflected from the ground with the reduced mass of Q_k'.

The wet deposition amounts of radionuclides are determined by the precipitation rate and the averaged concentration in cloud water estimated by the sub-grid cloud scheme followed by the diagnostic cloud model in ADAM2 [6] and the Regional Acid Deposition Model (RADM) version 2.6 [20-22]. The below cloud scavenging process is also included [6].

3. Simulation Results of 137Cs and 131I Concentrations and Depositions

^{131}I is assumed to be in the gas phase with the half-life time of 8.07 days whereas ^{137}Cs is to be in the aerosol phase with the aerodynamic mean diameter of 0.4 m, a logarithmic standard deviation of 0.3 and the half-life time of 30.2 years.

The emission rate of ^{131}I and ^{137}Cs from the Fukushima nuclear power plant accident estimated by [4,5,23] for the period from 05:00 JST 12 March to 24:00 JST 30 April 2011 are used in this study and given in **Figure 2**.

The first emission peak of 834 GBq s^{-1} of ^{131}I and 83.4 GBq s^{-1} of ^{137}Cs from 15:30 JST to 16:00 JST 12 March [1] was reported to be related to the hydrogen explosion in reactor unit 1. The highest emission of up to 1,110 GBq s^{-1} of ^{131}I and 110 GBq s^{-1} of ^{137}Cs during the period of 11:00 JST 14 March to 17:00 JST 15 March 2011 was reported to be relate to the hydrogen explosion in unit 4 and together with the hydrogen explosion in unit 2.

3.1. Simulated 131I and 137Cs Concentrations and Depositions by LPDM

The LPDM model has been employed to simulate radionuclide concentrations and depositions in the domain in **Figure 1(a)**. The Lagrangian particles are released at the rate of one particle per minute at the height of the first σ level (about 18-20 m above the ground) with the apportioned mass concentration equivalent to the emission rate in **Figure 2**.

The particle is released starting at 09:00 JST (00:00 UTC) 12 March and ending at 09:00 JST 30 April 2011 for the whole emission period from the Fukushima Dai-ichi nuclear power plant.

The concentration is calculated hourly at each level with the horizontal distance of 1,500 m. The hourly total deposition (wet and dry) is estimated at each grid with the horizontal distance of 1,500 m.

Figure 3 shows time variations of model simulated daily mean concentrations of ^{137}Cs and ^{131}I with the measured concentrations at 3 sites given in **Figure 1(a)**. Both nuclides are quite well simulated at all sites.

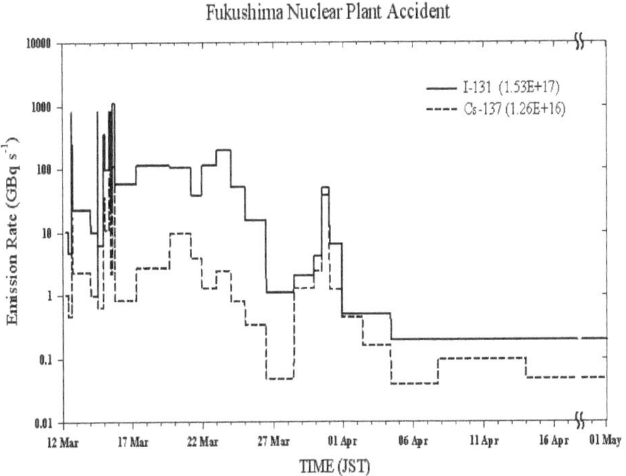

Figure 2. Time variations of emission rate of I-131 (solid line) and Cs-137 (dashed line) from Fukushima nuclear power plant from 12 March to 01 May 2011.

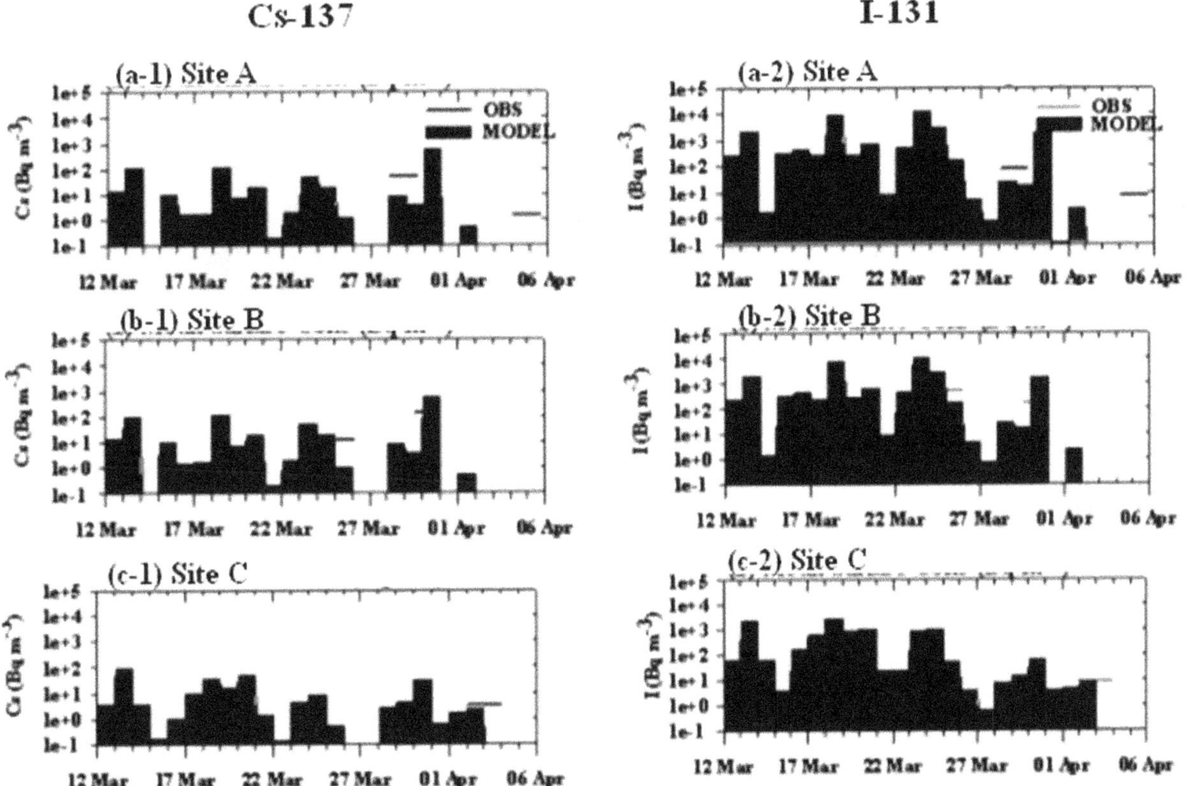

Figure 3. Time variations of model simulated by LPDM daily mean surface concentration (Bq m^{-3}) of (1) Cs-137 and (2) I-131 at (a) site A, (b) site B, and (c) site C for the period from 12 March to 5 April 2011. Observed concentration at each site is shown in with red bars.

Figure 4 shows the horizontal distributions at model simulated mean surface concentration, column integrated mean concentration and total deposition (wet+dry) of ^{137}Cs for the period from 12 March to 30 April 2011. The zone of the mean surface ^{137}Cs concentration exceeding 50 Bq m^{-3} extends northward from the power plant to 38.5N and southward to 35.5N with relative high ^{137}Cs concentrations along the coastline. The maximum surface mean ^{137}Cs concentration of 2.975 × 10^{2} Bq m^{-3} occurs near the power plant (**Figure 4(a)**). The horizontal distribution pattern of the column integrated mean ^{137}Cs concentration (**Figure 4(b)**) is quite similar to that of the surface mean concentration (**Figure 4(a)**) with the maximum value of 3.7 × 10^{7} Bq m^{-2} near the power plant. A similar horizontal distribution pattern is also seen in the horizontal distribution of the total deposition of ^{137}Cs. The maximum deposition of 1.78 × 10^{14} Bq m^{-2} occurs near the power plant (**Figure 4(c)**).

Figure 5 shows the horizontal distributions of the model simulated surface mean concentration, the column integrated mean concentration, and the total deposition of ^{131}I for the period from 12 March to 30 April 2011. The horizontal distribution pattern of the mean surface ^{131}I concentration (**Figure 5(a)**) is quite similar to that of ^{137}Cs (**Figure 4(a)**) but ^{131}I concentration is much higher than ^{137}Cs due to high emission rate of ^{131}I (**Figure 2**). The area enclosed by the isoline of surface mean ^{131}I concentration of 100 Bq m^{-2} is nearly the same as that of the surface mean ^{137}Cs concentration of 50 Bq m^{-2} (**Figure 4(a)**). The maximum surface mean ^{131}I concentration, and the maximum mean column integrated ^{131}I concentration (**Figure 5(b)**) are, respectively 1.96 × 10^{7} Bq m^{-3} and 2.44 × 10^{9} Bq m^{-2} that are 100 times higher than the corresponding values of ^{137}Cs (**Figures 4(a)** and **(b)**). However, the total deposition of ^{131}I with the maximum value of 5.96 × 10^{14} Bq m^{-2} (**Figure 5(c)**) is 4 times greater than that of ^{137}Cs, suggesting more effectiveness of the aerosol than the gas for the deposition.

3.2. Simulated 131I and 137Cs Concentrations and Depositions by the Eulerian Transport Model (ETM)

Figure 6 shows the model (ETM) simulated surface mean concentration, the column integrated mean concentration and the total deposition (wet+dry) of ^{137}Cs for the period from 12 March to 30 April 2011.

The emitted ^{137}Cs from the power plant affects all the downwind region of the model domain with some extension toward the upwind region during the analysis period (12 March to 20 April). The high surface mean ^{137}Cs concentration region extends southwestward from Alaska to Philippines with the surface mean maximum concentration of 20 Bq m^{-3} near the power plant (**Figure 7(a)**).

The further upwind extension up to 100°E of the atmospheric loading of ^{137}Cs is seen in **Figure 7(b)** with the maximum column integrated mean ^{137}Cs concentration of 2.78 × 10^{4} Bq m^{-2} near the power plant.

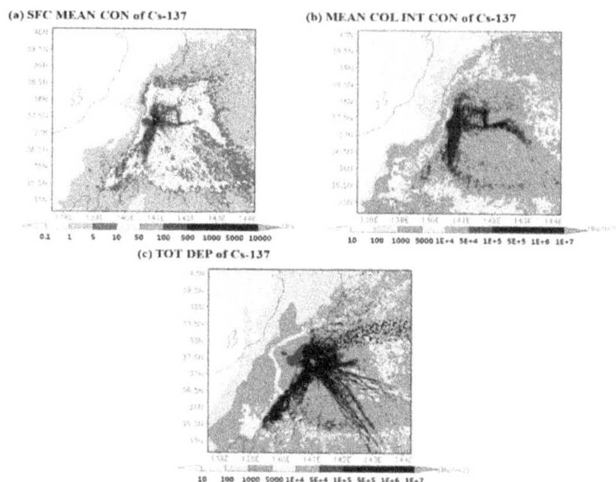

Figure 4. Horizontal distributions of (a) the near surface mean concentration (Bq m^{-3}), (b) the column integrated mean concentration (Bq m^{-2}), and the total deposition (Bq m^{-2}) of ^{137}Cs for the period from 12 March to 30 April 2011.

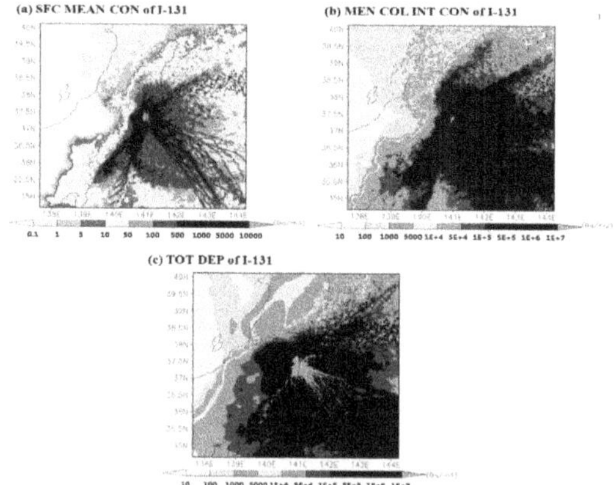

Figure 5. The same as in Figure 4 except for I-131.

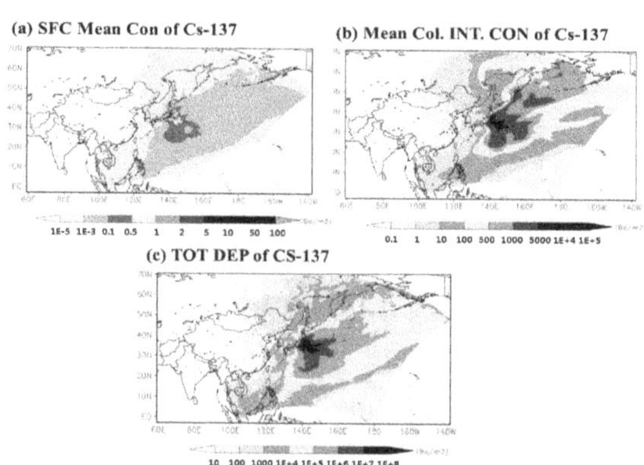

Figure 6. Horizontal distributions of (a) the near surface mean concentration (Bq m^{-3}), (b) column integrated mean concentration, and (c) total deposition of Cs-137 for the period from 12 to 20 April 2011.

Figure 7 shows the model (ETM) simulated surface mean concentration, column integrated mean concentration and total deposition (wet+dry) of ^{131}I for the period from 12 March to 30 April 2011.

The horizontal distribution patterns of these quantities of ^{131}I are quite resemble to those corresponding quantities of ^{137}Cs (**Figure 6**) but the maximum values are much higher than those of ^{137}Cs; The maximum value of the surface mean concentration (**Figure 7(a)**) the column integrated concentration (**Figure 7(b)**) and the total deposition (wet+dry) (**Figure 7(c)**) for ^{131}I is 851 Bq m^{-3}, 1.4×10^6 Bq m^{-2} and 8.88×10^5 Bq m^{-2}, respectively.

To understand the long-range transport process of the radionuclide plume, the daily averaged column integrated ^{131}I concentration is calculated and presented in **Figure 8** at other day interval from 12 March to 12 April 2011.

The radionuclide plume emitted from the power plant is transported to the downwind region of Alaska and the eastern boundary of the model domain within 4 days (on 15 March).

A well developed low pressure center located at the Bering Sea on 17 March makes the plume to be converged toward the low pressure center and then pushs the plume toward westward over Russia in association with the circulation (21 March). Thereafter the northerlies in association with the developing high pressure system over Siberia push the plume south and southeastward to Mongolia, East China and Korea (23-27 March).

On the mean time a high pressure system located to the south of the power plant on 19 March makes the emitted radionuclide plume from the power plant to be diverged toward southward to the easterly zone of the subtropical high pressure results in high atmospheric loading of the radionuclide (^{131}I) zone the along the northern boundary of the subtropical high pressure system that extends southwestward from the northwestern Pacific Ocean to Philippines and Vietnam (19-31 March).

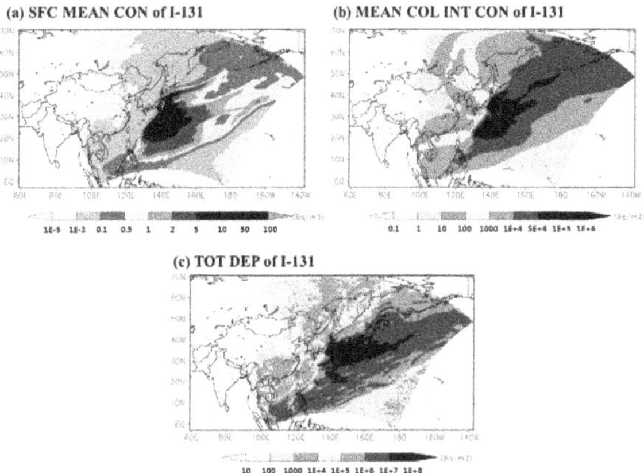

Figure 7. The same as in Figure 6 except for I-131.

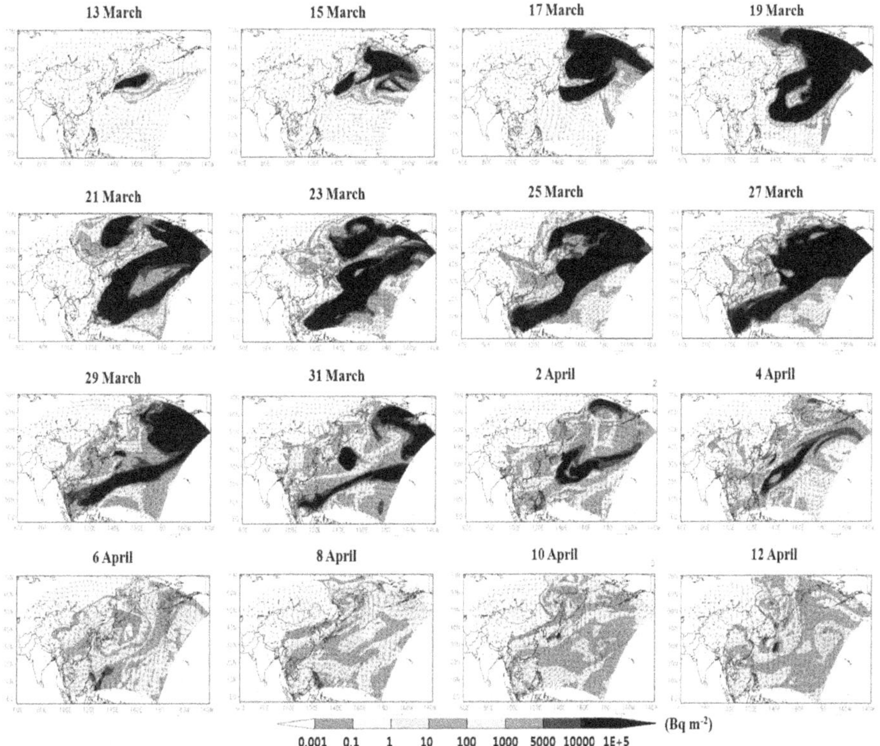

Figure 8. Horizontal distributions of the model simulated daily mean column integrated concentration (Bq m^{-2}) of I-131 for every two days starting from 12 UTC 12 March to 12 UTC 12 April 2011.

4. Conclusions

The Lagrangian Particle Dispersion Model (LPDM) in the 594 km × 594 km model domain centered at a power plant with the horizontal grid distance of 3 × 3 km^2 for meteorological fields and the Eulerian Transport Model (ETM) modified from the Asian Dust Aerosol Model 2 (ADAM2) in the model domain of 140° LON × 70° LAT with the horizontal grid distance of 27 × 27 km^2 co-centered with LPDM have been developed.

It is found that both models are able to produce the observed concentrations of ^{137}Cs and ^{131}I near the power plant reasonably. The LPDM model yields better results than those of ETM near the power plant. However, the long-range transport processes of the radionuclides are well simulated by the ETM model. Therefore, the presently developed two models are found to be useful for the concentration estimation in the near power plant area by the LPDM model and in the wide region by the ETM model.

The estimated maximum mean surface concentration, mean column integrated concentration and the total deposition by LPDM for the period from 12 March to 30 April are, respectively found to be 2.975×10^2 Bq m^{-3}, 3.7×10^7 Bq m^{-2} and 1.78×10^{14} Bq m^{-2} for ^{137}Cs and 1.96×10^4 Bq m^{-3}, 2.24×10^9 Bq m^{-2} and 5.96×10^{14} Bq m^{-2} for ^{131}I. The ETM model result indicates that the radionuclide plume released from the Dai-ichi power plant can affect wide regions not only the whole downwind region of the power plant but the upwind regions including Russia, Mongolia, Korea, the eastern part of China, Philippines and the parts of South East Asia.

The present study mainly pertains to the development of the emergency response modeling system that will be used for the operational model for the accidental releases. Further verification of the model with measured data is required to be used as an operational model.

5. Acknowledgements

This work was funded by the Korea Meteorological Administration Research and Development program uder Grant CATER 2012-2050.

REFERENCES

[1] K. Tanaka, Y. Takahashi, A. Sakaguchi, M. Umeo, S. Hayakawa, H. Tanida, T. Saito and Y. Kanai, "Vertial Profiles of Iodine-131 and Cesium-137 in Soils in Fukushima Perfecture Related to the Fukushima Daiichi Nuclear Power Station Accident, " *Geochemical Journal*, Vol. 46, 2012, pp. 73-76.

[2] A. Stohl, P. Seibert, G. wotawa, D. Arnold, J. F. Burkhart, S. Eckhardt, C. Tapia, A. Vargas and T. J. Yasunary, "Xenon-133 and Caesium-137 Releases into the Atmosphere from the Fukushima Dai-ichi Nuclear Power Plant: Determination of the Source Term, Atmospheric Dispersion, and Deposition, *Atmospheric Chemistry and Physics*, Vol. 12, 2011, pp. 2313-2343.

[3] G. Katata, H. Terada, H. Nagai and M. Chino, "Numerical Reconstruction of High Dose Rate Zones Due to the Fukushima Dai-Ichi Nuclear Power Plant Accident", *Journal of Environment Radioactivity*, Vol. 111, 2012, pp. 2-12.

[4] G. Katata, M. Ota, H. Terada, M. Chino and H. Nagai, "Atmospheric Discharge and Dispersion of Radionuclides during the Fukushima Dai-Ichi Nuclear Power Plant Accident. Part I: Source Term Estimation and Local-Scale Atmospheric Dispersion in Early Phase of the Accident," *Journal of Environment Radioactivity*, Vol. 109, 2012, pp. 103-113.

[5] M. Chino, H. Nakayama, H. Nagai, H. Terada, G. Katata and H. Yamazawa, "Preliminary Estimation of Release Aounts of 131I and 137Cs Accidentally Discharded from the Fukushima Daiichi Nuclear Power Plant into the Atmosphere, "*Journal of Nuclear Science and Technology*, Vol. 48, 2011, pp. 1129-1134.

[6] S.-U. Park, "Effects of Dry Deposition on Near-Surface Concentrations of SO2 during Medium-Range Transport," *Journal of Applied Meteorology*, Vol. 37, 1998, pp.486-496.

[7] S.-U. Park, A. Choe, E.-H. Lee, M.-S. Park and X. Song, "The Asian Dust Aerosol Model 2 (ADAM2) with the Use of Normalized Difference Vegetation Index (NDVI) Obtained from the Spot4/Vegetation Data," *Theoretical and Applied Genetics*, Vol. 101, 2010, pp. 191-208.

[8] D. A. Grell, J. Dudhia, and D. R. Stauffer, "A Description of the 5th Generation Penn State/NCAR Mesoscale Model (MM5), NCAR TECH. Note NCAR/TN-398, p. 117.

[9] J. Dudhia, D. Grell, Y.-R. Guo, D. Hausen, K. Manning, and W. Wang, "PSU/NCAR Mesoscale Modeling System Tutorial Class Note (MM5 Modeling System Version 2).

[10] S.-U. Park and H.-J. In, "Parameterization of Dust Emission for the Simulation of the Yellow Sand (Asian dust) Observed in March 2002 in Korea," *Journal of Geophysical Research*, Vol. 108, No. D19, 2003, p. 4618.

[11] S.-U. Park and E.-H. Lee, "Parameterization of Asian Dust (Hwangsa) Particle-Size Distributions for Use in Dust Emission Model," *Atmospheric Environment*, Vol. 38, 2004, pp. 2155-2162.

[12] R. A. Pielke, M. Arritt, M. Segal, M. D. Moran and R. T. McNider, "Mesoscale Numerical Modeling of Pollutant Transport in Complex Terrain", *Bound-Layer Meteor*, Vol. 41, 1987, pp. 59-74.

[13] R. T. McNider, "Investigation of the Impact of Topographic Circulations on the Transport and Dispersion of Air Pollutions," Ph.D. dissertation, University of Virginia, 1981, p. 195.

[14] T. Yamada, J. Kao, and S. Bunker, "Airflow and Air Quality Simulations over the Western Mountaineous Region with a Four-Dimensional Data Assimilation Technique," *Atmospheric Environment*, Vol.23, 1989, pp.539-554.

[15] W. L. Physick, and D. J. Abbs, "Modeling of Summertime Flow and Dispersion in the Coastal Terrain of Southeastern Asutralia," *Monthly Weather Review*, Vol.119, 1991,pp.1014-1030.

[16] F. B. Smith, "Conditioned particle motion in a homogenous turbulent field," *Atmospheric Environment*, Vol.2, 1968,pp.491-508.

[17] B. J. Legg, and M. R. Raupach, "Markov-Chain Simulations of Particle Deposition in Homogeneous Flows: The Mean Drift Velocity Induced by a Gradient in Eulerian Velocity Variance", *Bound.-Layer Meteor*, Vol.24, 1982, pp.3-13.

[18] M. L. Wesely, "Parameterization of Surface Resistances to Gaseous Dry Deposition in Regional-Scale Numerical Models," *Atmospheric Environment*, Vol.23, 1989, pp.1293-1304.

[19] M. L. Wesely and B. B. Hicks, "Some Factors that Affect the Dispersion Rates of Sulfur Dioxide and Similar Gases on Vegetation", *Journal of Air Pollution Control Association*,Vol.27,1977,pp.1110-1116.

[20] C. J. Walcek, and G. R. Taylor, "A Theoretical Method for Computing Vertical Distributions of Acidity and Sulfate Production within Cumulus Clouds," *J. Atmos. Sci.*, 43,1986,pp.339-355.

[21] J. S. Chang, R. A. Brost, I. S. A. Isaksen, S. Madronich, P. Middleton, W. R. Stockwell and C. J. Walcek, "A Three-Dimensional Eulerian Acid Deposition Model: Physical Concepts and Formulation," *Journal of Geophysical Research*, Vol. 92, 1987, pp. 14681-14700.

[22] R. L. Dennis, J. N. McHenry, W. R. Barchet, F. S. Binkovski and D. W. Byun, "Correcting RADM's Sulfate Underprediction: Discovery and Correction of Model Errors and Testing the Corrections through Comparisons against Field Data," *Atmospheric Environment*, Vol. 27A, No. 6, 1993, pp. 975-997.

[23] S. Furuta, S. Sumiya, H. Watanabe, M. Nakano, K. Imaizumi, M. Takeyasu, A. Nakada, H. Fujita, T. Mizutani, M. Morisawa, Y. Kokubun, T. Kono, M. Nagaoka, Y. Hiyama, T. Onuma, C. Kato and T. Kurachi, "Results of the Environmental Radiation Monitoring Following the Accident at the Fukushima Daiichi Nuclear Power Plant," *JAEA-Review*, Vol. 035, 2011, p. 89.

19

Effect of pH and Dissolved Silicate on the Formation of Surface Passivation Layers for Reducing Pyrite Oxidation

Shengjia Zeng[1], Jun Li[1], Russell Schumann[2], Roger Smart[1]
[1]Minerals and Materials Science & Technology, Mawson Institute
[2]Levay& Co.; University of South Australia, Adelaide, South Australia, Australia

ABSTRACT

Acid mine drainage (AMD)and toxic metal release generated by oxidation of sulphide minerals, particularly pyrite, in mine wastes, are a critical environmental issue worldwide. Currently, there are many options to diminish sulphide oxida- tion including barrier methods that isolate pyrite from oxygen or water, chemical additives and inhibition of iron-oxidizing bacteria. This study focuses on understanding the role that silicate and pH conditions play in the formation and stabilisation of pyrite surface passivation layers found in lab and field studies. The results from pyrite dissolution tests under various conditions showed that the pyrite oxidation rate has been reduced by up to 60% under neutral pH with additional soluble silicate. Solution speciation calculation predicted that crystalline goethite is formed in the experiment without silicate additionbutan amorphous iron hydroxide surface layer is stabilized by the addition of the silicate, inhibiting goethite formation and continuing pyrite oxidation. This coherent, continuous amorphous layer has been verified in SEM.

Keywords: Acid Mine Drainage; Pyrite; Passivation Coating; Silicate

1. Introduction

Acid mine drainage (AMD) is a critical global environmental issue from many mine wastes. The discharge of AMD with associated toxic metal release has caused acidification of water environment, leading to extermination of aquatic life and compromised water quality for agriculture and drinking. Sulphide oxidation of sulfidic minerals dominantly contributes to most of AMD issues at mining sites, and pyrite, the most abundant sulphide mineral in the earth's crust [1], is widely found in mining activities. Itsoxidation process under various conditions has been therefore widely studied to explore effective approaches and controlling factors to retard the oxidation process [2-5]. It is known that the oxidant (ferric ion or dissolved oxygen), water content and catalyst (i.e. iron oxidizing bacteria) have significant impact on the pyrite oxidation rate [6]. Reduction of any of these factors can contribute a remarkable reduction of the oxidation rate. The oxidant can be inhibited at pyrite particle surfaces by a surface barrier through the formation of passivation layers [2,3,7-10] and some coating agents, e.g. acetyl acetone, humic acids, ammonium lignosulfonates, oxalic acid and sodium silicate, were reported to be employed-for pyrite surface treatmentbut only minor reductions in pyrite oxidation rates resulted [1,3,9,11]. Huminicki and Rimstidt [12] reported that precipitation of iron hydroxide particle reduced oxidant's diffusion coefficient by more than five orders of magnitude. Evangelou [8] introduced a more stableiron hydroxide/silica coating generated via precipitation of an Fe-Si complex. Smart et al. [13] and Miller et al. [14] found from their long term column leaching and field studies at the Grasberg mine Indonesia that silicate may be responsible for stabilising the iron oxy-hydroxide passivation layer on pyrite surfaces. It was also reported from some of fund amental studies that silicate in hibits the trans formation ofamorphous iron hydroxide $Fe(OH)_3$ to crystalline goethite (FeOOH) in synthesis processes [4,15,16]. However, little has been known on the reaction path ways of the surface coating associated with silicate or the trans formation process of iron hydroxideprecipitated and coating stability. In this study, pyrite dissolution tests under various pH conditions in the presence of low concentrations of Na_2SiO_3(10 and 20 mg/Lsilicate as Si),were performedto investigate the influence of pH and dissolved silicate on the stability of iron oxyhydroxide surface layers formed during oxidative dissolution of pyrite at circum-neutral pH.

2. Method and Materials

Pyrite(particle size 38 - 75 μm)obtained from Geo Discoveries (NSW, Australia) was leached under conditions shown in **Table 1** Dissolved silicate, shown as Si concentrations from assays, has been chosen based on concentrations in effluent of column leach tests. Experiments were conducted at four different solution pHs (3.5, 4.5, 5.5 and ~8) and with three concentrations of dissolved Si (none added, 10 mg/L and 20 mg/L added as sodium silicate).Solution samples were periodically taken for Fe, S and Si concentrations by ICP analysis as well as for solution Eh and pH measurements. Solution modelling software, PHREEQC [17], was used to predict the precipitation of possible secondary minerals based on measured Eh, pH and the dissolved element concentrations. This is a thermodynamic prediction of precipitates and solution species and requires experimental verification where possible. Scanning electron microscopy (SEM) was introduced to investigate the surface coating.

3. Results and Discussion

3.1. Effect of pH and Silicate Concentrationonpyrite Oxidation

Figure 1 shows pyrite oxidation rates plotted as a function of reaction time for each of the different solution pH and silicate concentrations examined. Oxidation rates were calculated from the total solution concentration (measured by ICP) in samples removed from the reactions throughout 160 days of dissolution. Initial oxidation rates were between 3 and 4 × 10^{-10} mol/m^2/s for all experiments, which decreased to around 1 × 10^{-10} mol/m^2/s during about 100 days of oxidation for solution pH between 3.5 and 5.5 irrespective of how much silicate had been added. At the higher pH (near 8) of the saturated calcite solutions there was a difference in pyrite oxidation rates as a result of silicate addition. With no silicate added, pyrite oxidation rates in calcite saturated solution were about 60%fasterthan those at lower pH after about 40 days, while in the presence of 10 - 20 mg/L Si the oxidation rates were slower at the higher pH of the calcite saturated solutions after about 40 days. The pyrite oxidation rate after about 100 days at circum-neutral pH and with 20 mg/L Si added, was about half of the rate in similar solutions at lower pH (3.5 - 5.5).

Table 1. Experimental details of pyrite dissolution tests.

Object	Description
Pyrite	38-75 μm, BET surface area: 0.539 m^2/g , usage: 2 g for each sample
Calcite	38-75 μm, usage: 1.66 g (mole ratio: 1:1 with pyrite),
Solution A	1L of 0.01 mol/l KCl, **No Si** addition
Solution B	1L of 0.01 mol/l KCl, **10** mg/L**Si** (Na_2SiO_3)
Solution C	1L of 0.01 mol/l KCl, **20**mg/L**Si** (Na_2SiO_3)
pH	Manually controlled at **3.5**, **4.5** and **5.5** or buffered around**8**by calcite

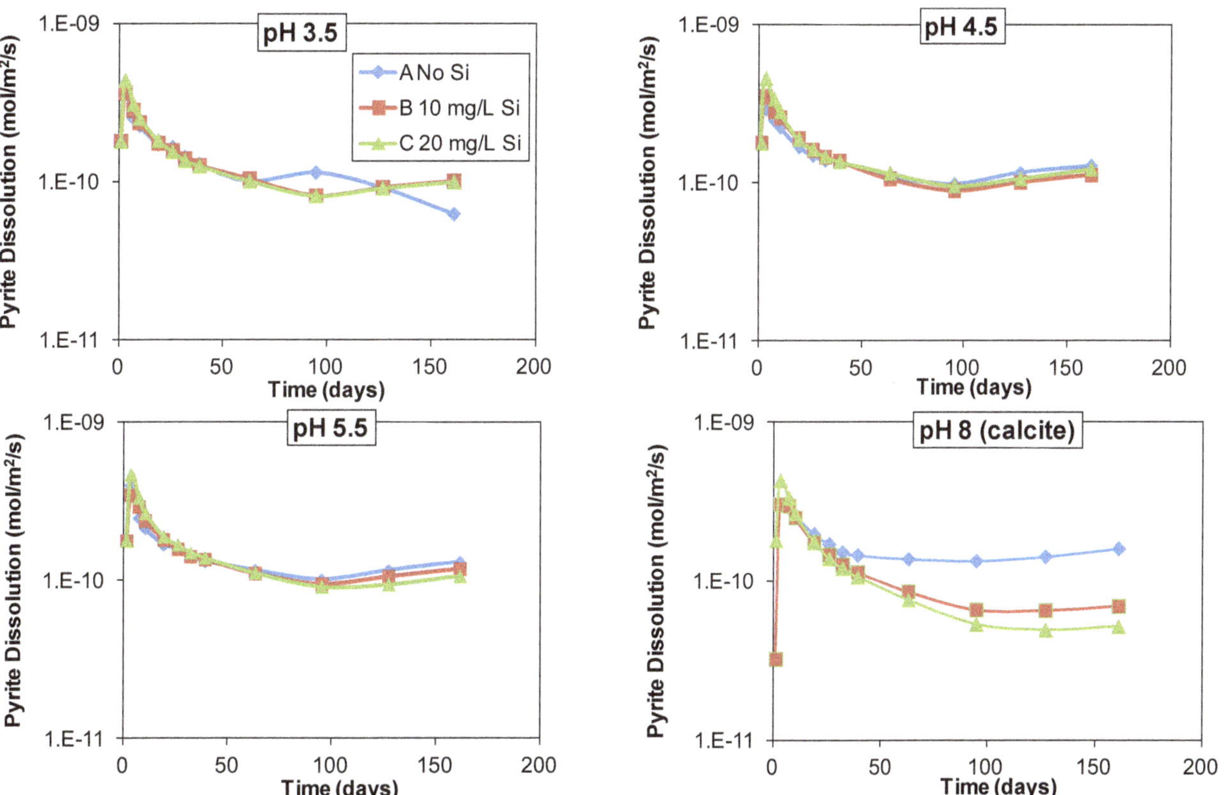

Figure 1. Pyrite dissolution rates as function of time at different solution pH and silicate concentrations.

The decreasing oxidation rate as a function of time is indicative of passivating iron oxy-hydroxide layer formation found in SEM studies of lab and field samples [13, 14]. These results suggest that across the pH range 3.5 to 5.5, neither the pH nor the concentration of silicate affect the nature of the iron oxy-hydroxide layer that forms as oxidation proceeds. However at the higher pH of calcite saturated solution there appears to be a clear influence of silicate on the oxidation rate, suggesting that increasing solution silicate concentration results in surface layers that are less permeable to oxygen and more stable.

3.2. Change in Solution Fe concentrations

Figure 2 shows solution iron concentrations as a function of time for all of the pyrite oxidation experiments. Not unexpectedly, there is a strong correlation between the solution pH and solution iron concentration, with higher concentrations in the lower pH solutions. During the first 10 days of pyrite oxidation at pH 3.5 and 4.5 there was stoichiometric dissolution of iron and sulphur. After this time, the ratio of iron to sulphur decreased below two, indicative of iron oxy-hydroxide precipitation [12]. At pH 5.5 and in saturated calcite solution the ratio of iron to sulphur in solution was less than two from the start, indicating formation of passivating layers from the beginning. Despite these differences, as discussed above, there appears to be no difference in pyrite oxidation rates with silicate addition except for oxidation in calcite saturated solution after about 40 days. While these results may suggest differences in the amount of precipitated iron oxy-hydroxide, this does not appear to strongly influence the oxidation rate in solutions with pH between 3.5 and 5.5.

After about 100 days, there was a small increase in the concentration of iron in the solutions at pH 3.5 and 4.5. This corresponds with a slight increase in the oxidation rate and may suggest some dissolution of the iron oxy-hydroxide coating at lower pH values [12].

The data shown in **Figure 2** also indicate a correlation between the concentration of iron in solution and that of silicate. They suggest that iron may be stabilized by complexation with silicate enhancing its solubility at the lower pH values. Other investigations on the influence of silicate on pyrite oxidation have shown that, at higher silicate concentrations than those used here and at lower pH, pyrite oxidation is actually enhanced in the presence of silicate, possibly due to stabilization of Fe (III) by complexation with silicate [4].

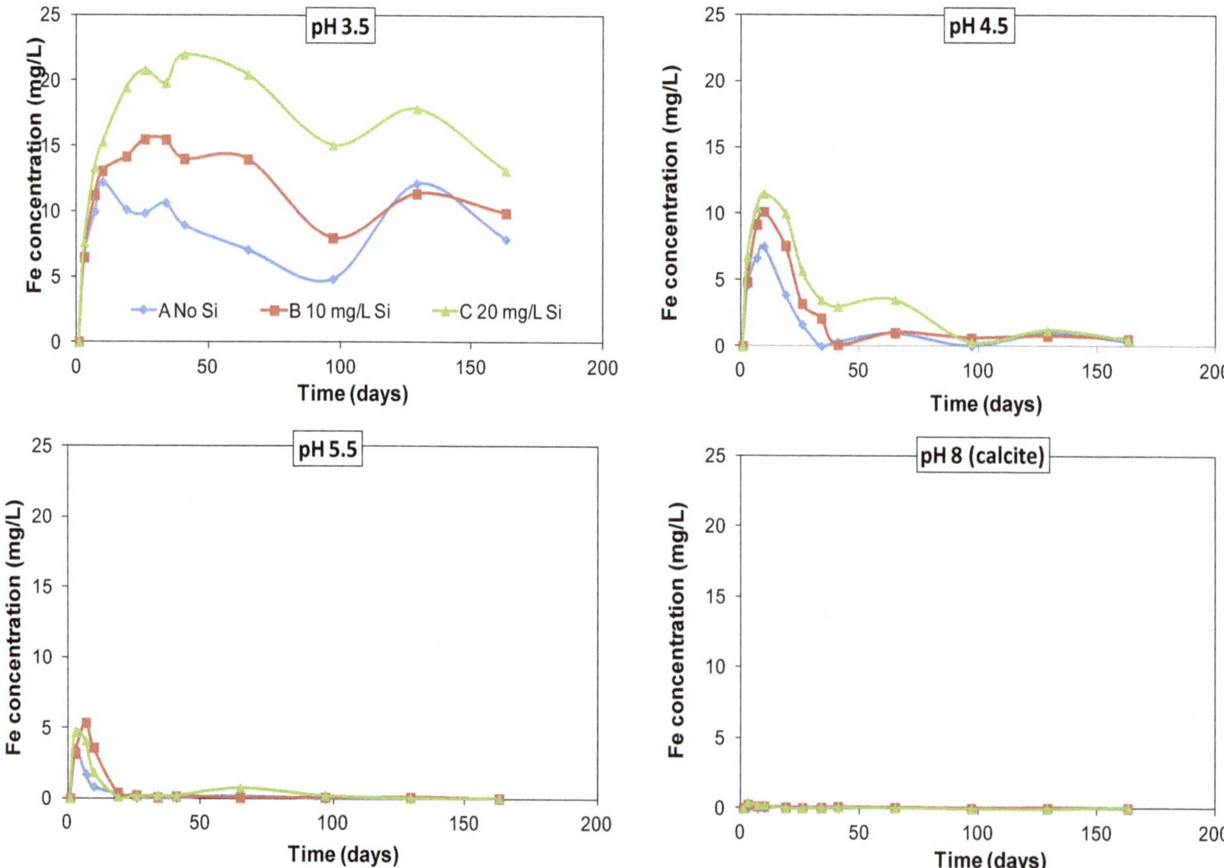

Figure 2. Iron concentration as function of time and pH.

3.3. Change in Solution Si Concentrations

Figure 3 shows solution Si concentrations as a function of time for all of the pyrite oxidation experiments except series A where no silicate was added and silicon concentrations were generally non-detectable (<0.05 mg/L). There appears to be a clear correlation between solution pH and the concentration of silicon in solution. At pH 3.5 there is little decrease in the concentration of Si during the 160 days of the experiments. At the circum-neutral pH conditions of oxidation in calcite saturated solution there is close to a 50% decrease in the concentration of silicate, suggesting adsorption and/or precipitation of silicate with iron oxy-hydroxide. These results suggest that as the pH increases there is likely to be an increase in the amount of silicate in the iron oxy-hydroxide coating passivating the pyrite surface. However, as discussed above this does not appear to have influenced oxidation rates except in the case of oxidation at circum-neutral pH. It therefore appears that a combination of both relatively high silicate concentration and neutral to high pH is required to have a significant effect on the oxygen permeability of iron oxy-hydroxide layers formed during pyrite oxidation.

3.4. Speciation Calculations

Speciation calculations for the pyrite dissolution experiments conducted in calcite saturated solutions predict that goethite is saturated both in the presence and absence of added silicate (**Table 2**). In contrast, amorphous iron hydroxide is predicted to be unsaturated in the absence of added silicate suggesting that conversion to goethite via dissolution and re-precipitation is thermodynamically likely. However, in the solution to which 20 mg/L silicate (as Si) has been added, speciation calculations predict that amorphous iron hydroxide is saturated indicating increased stability of the amorphous phase in the presence of silicate. These results are consistent with the observations of silicate -stabilized amorphous iron hydroxide and retarded the crystallization of goethite reported in literature [4,15,18,19].

3.5. Pyrite Surface Analysis

Scanning electron microscopy (SEM) was applied to identify whether iron ox-hydroxide coatings formed on pyrite particles which have slowed pyrite oxidation as the reaction has proceeded, suggested by the oxidation rate data and solution analysis. **Figure 4** shows SEM images of pyrite particles sampled from dissolution experiments conducted in calcite-saturated solution. The top images display the surface of pyrite after 160 days dissolution in a solution where no silicate has been added. An iron ox-hydroxide coating is clearly visible with a needle-like morphology suggestive of a crystalline goethite structure. The bottom images show pyrite particles taken after 160 days from a saturated calcite solution to which 20 mg/L silicate (as Si) had been added. Again an iron ox-hydroxide coating is obvious on the pyrite surface, however, with very different morphology to that on pyrite from the solution in which no silicate had been added. In this instance the coating appears to be of a more amorphous nature. The Energy dispersed spectroscopy (EDS) spectra of the pyrite particles shown in **Figure 5** confirms that

Table 2. Saturation indices (SI) of various mineral phases calculated using PHREEQC for saturated calcite solutions from pyrite dissolution with and without added silicate.

Species	SI No added silicate	SI 20 mg/L (as Si) silicate added
Fe $(OH)_3$ (amorphous)	-0.2	0.5
Goethite	4.9	5.6

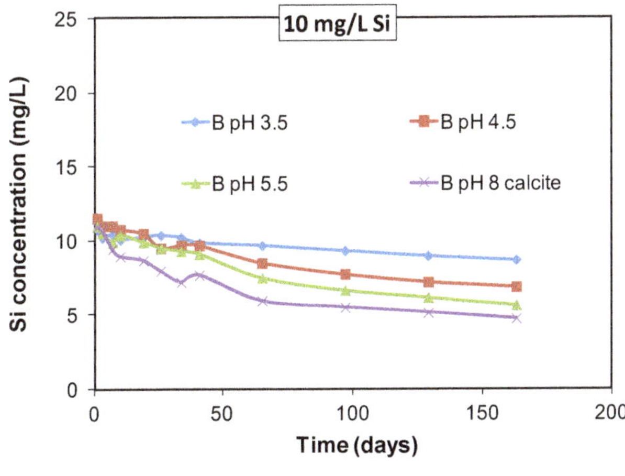

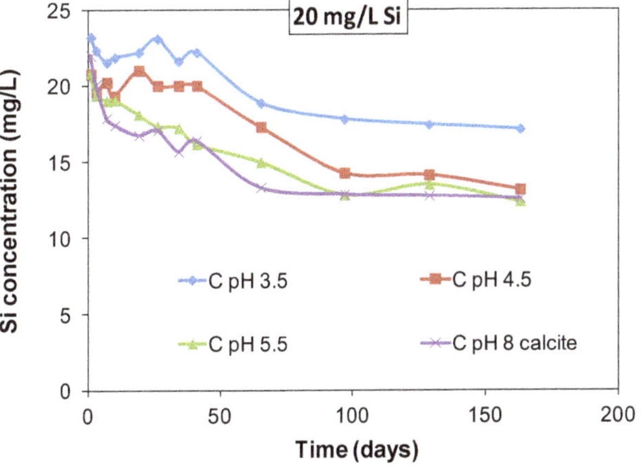

Figure 3. Silicate concentrations as function of time and pH.

Figure 4. SEM images of pyrite surface coatings after dissolution in saturated calcite solution for 160 days with no added silicate (top images) and 20 mg/L (as Si) added silicate (bottom images).

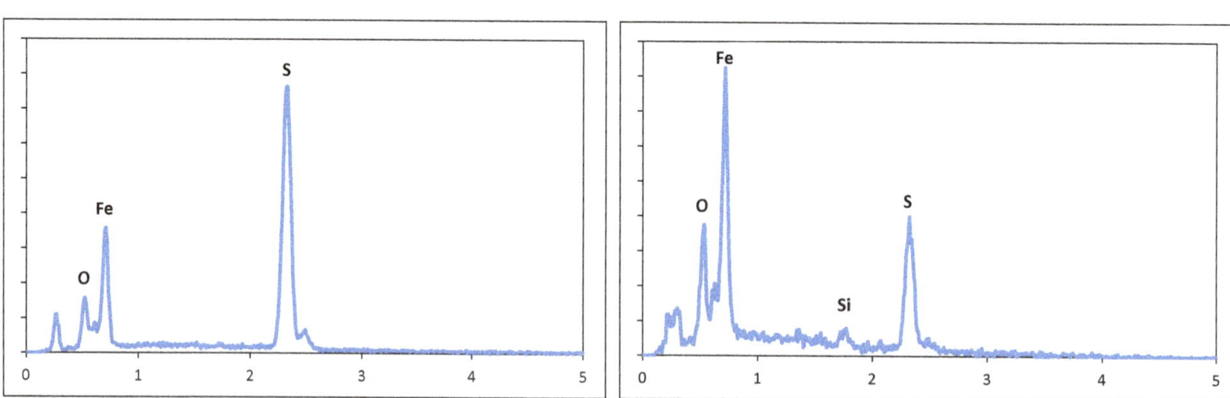

Figure 5. EDS spectra of pyrite surface coatings after dissolution in saturated calcite solution for 160 days with no added silicate (left) and 20 mg/L (as Si) added silicate (right).

both coatings contain Fe, S and O. Si is only found in the less crystalline coating (amorphous-like coating) formed in the presence silicate. A number of studies examining the transformation from ferrihydrite (an amorphous phase) to goethite (a crystal phase) suggest that the addition of soluble silicate inhibits the crystallization process [4,15, 16,19]. Our experimental results are consistent with these studies indicating that in solutions with added silicate, the conversion of the initially formed amorphous iron hydroxide is inhibited, while in the silicate free solution transformation to goethite is more pronounced. It appears that the presence of silicate inhibits transformation of amorphous iron oxy-hydroxide to a more crystalline goethite-like phase. The former phase appears to be less per- meable to oxygen and therefore pyrite oxidation is reduced more at circum-neutral pH when the solution concentration of silicate is higher.

4. Conclusions

Based on the results from the pyrite dissolution tests and the surface analyses, a possible mechanism for the formation and stabilization of iron oxy-hydroxide layers on pyrite during oxidation at pH > 5.5 was proposed. In the presence of dissolved silicate an amorphous layer with low oxygen permeability may form via:

- Ferrous ion is oxidized into ferric ion by dissolved oxygen;
- Ferric ion hydrolyses forming colloidal ferric hydroxide. Colloidal particles are aggregated onto the pyrite surface via adsorption eventually resulting in complete surface coverage;
- Silicate is absorbed into the colloidal iron hydroxide

particles displacing hydroxyl groups and forming a Si-O-Fe bond during the formation of colloid particles;

- A layer with low oxygen permeability is established which reduces oxidation rates significantly. The transformation to goethite via dissolution and re-precipitation is retarded.

5. Acknowledgements

This research has been funded by an Australian Postgraduate Award Industry (APAI) through an Australian Research Council Linkage Project Grant with AMIRA International (Mr Gray Bailey). Sponsors of the Savage River Rehabilitation Project (SRRP), Hidden Valley Services, BHP Billiton Iron Ore, Rio Tinto Ltd. and Teck Ltd. to support the Project are gratefully acknowledged.

REFERENCES

[1] J. Satur, N. Hiroyoshi, M. Tsunekawa, M. Ito and H. Okamoto, "Carrier-Microencapsulation for Preventing Pyrite Oxidation," *International Journal of Mineral Processing*, Vol.83, No. 3–4, 2007, pp. 116-124.

[2] E. Ahlberg and A. E. Broo, "Oxygen Reduction at Sulphide Minerals. 3. The Effect of Surface Pre-Treatment on the Oxygen Reduction at Pyrite," *International Journal of Mineral Processing*, 1996, Vol. 47, No. 1–2, pp. 49-60.

[3] N. Belzile, S. Maki, Y. W. Chen and D. Goldsack, "Inhibition of Pyrite Oxidation by Surface Treatment," *Science of The Total Environment*, Vol. 196, No. 2, 1997, pp. 177-186.

[4] R. Cornell, R. Giovanoli and P. Schindler, "Effect of Silicate Species on the Transformation of Ferrihydrite into Goethite and Hematite in Alkaline Media," *Clays and Clay Minerals,* Vol. 35, No. 1, 1987, pp. 21-28.

[5] D. B. Johnson and K. B. Hallberg, "Acid Mine Drainage Remediation Options: A Review," *Science of The Total Environment*, Vol. 338, No. 1–2, 2005, pp. 3-14.

[6] A. P. Chandra and A. R. Gerson, "The Mechanisms of Pyrite Oxidation and Leaching: A Fundamental Perspective," *Surface Science Reports,* Vol. 65, No. 9, 2010, pp. 293-315.

[7] C. L. Caldeira, V. S. T. Ciminelli, A. Dias and K. Osseo-Asare, "Pyrite Oxidation in Alkaline Solutions: Nature of the Product Layer," *International Journal of Mineral Processing*, Vol. 72, No. 1–4, 2003, pp. 373-386.

[8] V. P. Evangelou, "Pyrite Microencapsulation Technologies: Principles and Potential Field Application," *Ecological Engineering*, Vol.17, No. 2–3, 2001, pp. 165-178.

[9] Lalvani, S. B., B. A. DeNeve and A. Weston, "Passivation of Pyrite Due to Surface Treatment," *Fuel*, Vol. 69, No. 12, 1990, pp. 1567-1569.

[10] K. Nyavor, N. O. Egiebor and P. M. Fedorak, "Suppression of Microbial Pyrite Oxidation by Fatty Acid Amine Treatment, *Science of The Total Environment*, Vol. 182, No. 1–3, 1996, pp. 75-83.

[11] C. L. Jiang, X. H. Wang and B. K. Parekh, "Effect of Sodium Oleate on Inhibiting Pyrite Oxidation," *International Journal of Mineral Processing*, Vol. 58, No. 1–4, 2000, pp. 305-318.

[12] D. M. C. Huminicki and J. D. Rimstidt, "Iron Oxyhydroxide Coating of Pyrite for Acid Mine Drainage Control. Applied Geochemistry," Vol. 24, No. 9, 2009, pp. 1626-1634.

[13] R. S. C. Smart, S. D. Miller, W. S. Stewart, Y. Rusdinar, R. E. Schumann, N. Kawashima and J. Li, "In Situ Calcite Formation in Limestone-Saturated Water Leaching of Acid Rock Waste," *Science of The Total Environment*, Vol. 408, No. 16, 2010, pp. 3392-3402.

[14] S. Miller, R. Schumann, R. Smart and Y. Rusdinar, "Ard Control by Limestene Induced Armouring and Passivation of Pyrite Mineral Surfaces, in *ICARD*," Skellefta Sweden, 2009, p. 11.

[15] L. Dyer, P. D. Fawell, O. M. G. Newman and W. R. Richmond, "Synthesis and Characterisation of Ferrihydrite/Silica Co-Precipitates," *Journal of Colloid and Interface Science*, Vol. 348, No. 1, 2010, pp. 65-70.

[16] U. Schwertmann and R. M. Cornell, "Iron Oxides in the Laboratory: Preparation and Characterization, " 1991, Weinheim: VCH.

[17] S. Santomartino and J. A. Webb, "Estimating the Longevity of Limestone Drains in Treating Acid Mine Drainage Containing High Concentrations of Iron," Applied Geochemistry,Vol. 22, No. 11, 2007, pp. 2344-2361.

[18] D. Mohapatra, P. Singh, W. Zhang and P. Pullammanappallil, "The Effect of Citrate, Oxalate, Acetate, Silicate and Phosphate on Stability of Synthetic Arsenic-Loaded Ferrihydrite and Al-Ferrihydrite, " *Journal of Hazardous Materials*, Vol. 124, No. 1–3, 2005, pp. 95-100.

[19] P. J. Swedlund, G. M. Miskelly and A. J. McQuillan, "An ATR-FTIR Study of Silicate Adsorption onto Ferrihydrite," *Geochimica et Cosmochimica Acta,* Vol. 70, No. 18,(Supplement), 2006, pp. A632.

20

Growing Public Health Concerns from Poor Urban Air Quality: Strategies for Sustainable Urban Living

Bhaskar Kura, Suruchi Verma, Elena Ajdari, Amrita Iyer
Civil & Environmental Engineering, University of New Orleans, Louisiana, U.S.A

ABSTRACT

Urban areas around the world, particularly in emerging nations such as China, India, and Brazil are experiencing high levels of air pollution due to increased population, economy, spending, and consumption, all of which contribute to deterioration in environmental and public health conditions in urban areas. This paper briefly discusses important sources of air pollution, air pollutants of concern, public health impacts, and proposed strategies to combat urban air pollution and promote sustainable urban living. A team of researchers under the mentorship of the main author is working on a number of air quality projects that involve air quality monitoring (sources, ambient, indoor, and occupational), emissions modeling, atmospheric dispersion modeling, air pollution control, and development of knowledge-based systems to manage air quality. This paper presents potential strategies that could help address the growing public health concerns in urban areas and promote sustainable and healthy living.

Keywords: Urban Air Quality; Exposures to Air Pollutants; Risk Assessment; Public Health

1. Introduction

The populations of the rapidly expanding cities of Asia, Africa, and Latin America are increasingly exposed to ambient air pollutants by several folds compared to the levels of exposures in developed countries. It is estimated that more than 1 billion people are exposed to outdoor air pollution annually. Urban air pollution is linked to about 1 million premature deaths and 1 million pre-native deaths each year [1]. Urban air pollution is estimated to cost approximately 2% of GDP in developed countries and 5% in developing countries [1]. Rapid urbanization has resulted in increasing urban air pollution in major cities, especially in developing countries.

The potential for serious consequences of exposure to high levels of ambient air pollution was made clear in the mid-20th century when cities in Europe and the United States experienced episodes of air pollution, such as the incidents of London Fog of 1952 [2] and Donora Smog of 1948 [3], which resulted in large numbers of excess deaths and hospital admissions. In more recent years, the "gray sky" phenomenon has been the subject of growing public concern. Research shows that high levels of ambient fine particles ($PM_{2.5}$) lead to poor visibility [4].

Urban air pollution occurs when there are continuous or large emissions of air pollutants. The level of pollution depends on the meteorological (wind conditions and atmospheric stability) conditions as well. Urban air pollution is usually associated with high quantities of sulfur dioxide, oxides of carbon (dioxide and monoxide), oxides of nitrogen, and particulate matter (PM_{10} and $PM_{2.5}$) present in the atmosphere. Depending on other emission sources in cities, heavy metals and volatile organic compounds (VOCs) are also present. In the United States air pollutants are broadly categorized into criteria pollutants (SO_2, NO_x, CO, O_3, PM_{10} and Pb) and hazardous air pollutants (HAPs). Criteria pollutants have federal standards as promulgated by the United States Environmental Protection Agency (U.S. EPA) which are shown in **Table 1**. Whereas HAPs do not have national ambient standards and the goal is to eliminate them from the air environment to the extent possible as they are considered to be hazardous/toxic to human health at any levels.

Data from the World Health Organization (WHO) indicates that the world's average PM_{10} levels by region ranges from 21 to 142 $\mu g/m^3$, with a world average of 71 $\mu g/m^3$[5]. Scientific studies suggest that fine particulate matter – less than 10 microns in diameter (PM_{10}) – is likely to be the most dangerous because such fine particles can be inhaled deeply into the lungs where the clearance time of deposited particles is much longer, increasing the potential for adverse health effects.

Countries like China, India and Pakistan have recently become highly polluted in terms of air pollution. By the end of 2011, the mainland of the People's Republic of China had a total urban population of 691 million or 51.2% of the total population, rising from 26% in 1990.

In the 11-yr period of 1990—2001, the urban population increased 10%, from 26% to 36% of the population as seen in **Figure 1**. Similarly, GDP changes are shown in **Figure 2**[6].

Table 1. National ambient air quality standards (NAAQs) for criteria pollutants [Adapted from 17].

Pollutant		Primary/Secondary	Average Time	Level	Urban Sources	Health Impacts
Carbon Monoxide		primary	8-hour	9 ppm (10.31 mg/m^3)	Vehicles	Fatigue, impaired vision, CNS and cardiovascular defects
			1-hour	35 ppm (40.09 mg/m^3)		
Lead		primary and secondary	Rolling 3 month average	0.15 μg/m^3	Vehicles	Lead poisoning in children, CNS defects, renal and reproductive malfunctions
Nitrogen Dioxide		primary	1-hour	100 ppb (188.18 μg/m^3)	Vehicles	Increased bronchial sensitivity and reduced lung capacity
		primary and secondary	Annual	53 ppb (99.73 μg/m^3)		
Ozone		primary and secondary	8-hour	0.075 ppm (147 μg/m^3)	Secondary pollutant; Photochemical reactions of vehicle emission	Respiratory defects and reduced lung capacity
Particle Pollution	$PM_{2.5}$	primary	Annual	12 μg/m^3	Vehicles, power plants, industry, agriculture, surface dust, sea spray	Respiratory and cardiovascular defects. Also, an increase in short term mortality rate. Increased cardiopulmonary and lung cancer mortality rate.
		secondary	Annual	15 μg/m^3		
		primary and secondary	24-hour	35 μg/m^3		
	PM_{10}	primary and secondary	24-hour	150 μg/m^3		
Sulfur Dioxide		primary	1-hour	75 ppb (196.5 μg/m^3)	Power plants, vehicles	Changes in the pulmonary function and increase in mortality rate.
		secondary	3-hour	0.5 ppm (1.31 mg/m^3)		

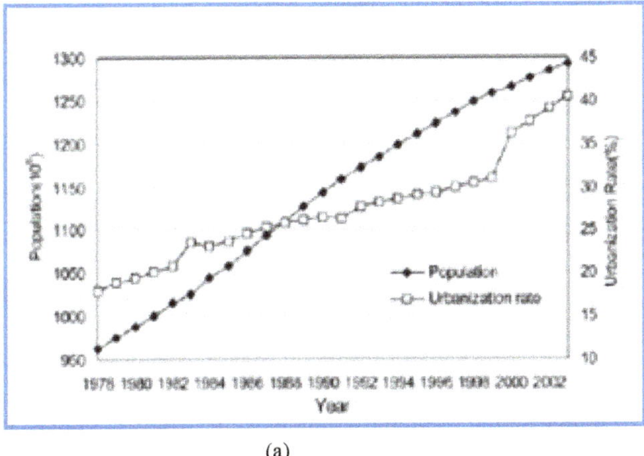

(a)

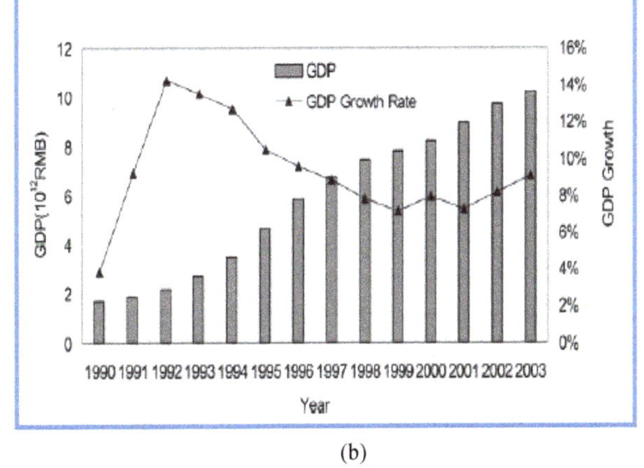

(b)

Figure 1. (a) Population and urbanization rate increase in China and (b) Gross domestic product growth in China from 1990 to 2003 [6].

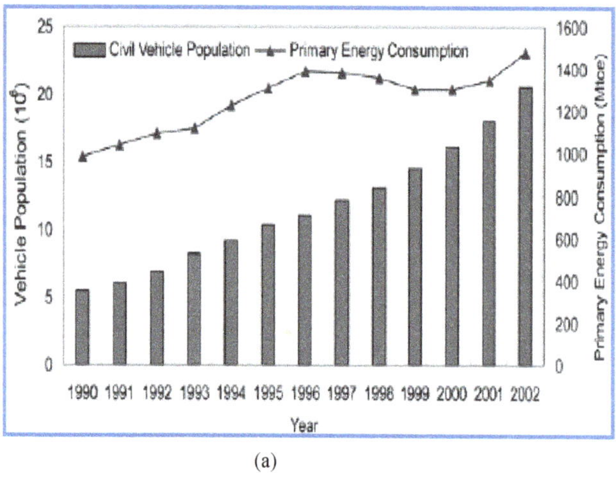

(a)

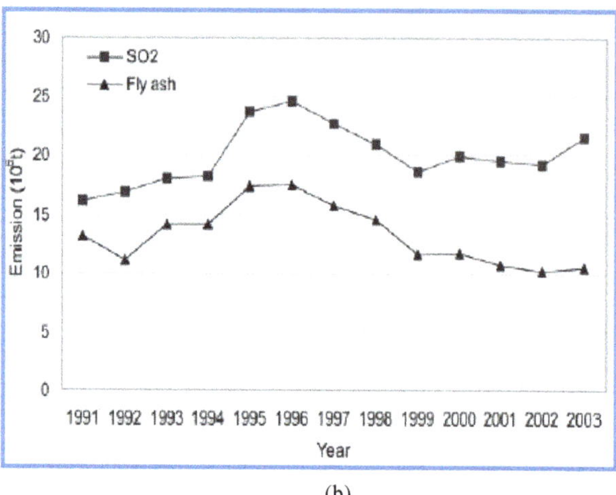

(b)

Figure 2. (a) Energy consumption and vehicle population growth in China and (b) SO2 and fly ash emission in China [6].

More than 450 million of China's 1.3 billion people are now living in urban areas where over 70 percent of China's gross domestic product (GDP) is generated [7]. Two out of three cities in China failed to meet the residential ambient air quality standard, resulting in large population exposure to health risks such as chronic bronchitis, pulmonary heart disease, and lung cancer. Respiratory diseases are a leading cause of premature deaths in China. In the long term, China faces increasing urbanization; according to predictions, nearly 70% of the population will live in urban areas by 2035[8].

In addition to PM_{10} and $PM_{2.5}$, China has large emissions of sulfur dioxide and total suspended particles (TSP) [9]. In 2001, the concentration of $PM_{2.5}$ in Beijing averaged 110 μg/m^3, more than seven times the ambient air quality standard established for the United States by the U.S. EPA for fine particulate matter [10].

According to the 2001—2008 "State of the Environment in China" reports, the average daily concentration of SO_2 in urban air was between 1 - 389 μg/m^3 while the average daily NO_2 urban air concentration in China fluctuated between 2 and 77μg/m^3. At present, the particulate pollution in China's ambient air is very serious, with not only high levels of TSP concentrations but also comparatively high levels of PM_{10} and $PM_{2.5}$ concentrations [11].

Figures 3-5 depict the ambient air quality in some of the cities of China [9].

In addition to air pollution, water pollution and other forms of pollution also pose serious threat to human health. Clean water is very essential for consumption by an individual. However, in most of the developing cities (or countries), clean water is readily available via water treatment plants set up by various governmental and non-governmental agencies. Air pollution is one thing that can only be treated at the source of the pollution. Once the pollutant is emitted into the atmosphere, it is very difficult to clean up.

2. Sources of Urban Air Pollution

Several sources were identified that cause urban air pollution. The major source of deterioration of urban air quality is transportation. Vehicles like cars or trucks run on internal combustion engines which use gasoline or other fossil fuels. This process of burning the fuel to power vehicles contributes to air pollution by releasing air pollutants such as particulates (TPM, PM_{10}, $PM_{2.5}$), SO_2, NO_x, CO, and unburned hydrocarbons. These emissions cause cancer and other serious health issues.

The American Lung Association reports that 30,000 people are killed by car emissions annually in the United States alone [12]. A study by the World Health Organization reported in National Geographic News states that air pollution in China is related to around 656,000 deaths every year throughout the country [13]. China currently has one car for every 17.2 people. In the United States, there's one car for every 1.3 people. If China were to catch up with the U.S. car ownership rate, the country would field a billion vehicles all by itself [14].

Listed below are the specific air pollutants emitted from transportation related sources. Percentages mentioned indicate contribution of these transportation sources to the overall ambient concentration of that particular air pollutant.

- **Carbon monoxide (CO)** [70% to 90%]
- **Hydrocarbons** likemethane (CH_4), gasoline (C_8H_{18}) and diesel vapors, benzene (C_6H_6), formaldehyde (CH_2O), butadiene (C_4H_6) and acetaldehyde (CH_3CHO). [50%]
- **Oxides of nitrogen (NO_x)** - [45% to 50%]
- **Greenhouse gases** like carbon dioxide [30%] in developed countries and [15%] worldwide. China emits 6,018 million ton of greenhouse gases each year.
- **Particulate matter (PM_{10}/$PM_{2.5}$)** - [25%]. In the United States particulate matter (PM_{10}) levels above 50 micrograms are considered unsafe. In Europe, the average levels for $PM_{10}$is approximately 40 μg/m^3, and in Beijing the average level is 141 μg/m^3.

- **Sulfur dioxide (SO_2)** - [5%]
- **Lead**– Lead is a toxic metal mainly used as an anti-knocking agent in gasoline (Lead tetraethyl - $Pb(C_2H_5)_4$) and is also used in batteries (lead dioxide as an anode and lead as a cathode).
- **Odors**- Diesel and gasoline engines are the major sources of odors. [15]

Gas stations also contribute to urban air pollution, particularly because of the evaporation of gasoline during filling of gas station storage tanks and vehicles. The main compounds are: benzene, toluene, ethylbenzene, xylenes, lead, methyl tertiary butyl ether (MTBE; MTBE is currently classified as a potential human carcinogen), ethylene dichloride (EDC) and naphthalene. [16] Dry cleaning shops are common sources present in the urban environment which pose significant threat to human health. These facilities emit volatile organic compounds (VOC), most of which are considered to be HAPs. Dry cleaning units emit perchloroethylene (Perc) and petroleum solvents[18]. Perc may cause cancer in human and is identified as HAP. Perchloroethylene can be measured in the breath, and breakdown products of it can be measured in the blood and urine [19]. Petroleum solvents used in this industry also emit HAPs and VOCs. Specific VOCs include: Stoddard solvent which is a mixture of C5-C12 petroleum hydrocarbons containing 30% - 50% straight-and branched-chained alkanes, 30% - 40% cycloalkanes, and 10% - 20% alkyl aromatic compounds[20].Other contaminants are carbon tetrachloride, trichloroethylene, perchloroethylene, 1,1,2-trichlorotrifluoroethane, 1,1,1-trichloroethane, glycol ethers, decamethylcylcopentasiloxane, n-propyl bromide, pure dry™[20].

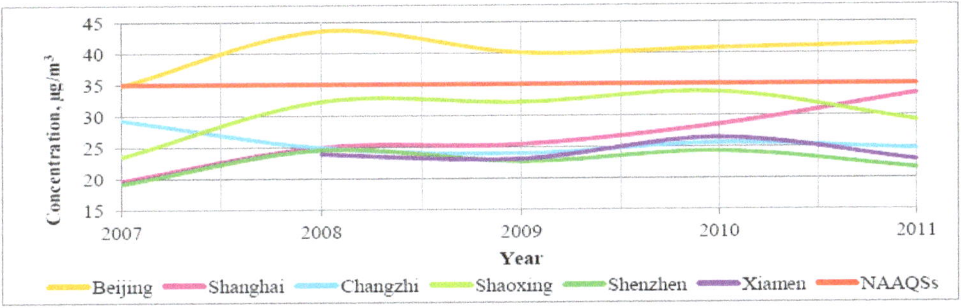

Figure 3. PM_{10} yearly ambient concentration in some cities of China [9].

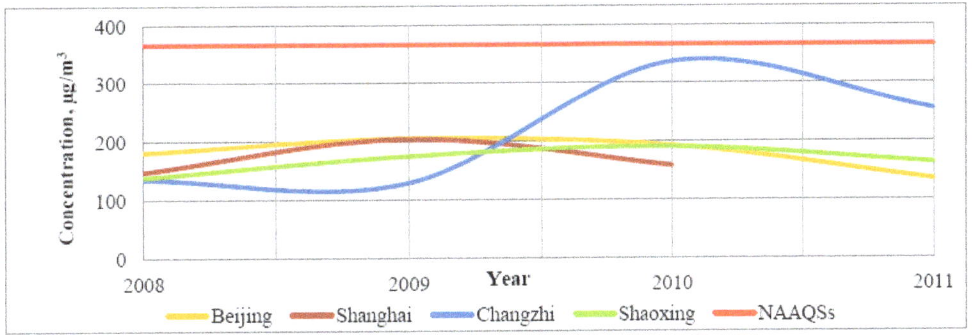

Figure 4. SO_2 yearly ambient concentration in some cities of China [9].

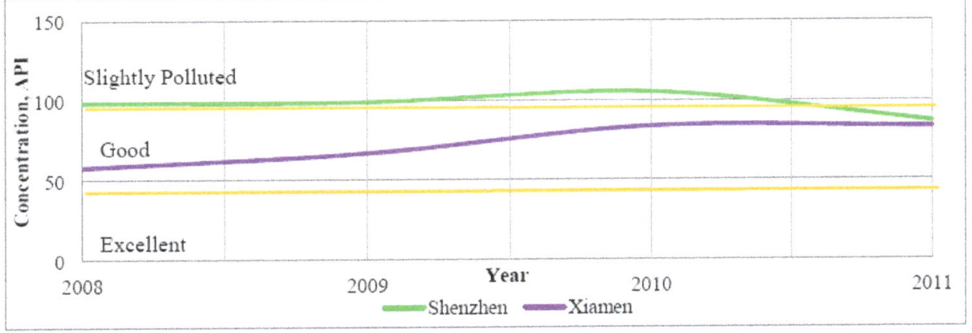

Figure 5. NO_2 yearly ambient concentration in some cities of China [9].

Compounds in VOCs can react with sunlight and contribute to ground level ozone which can cause respiratory disease. In 1991, California wasthe first state which declared perchloroethylene as a toxic chemical.As a result of thisban will be implemented on using this material by thedry cleaners in that state from 2023 [21]A recent study at Georgetown University shows that perc remains in dry-cleaned clothes and the level of this chemical increases with repeated dry cleanings [22].

3. Health Impacts from Exposure to Air Pollutants

The health impacts of a particular air pollutant depends on the pathway that it enters a human body and the type of interactions it goes through once it is inside the body. This is depicted in **Figure 6**. Usually, the pathway of exposure to an air pollutant is inhalation of the pollutant. Another route of exposure could be dermal where the skin comes in contact with the pollutant. Skin offers a good resistance to any air pollutant, but in case of any cracks or aberrations, it is highly possible that the pollutant might enter the body through the skin.

The extent of exposure depends on the dose that the individual is exposed to and the time that the individual was exposed for. Even when the individual goes through a very small dosage of the pollutant, if the time period of exposure is long enough, the person can suffer a fatality in some cases. Based on this, the health impacts are classified as acute and chronic.

Air pollutants majorly affect the respiratory system that mainly constitutes the lungs. The various effects on the respiratory system are increased respiratory (lung) illnesses, asthma exacerbations (makes asthma worse), decreased lung function in children, and chronic respiratory illnesses. In some cases with high levels of exposure, air pollutants can cause cancer and premature death.

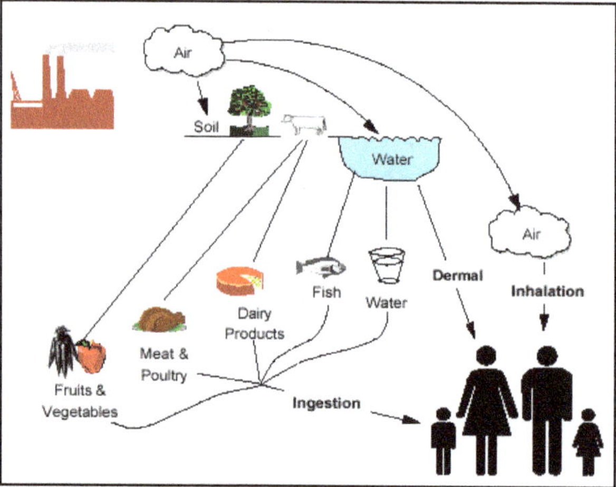

Figure 6. Exposure pathways for air pollutants (Source: EPA).

Once inside the lungs, the air pollutants get picked up by the bloodstream and can cause a number of cardiovascular impacts. These include, but not limited to, myocardial infarction (heart attack), stroke, hypertension (high blood pressure), atherosclerosis (artery disease), arrhythmia (irregular heartbeat), and thrombosis (abnormal clotting).

In several cases, certain reproductive (birth) defects have been observed. These birth defects are preterm babies (born earlier than they should be), low birth weight, slow growth in the womb, miscarriage, still birth, premature birth, infant mortality. Studies have found that a 10% increase in PM and SO_2 pollution are associated with a 1% increase in infant deaths. Also, breathing high levels of urban air pollution almost tripled a mother's chances of having a low birth weight baby.

More in-depth health impacts are shown in **Tables 1-3**.

Quantifying the magnitude of these health impacts in cities worldwide, however, presents considerable challenges owing to the limited availability of information on both effects on health and on exposures to air pollution in many parts of the world. Man-made urban air pollution is a complex mixture with many toxic components.

4. An Integrated Approach to Risk Assessment

An air quality research team at the University of New Orleans (UNO) is involved in research that promotes integrated air quality management using health risk as an important criteria. As part of integrated air quality management, the team typically uses the following strategies and research methods to achieve the overall air quality management goals:

- **Air quality monitoring**: Monitoring of ambient air quality, source emissions, indoor air quality, and occupational exposures are monitored as part of the research methodology.
- **Modeling**: Modeling strategies used include, emissions modeling, atmospheric dispersion modeling, exposure modeling (both ambient conditions and indoor conditions).
- **Clean/green technologies**: As part of this initiative, research efforts are made to optimize industrial and municipal processes with the goal of (a) promoting/increasing efficiency, (b) minimize consumption of natural resources, input materials, fuel, and energy, (c) minimize life-cycle costs, (d) reduce waste quantities, and (e) minimize health risks (cancer and non-cancer health risks). UNO has an emissions test facility which can be used to simulate industrial and other processes to evaluate existing manufacturing technologies and recommend process changes to achieve environmental friendliness.

Table 2. Health Effects of Airborne VOCs.

Air Pollutant	OSHA PEL ($\mu g/m^3$)	NIOSH REL ($\mu g/m^3$)	Toxicity/Health Effects	URE $(\mu g/m^3)^{-1}$	RFC (mg/m^3)	Urban Sources
1,1,1-Trichloroethane / Methyl Chloroform (CH_3CCl_3)	1,900,000	1,900,000	Irritation eyes, skin; headache, lassitude (weakness, exhaustion), central nervous system depression, poor equilibrium; dermatitis; cardiac arrhythmias; liver damage	N/A	5	Dry Cleaning
1,1,2-Trichlorotrifluoroethane/CFC-113 (CCl_2FCClF_2)	7,600,000	7,600,000	Irritation skin, throat, drowsiness, dermatitis; central nervous system depression; in animals: cardiac arrhythmias, narcosis	N/A	N/A	Dry Cleaning
1,3-Butadiene (CH_2=CHCH=CH)	2,200	A	Irritation eyes, nose, throat; drowsiness, dizziness; liquid: frostbite; teratogenic, reproductive effects; [potential occupational carcinogen]	3.00E-05	2.00E-03	Traffic; Gas Stations;
Benzene (C_6H_6)	F	320	Irritation eyes, skin, nose, respiratory system; dizziness; headache, nausea, staggered gait; anorexia, lassitude (weakness, exhaustion); dermatitis; bone marrow depression; [potential occupational carcinogen]	7.80E-06	3.00E-02	Traffic; Gas Stations;
Carbon tetrachloride (CCl_4)	63,000	12,600	Irritation eyes, skin; central nervous system depression; nausea, vomiting; liver, kidney injury; drowsiness, dizziness, incoordination; [potential occupational carcinogen]	6.00E-06	1.00E-01	Dry Cleaning
Ethylbenzene ($CH_3CH_2C_6H_5$)	435,000	435,000	Irritation eyes, skin, mucous membrane; headache; dermatitis; narcosis, coma	2.50E-06	1	Gas stations
Methyl-tert-butyl ether (MTBE) ((CH3)3COCH3)	N/A	N/A	Headache, nausea, dizziness, irritation of the nose or throat, and sense of confusion	2.60E-07	3	Gas stations
Toluene ($C_6H_5CH_3$)	760,000	375,000	Irritation eyes, nose; lassitude (weakness, exhaustion), confusion, euphoria, dizziness, headache; dilated pupils, lacrimation (discharge of tears); anxiety, muscle fatigue, insomnia; paresthesia; dermatitis; liver, kidney damage	N/A	5	Gas stations
Formaldehyde (HCHO)	922.5	19.68	Irritation eyes, nose, throat, respiratory system; lacrimation (discharge of tears); cough; wheezing; [potential occupational carcinogen]	1.30E-05	9.80E-03	Traffic
Acetaldehyde (CH_3CHO)	360,000	N/A	Irritation eyes, nose, throat; eye, skin burns; dermatitis; conjunctivitis; cough; central nervous system depression; delayed pulmonary edema; in animals: kidney, reproductive, teratogenic effects; [potential occupational carcinogen]	2.20E-06	9.00E-03	Traffic
o- Xylene ($C_6H_4(CH_3)_2$)	435,000	435,000	Irritation eyes, skin, nose, throat; dizziness, excitement, drowsiness, incoordination, staggering gait; corneal vacuolization; anorexia, nausea, vomiting, abdominal pain; dermatitis	N/A	1.00E-01	Gas stations
p- Xylene ($C_6H_4(CH_3)_2$)	435,000	435,000	Irritation eyes, skin, nose, throat; dizziness, excitement, drowsiness, incoordination, staggering gait; corneal vacuolization; anorexia, nausea, vomiting, abdominal pain; dermatitis	N/A	1.00E-01	Gas stations
m –Xylene ($C_6H_4(CH_3)_2$)	435,000	435,000	Irritation eyes, skin, nose, throat; dizziness, excitement, drowsiness, incoordination, staggering gait; corneal vacuolization; anorexia, nausea, vomiting, abdominal pain; dermatitis	N/A	1.00E-01	Gas stations
Ethylene dichloride ($ClCH_2CH_2Cl$)	202,500	4,000	Irritation eyes, corneal opacity; central nervous system depression; nausea, vomiting; dermatitis; liver, kidney, cardiovascular system damage; [potential occupational carcinogen]	2.60E-05	2.40E+00	Gas stations
Trichloroethylene ($ClCH=CCl_2$)	537,000	N/A	Irritation eyes, skin; headache, visual disturbance, lassitude (weakness, exhaustion), dizziness, tremor, drowsiness, nausea, vomiting; dermatitis; cardiac arrhythmias, paresthesia; liver injury; [potential occupational carcinogen]	2.00E-06	6.00E-01	Dry Cleaning
Perchloroethylene ($Cl_2C=CCl_2$)	678,000	N/A	Irritation eyes, skin, nose, throat, respiratory system; nausea; flush face, neck; dizziness, incoordination; headache, drowsiness; skin erythema (skin redness); liver damage; [potential occupational carcinogen]	5.90E-06	2.70E-01	Dry Cleaning

Table 3. Health effects of airborne metals.*

Air Pollutant	OSHA PEL (μg/m^3)	NIOSH REL (μg/m^3)	Toxicity/Health Effects	URE (μg/m^3)$^{-1}$	RFC (mg/m^3)	Urban Sources
Lead (Pb)	50	50	Lassitude, insomnia; facial pallor; anorexia, constipation, abdominal pain, colic; anemia; gingival lead line; tremor; paralysis wrist, ankles; encephalopathy; kidney disease; irritation eyes; hypertension	N/A++	1.5E-04 [2 times more potent that Hg]	Traffic-Brake Dust
Antimony (Sb)	500	500	Chronic poisoning, functional disorders of the heart, degeneration of the heart muscle	N/A	2.2E-04 [1.4 times more potent that Hg]	Traffic-Brake Dust
Copper (Cu)	1,000	1,000	Irritation-Eye, Nose, Throat; Cumulative Lung Damage	N/A	N/A	Traffic-Brake Dust
Zinc (Zn)	5,000	5,000	Irritation-Eye, Nose, Throat, Skin---Marked Respiratory Effects---Acute lung damage/edema Chronic (Cumulative) Toxicity-Suspect Carcinogen or mutagen	N/A	N/A	Vehicular tires
Mercury (Hg)	100	50	Irritation eyes, skin; cough, dyspnea, bronchitis, pneumonitis; tremor, insomnia, lassitude ; stomatitis, salivation; gastrointestinal disturbance, anorexia, weight loss; proteinuria	N/A	3.00E-04	Various
Chromium (Cr)	1,000	500	Cancer; irritation eyes, skin; lung fibrosis (histologic)	1.2E-02 [38 times more carcinogenic than Ni]	1E-04 [3 times more potent that Hg]	Various
Manganese (Mn)	5,000	1,000	Asthenia, insomnia, mental confusion; metal fume fever: dry throat, cough, chest tightness, dyspnea, rales, flu-like fever; low-back pain; vomiting; malaise ; lassitude ; kidney damage	N/A	0.5E-04 [6 times more potent that Hg]	Various
Nickel (Ni)	1,000	15	Sensitization dermatitis, allergic asthma, pneumonitis; [potential occupational carcinogen]	3.12E-04	0.9E-04 [3.3 times more potent that Hg]	Various
Cadmium (Cd)	5	N/A	Pulmonary edema, dyspnea, cough, chest tightness, headache; chills, muscle aches; nausea, vomiting, diarrhea; anosmia, emphysema, proteinuria, mild anemia; [potential occupational carcinogen]	0.18E-02 [5.8 times more carcinogenic than Ni]	0.1E-04 [30 times more potent that Hg]	Various
Arsenic (As)	10	2	Ulceration of nasal septum, dermatitis, gastrointestinal disturbances, peripheral neuropathy, hyperpigmentation of skin, [potential occupational carcinogen]	0.43E-02 [13.8 times more carcinogenic than Ni]	0.15E-04 [20 times more potent that Hg]	Various

*Note: Ambient Standards for the above air pollutants are not available. Some sources recommend 1/1000th value of OSHA PEL as ambient limit. N/A- Not Available

- **Engineering controls**: The research team identifies, evaluates, and develops air pollution control technologies appropriate for various industry sectors for both particulates and gaseous air pollutants.

Development of decision support systems: Kura's research team developed a number of decision support systems for industrial environmental management, occupational exposures and health management, health risk management system for public, and life cycle assessment and life cycle costing predictive model for industrial processes.

5. Proposed Research and Management Strategies for Urban Air Quality Management

As the urban air environment is contaminated with a variety of pollutants such as criteria pollutants (CO, SO_2, NO_x, PM_{10}, $PM_{2.5}$, Pb, O_3) and hazardous air pollutants (heavy metals, volatile organic compounds, and others) from a wide range of sources such as transportation, commercial, and industrial sources, urban air quality management requires a comprehensive and integrate approach. Kura's research team at UNO has significant experience in this area. A conceptual urban air quality management approach is described in this section:

- Identify **sources** of air pollution in the immediate urban area and the area of influence
- Knowing the sources of air pollution to identify the **specific pollutants** that may be released in the air environment
- Conduct **preliminary air quality monitoring** to identify the concentration ranges for air pollutants of concern
- Identify the best **monitoringtechniques, equipment, and resources** needed to understand the short term and long term trends of these pollutants
- Establish a permanent **network of air quality monitoring stations** to measure criteria pollutants and hazardous air pollutants
- Develop a **database** of air pollutant emission quantities from various sources that have been identified for

the urban area being investigated

- Evaluate the **relationships** between the emission quantities and the ambient concentrations under various meteorological conditions
- Prioritize**the sources of air pollution** for regulating as appropriate
- Use the data obtained from the network of air quality monitoring stations for **developing policies to control emissions** and also to make changes in public behavior which may be responsible for poor air quality (traffic; open burning; fuel combustion; others)
- Develop a **decision support system** to integrate the ambient air quality data to compute exposures and health risk probabilities (cancer and non-cancer)
- Use the data from the decision support system to **assist** the policy makers, scientists, and the public to achieve the required air quality management goals.

A proposed integrated air quality management plan for urban areas is illustrated in **Figure 7.**

6. Summary and Conclusions

This paper presents various urban air quality challenges experienced by many cities around the world. More specifically, types of sources present in urban areas, types of air pollutants, and health hazards associated with most common urban air pollutants. Due to the high concentration of population, recent growth in the economy and increased consumption resulted in deteriorated air quality. Information provided in this paper and some suggested strategies discussed should be helpful to initiate a dialogue among interested parties. While it is not practically feasible to discuss all aspects of integrated urban air quality management strategies, this paper should serve well in undertaking further actions by the stake-holders from the governmental agencies, the scientific community, and the public.

7. Acknowledgements

Principal author (who serves on the Board of Trustees of ITU) acknowledges and thanks the financial support received from the International Technological University (ITU), San Jose, CA towards conference travel.

REFERENCES

[1] Urban Air Quality. United Nations Environment Programme, Urban Environment Unit - http://www.unep.org/urban_environment/issues/urban_air.asp

[2] Stegeman, John J. & Solow, Andrew R,"A Look Back at the London Smog of 1952 and the Half Century Since A Half Century Later: Recollections of the London Fog," *Environmental Health Perspectives*, 2002.

[3] "Overview of the 1948 Donora Smog," Pennsylvania Department of Environmental Protection.

[4] Y.L. Sun, G.S. Zhuang, Y. Wang, L.H. Han, J.H. Guo, M. Dan, W.J. Zhang, Z.F.Wang and Z.P. Hao, "The Air-Borne Particulate Pollution in Beijing—Concentration, Composition, Distribution and Sources," 2004.

[5] Annual mean PM10 (Particulate matter with diameter of 10 µm or less), by city. World Health organization - http://www.who.int/phe/health_topics/outdoorair/databases/en/index.html

[6] J.M. Hao and L.T. Wang, "Improving Urban Air Quality in China: Beijing Case Study," 2005.

[7] Juan, S., "Census: Population hits 1.37b," *China Daily*, 2011. http://europe.chinadaily.com.cn/china/2011-04/29/content_12418282.htm

[8] Urban and Rural Areas 2011, United Nations, Department of economic and Social Affairs. http://esa.un.org/unup/Wallcharts/urban-rural-areas.pdf

[9] C. Blazo, B. Kura and N. Halageri, "China: The Fastest Growing Economy and the Associated Air Quality Challenges", Automatic Welding Machinery Association, 2012.

[10] J.L.Wang, Y.H. Zhang, M. Shao, et al. "The Chemical Composition and Quantitative Relationship between Meteorological Condition and Fine Particles in Beijing," *Jounal of Environmental Sciences*, Vol. 16, 2004, pp. 860-864.

[11] Air Quality Information Transparency Index, Institute of Public & Environmental Affairs - http://www.ipe.org.cn/upload/report-aqti-en.pdf

[12] Jacob, S., "Facts of Car Pollution", 2010. http://www.livestrong.com/article/156537-facts-of-car-pollution/

[13] National Geographic News: http://news.nationalgeographic.com/news/2007/07/070709-china-pollution.html

[14] Plumer, B., "Air Pollution now Kills more than High cholesterol," 2012. http://www.washingtonpost.com/blogs/wonkblog/wp/2012/12/20/air-pollution-now-kills-more-people-than-high-cholesterol/

[15] Jean-Paul Rodrigue, "The Geography of Transport Systems," 3rd Edition, 2013.

[16] Environmental Pollution Centers

[17] NAAQs for criteria pollutants - http://epa.gov/air/criteria.html

[18] EPA-outdoor Air-industry,business, and home: Dry Cleaning Operations - Additional information, available at: http://www.epa.gov/oaqps001/community/details/drycleaning_addl_info.html#activity2

[19] *EPA- Air Toxics Health Effects Notebook, available at* http://www.epa.gov/ttn/atw/hlthef/tet-ethy.html

[20] Chemicals used in drycleaning operations report, Bill Linn, Florida Department of Environmental Protection

(FDEP), Scott Stupak, North Carolina Superfund Section, provided technical support for database development, January 2002, Revised July 2009

[21] NBC news, available at: http://www.nbcnews.com/id/16816627/ns/us_news-environment/t/calif-air-regulators-ban-dry-cleaning-chemical/#.USQcPB0skuA

[22] Prof. Paul Roepe, GeorgetownUniversity, available at: http://www.georgetown.edu/story/dry-cleaning-study.html

Feasibility Study for Power Generation during Peak Hours with a Hybrid System in a Recycled Paper Mill

Adriano Beluco[1], Clodomiro P. Colvara[2], Luis E. Teixeira[3], Alexandre Beluco[3]
[1]Universidade Ritter dos Reis, Porto Alegre, Brazil
[2]Genesys Participações Ltd., Porto Alegre, Brazil
[3]Institution Pesquisas Hidráulicas, Universidade Federal do Rio Grande do Sul, Porto Alegre, Brazil

ABSTRACT

The differential pricing for peak hours encourages industrial consumers to look for independent power supplies for the period from 19 to 22 hours. This paper presents a study to identify the optimal solution for a recycled paper mill that also intends to work in that period. The factory is located in Rio Grande do Sul, in southern Brazil, and considers the use of a diesel gen set, a micro hydro power plant and possibly PV modules. Two micro hydro power plants were considered in the study, an old plant to be renewed and another to be fully implemented. The software Homer was used as a tool to determine the most feasible combination of components considered in the study. The sale of surplus power to the energy system appears as a key to viability of alternatives that are not based solely on diesel generators. The optimal solution consists of a combination of diesel generators and micro hydro power plant, in one case, and only on hydroelectric power plant in another, with a significant penetration of PV modules if its cost is reduced to 12% of the current price, selling an amount of energy equal to that which is bought. The annual water availability in one of the sites requires diesel supplement, while the other, more abundant, this supplement is not necessary.

Keywords: Hybrid Energy Systems; Micro Hydro Power; PV Modules; Energetic Complementarity; Feasibility Study; Computational Simulation; Software Homer

1. Introduction

The peak time is three to four hours of the end of the day on which the power consumption increases dramatically by matching habits of various types of consumers.

This increase in consumption is traditionally fought with distinction in energy tariffs for that time, typically for the period from 19 hours to 22 hours.

The reversible hydropower plants provide a greater supply of energy at peak times, made possible by the difference in prices of energy at peak times and off-peak.

In Brazil, residential consumers still do not pay for more expensive energy at peak times, but the industrial consumers pay charges that may be more than six times higher than rates outside of peak hours. This differentiation forces industrial consumers to seek solutions for independent power supplies during peak hours.

Differences in tariffs are a tool for managing loads of different profiles and power that can be delivered to supply these loads [1,2]. This is a topic of ongoing study and the specific case of Brazil, for its size, the variety of loads and the pressure for economic development, awakens many discussions on applicable models for pricing [3].

The search for alternative energy supplies that may be competitive with electricity tariffs valid during peak hours are an obvious opportunity for development of renewable resources. Both the energies already established, such as solar and wind power and biomass fuels, as those still in development, can be employed.

This paper presents a study to identify the optimal solution for a recycled paper mill that also intends to work during the period from 19 to 22 hours without relying solely on the power supply provided by the grid.

2. The Recycled Paper Mill and the Problem to Be Solved

The factory is located near the city of Caxias do Sul[1], Rio Grande do Sul, and produces recycled paper for cleaning,

[1]The city of Caxias do Sul can be located on Google Maps [4] at http://goo.gl/maps/nQIiZ.

marketed in the southern states of Brazil. The plant requires about 60 employees and operates 21 hours a day, stopping only at peak times.

The mill has two paper making machines, each with a capacity to produce 900 tons per month, amounting to a total capacity of 1800 tons per month. These machines can operate producing recycled paper and paper produced from virgin pulp.

The two papermaking machines are installed in a building with total covered area of 10,500 square meters. This building houses all the facilities needed to operate these machines. There is also an administrative building and a large area dedicated to the treatment of effluents from industrial process.

Two machines, their auxiliary equipment and wastewater treatment plant consume a total power of 300 kW. The offices, lighting and other auxiliary systems, such as security, consume 25 kW.

The big difference in pricing of electricity during peak hours, compared to the off-peak, requires the company to cease production for three hours a day. Thus, the company (capable of operating 24 hours per day) currently produces about 1500 tons of paper per month.

This study aims to find a solution to the energy supply during peak hours, enabling the operation of the company during the 24 hours of the day.

Obviously, the company reserves the peak hours to perform daily maintenance tasks. The intention is to follow the schedule allowing peak for these activities, but with the intention to proceed with the production of paper whenever possible.

The problem can be viewed as a simple extension of a system based on diesel, with additional modules, wind turbines and a plant [5,6]. The isolated system and the system connected to the grid must be compared [7,8]. Contribution to sustainable development is an additional motivation for everyone involved [9,10].

The study was performed in two stages, as described in the next section. At first, the acquisition of a disabled micro hydro power plant was considered. At the second timing, the construction of a new hydro power plant at another location has been considered.

The study was based on computer simulations with software Homer, described in a later section. Homer simulates the operation of a power system and provides the optimal combination of components considered, receiving input data like component costs and their performances. **Figure 1** shows schematically the system was simulated, described in detail in the next section.

3. Components of the Energy System

The study was conducted in two phases, known hereafter as Alternative A and Alternative B. These alternatives

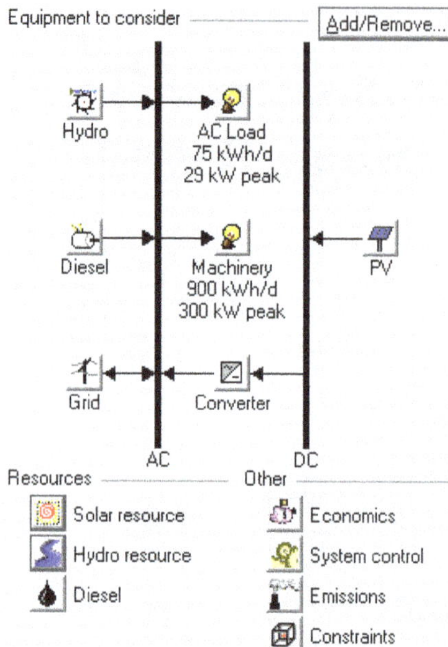

Figure 1. Hydro PV diesel hybrid system used to supply the recycled paper mill considered in the study.

are described in the following sections.

The first solution considered was a diesel generator set. Over time, it has become possible to acquire a disabled micro hydroelectric power plant with output of 240 kW and potential for expansion. A large covered area of the building that houses the two machines encouraged the inclusion of PV power generation in the study. These three possibilities were considered in Alternative A. A diesel generator set, another hydroelectric plant to be fully built and the same idea of using the covered area for photovoltaic generation is Alternative B.

These two alternatives take into account the sale of surplus power to the interconnected power system as an important factor in their development, both in the off-peak period and even at peak times.

In this study, the demand for electricity was divided into two loads, one for the office and the other corresponding to the papermaking machines and their auxiliary systems. **Figure 2** shows the demand profile for the first load and **Figure 3** shows the profile for the second load. These consumer loads only consider consumption to peak hours, since there is no intention to break the demand contract with the concessionaire of electric power distribution to the rest of the day.

Figures 2 and **3** show two graphs each. At first, the average electrical demand for each month, the deviations around these averages and maximum and minimum values are shown. The second graph consists of a diagram that summarizes the distribution of electrical demand for each hour of the day over all days of the year, according to the legend on the right. The step that appears in the months of November, December, January and February

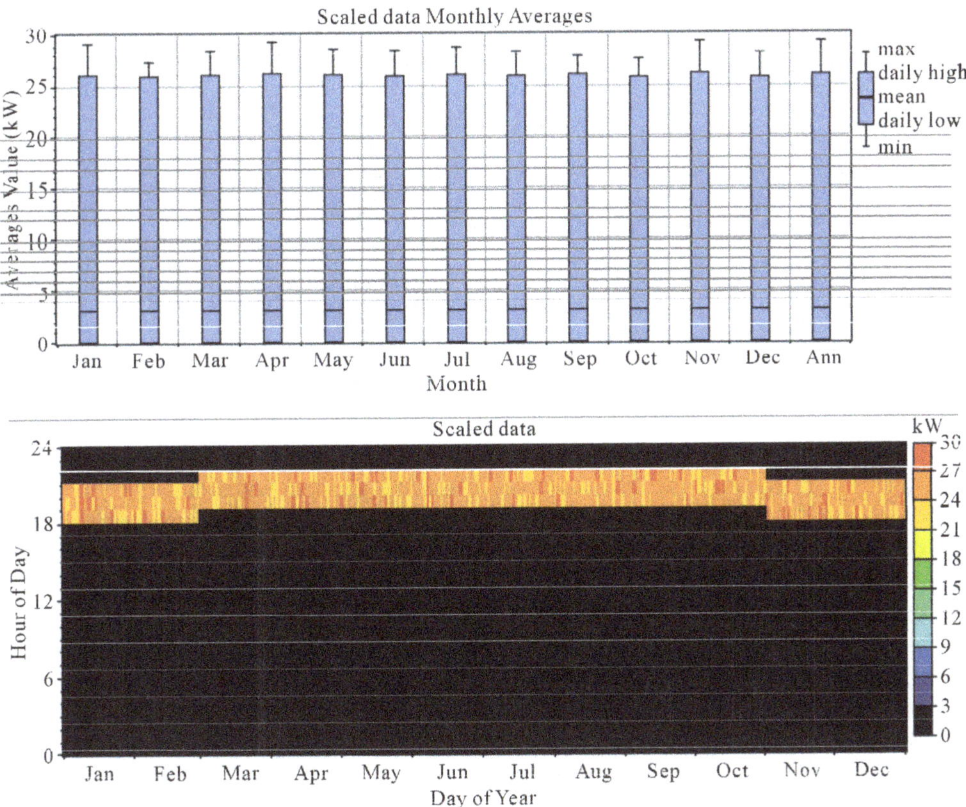

Figure 2. Average daily demand profile of the office of the paper mill.

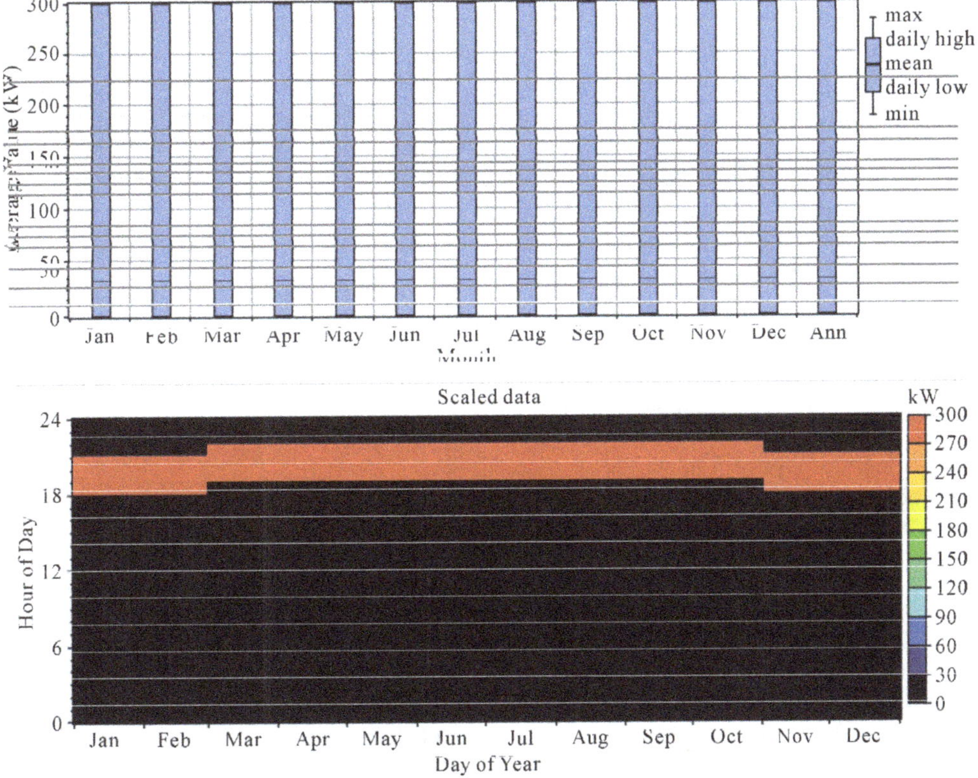

Figure 3. Profile of average daily demand for machinery of the paper mill.

matches the DST (daylight saving time).

3.1. Alternative A

The diesel generator set consists of two units with 150 kW each and possibly a third rig with 30 kW. The generators of 150 kW present initial cost of $36,000 and replacement cost at $32,400. Equipment with 30 kW presents initial cost of $8200. All have useful life of 15,000 hours.

The micro hydroelectric power plant considered in this alternative is the Trail 48, located in the municipality of Ivoti[2], Rio Grande do Sul, launched in 1914 with 240 kW of installed power as a private venture, operated for several years by the government and deactivated in 1971. The distance to the headquarters of the company is around 72 km.

The **Figure 4** shows the monthly average stream flow in the small dam that is part of this plant. The data set is synthetic and is compiled from data rainfall stations in the region. It is evident the wide variation between the months of highest and lowest availability, including energy restriction if the plant is expanded.

The cost of restoration of the facilities was not included in this study and may reverse in indirect benefits for a certain tourism potential in the region. The cost of refurbishing the plant was valued at $200,000 to modernize equipment and increase the installed capacity to 252 kW.

The dam is only 2.7 meters high and about 42 meters long. The annual average stream flow is 0.681 m^3/s. There is a strong commitment to the small volume of storage due to silting occurred mainly due to human occupation of their surroundings.

A large covered area of the building that houses the papermaking machines encouraged the consideration of PV modules in this study. The opportunity to purchase PV modules from countries with low production costs or obtaining government incentives directed insertion of the modules in this study seeking the cost that would allow greater penetration of PV energy.

The software Homer [11] allowed the acquisition of data from solar availability in the region of Caxias do Sul, where the factory is located. **Figure 5** shows the average incident solar radiation on a horizontal plane for each month, the deviations around these averages and maximum and minimum values are shown.

3.2. Alternative B

This alternative differs from Alternative A because it considers a new micro hydroelectric power plant. It's actually a site with potential newly identified and not yet cataloged, located about 10 km from the headquarters of the factory. The height is 88 meters, the dam is approximately 30 meters long and 12 meters high, the annual average stream flow is 0.718 m^3/s, with a total estimated potential of about 427 kW.

The **Figure 6** shows the monthly average stream flow for the location originally designed for the implementation of a small dam. The data set is also synthetic and was compiled from data of rainfall stations in the region without storage capacity.

3.3. Energetic Complementarity between Water Availability and Solar Radiation

The water availability, both in **Figure 4** as in **Figure 6**, and the incident solar radiation, in **Figure 5**, shows a clear and evident complementarity. Following the proposal of Beluco *et al.* [12], this complementarity can be assessed by the complementarity index κ.

Comparing **Figures 4** and **5**, the values 0.98 for complementarity in time, 0.82 for energy complementarity and 0.78 for complementarity between amplitudes are obtained. The overall index value is 0.63.

Comparing **Figures 5** and **6**, the values 0.92 for complementarity in time, 0.74 for energy complementarity and 0.89 for complementarity between amplitudes are obtained. The overall index value is 0.61.

The final values of the indices were low (0.63 and 0.61), but the complementarity that strikes the eye observing the figures are evaluated by complementarity in time, where values were higher (0.92 and 0.98).

References [13,14] present results that show the influence of different degrees of complementarity on the performance of hybrid energy systems. This subject is not well known and is still a matter of research.

4. Simulations with Homer

The software Homer [11] was developed by National Renewable Energy Laboratory (NREL) and is available for universal access in its version 2.68 beta. Homer simulates a system for power generation over the time period of 25 years at intervals of 60 minutes, presenting the results for a period of one year [15,16].

Simulations were performed for the following values for the optimization variables: 0.00 kW and 2000 kW for PV modules; 0 kW, 150 kW and 300 kW for diesel generation set; 0 kW and 380 kW for the purchase capacity from the grid; and 0.0 kW, 380 kW and 2000 kW for the converter power capacity.

Simulations were performed for the following values for the sensitivity inputs: 500 l/s, 900 l/s, 1,300 l/s and 1700 l/s for stream flow; $0.80, $1.10, $1.40 and $1.70 per liter of diesel; 1.00, 0.25, 0.15, 0.10, 0.05 and 0.01 for PV Capital Cost, PV Replacement Cost and PV

[2]The city of Ivoti can be located on Google Maps [4] at http://goo.gl/maps/lD30s.

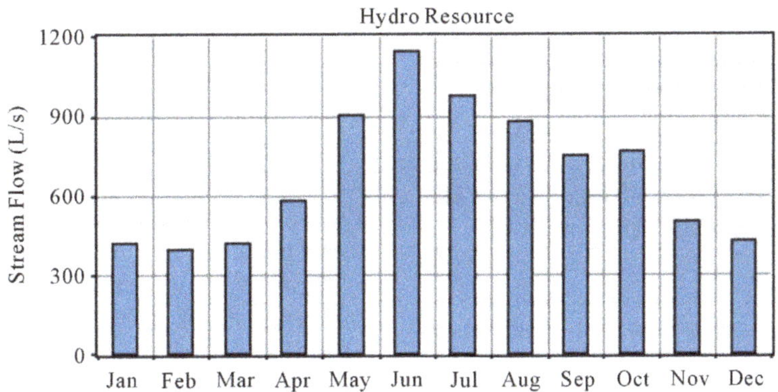

Figure 4. Trail 48 micro hydro characteristics.

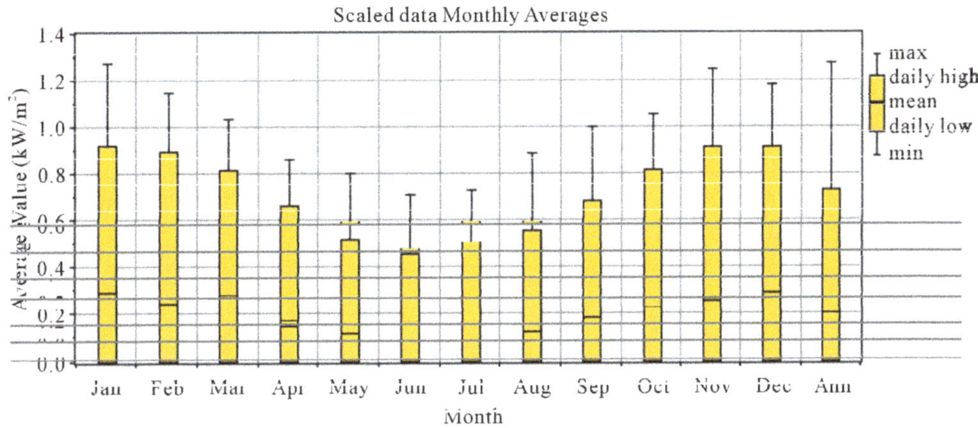

Figure 5. Incident solar radiation on a horizontal plane, obtained with software Homer.

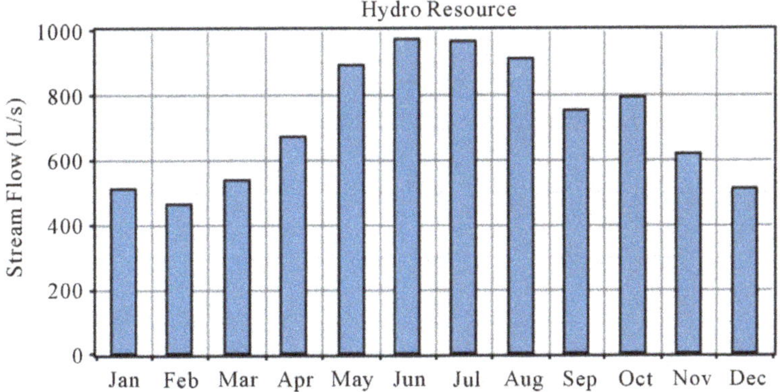

Figure 6. Trail 48 micro hydro characteristics.

Operating and Maintenance Costs multipliers, these latter two linked to the first.

$200,000, $320,000, $400,000 and $480,000 for hydro capital cost, $160,000, $256,000, $320,000 and $384,000 for hydro replacement cost, $4000, $6000, $8000 and $10,000 for hydro operation and maintenance costs, linked to 740 l/s, 860 l/s, 980 l/s and 1100 l/s for design flow rate, for Alternative A.

$1,075,000, $1,285,000, $1,500,000 and $1,700,000 for hydro capital cost, $800,000, $960,000, $1,200,000 and $1,400,000 for hydro replacement cost, $20,000, $24,000, $30,000 and $36,000 for hydro operation and maintenance costs, linked to 600 l/s, 720 l/s, 840 l/s and 960 l/s for design flow rate, for Alternative B.

A set of 72 simulations, with 1536 different values for the variables of sensitivity, for Alternative A, and other 72 simulations, with 1536 different values for the variables of sensitivity, for Alternative B, were performed. The results are presented and discussed in the next section.

5. Results and Discussion

Figures 7 and **8** show the main results for the Alternative A. The system of **Figure 7** is able to sell 2000 kW to the grid and the system of **Figure 8** has no capacity for sale to the grid. These two figures show the optimal combinations of components of the system of **Figure 1** for different values of average stream flow and diesel price.

It can be seen in **Figure 7** that the combination of grid connection and hydro power plant would prevail only for higher values of average stream flow. All combinations in brown in this figure correspond to the combination of grid and hydro, supported by diesel generator set.

For larger values of design flow rate and peak power price, in **Figure 8**, with no ability to sell energy to the grid, the area in the figure corresponding to the combination of grid supplies and hydro is smaller on the right side. It appears on the left, in black, a series of combinations of grid and diesel generator set only.

In **Figure** 7, the leftmost points of the optimization space have energy costs near \$0.2/kW·h. This value will decrease further to the right of the chart, passing situations highly profitable due to higher water availability. The sale of surplus power to the grid allows these combinations present operating profit. Equilibrium occurs approximately at points around the stream flow of 900 l/s.

In these two figures, the regions corresponding to combinations including diesel generator set, the leftmost points, with less water availability, include two generators of 150 kW. More to the right, with higher water availability, combinations include one generator.

Figures 9 and **10** show the main results for the Alternative B. The same manner as for A, the system of **Figure 9** is able to sell 2000 kW to the grid and the system of **Figure 10** has no capacity for sale. These two figures also show the optimal combinations of the system of **Figure 1** for different values of average stream flow and diesel price.

Figure 9 shows a much more favorable situation for the combination of grid connection and hydro power plant than that of **Figure** 7. The combinations which include diesel generator set, in brown, are now reduced to the lower right corner of the optimization space.

This difference is due in part to the best results obtained with a hydro power plant with higher available power and less stream flow variation. In Alternative A, the hydro power plant has a design flow rate of 740 l/s and an installed power of 240 kW, while in B, the design flow rate is 600 l/s and the power is equal to 380 kW.

In **Figure 10**, without possibility of selling energy to the grid and allowed to purchase energy, it arises in extreme lower left of optimization space a region corresponding to the combinations of grid connection and diesel generators only. All combinations include two generators of 150 kW, with cost of energy around \$0.400/kW·h.

The simulations with different values for the stream flow can be useful for two reasons. First, the time series used in the study were synthesized from data sets obtained from stations in neighboring river basins, providing

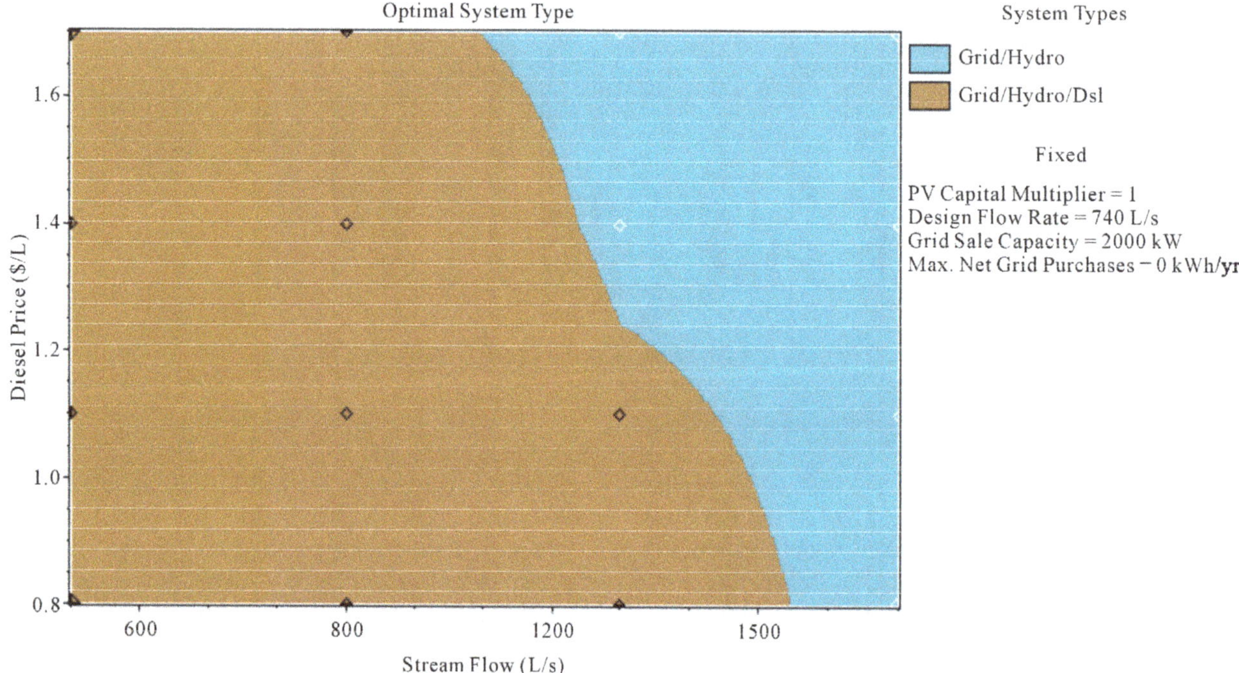

Figure 7. Results for the optimization space of the system of Figure 1, with components of Alternative A.

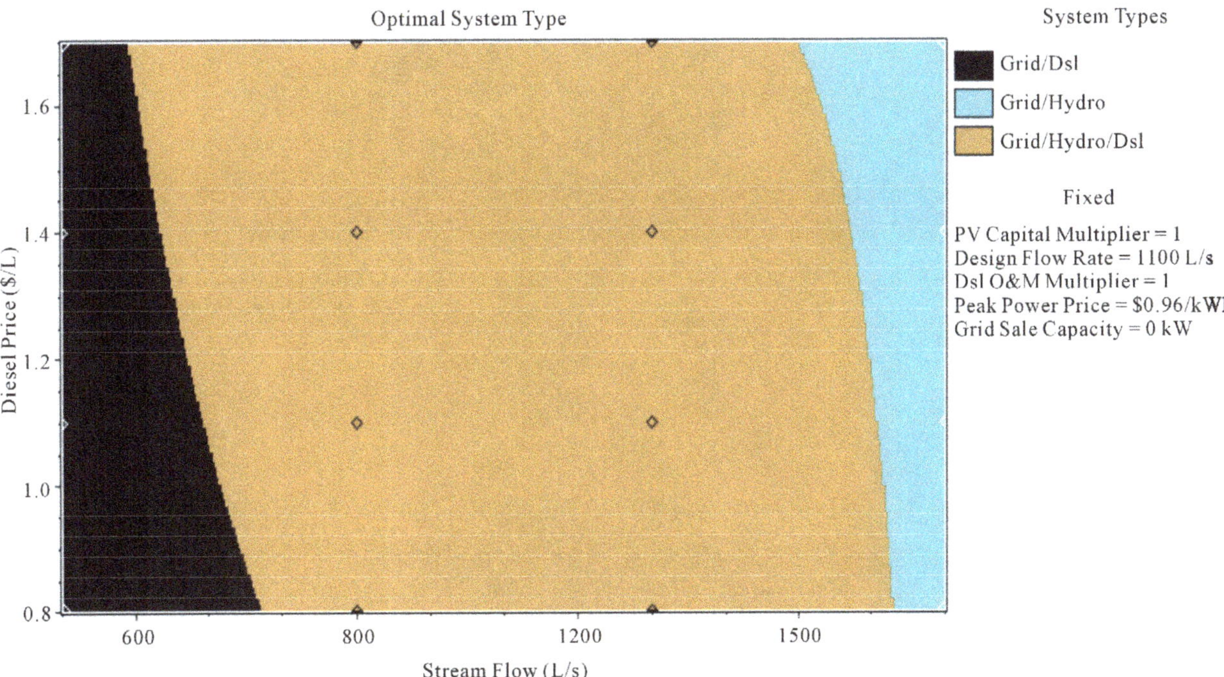

Figure 8. Results for the optimization space of the system of Figure 1, Alternative A, without grid sale capacity.

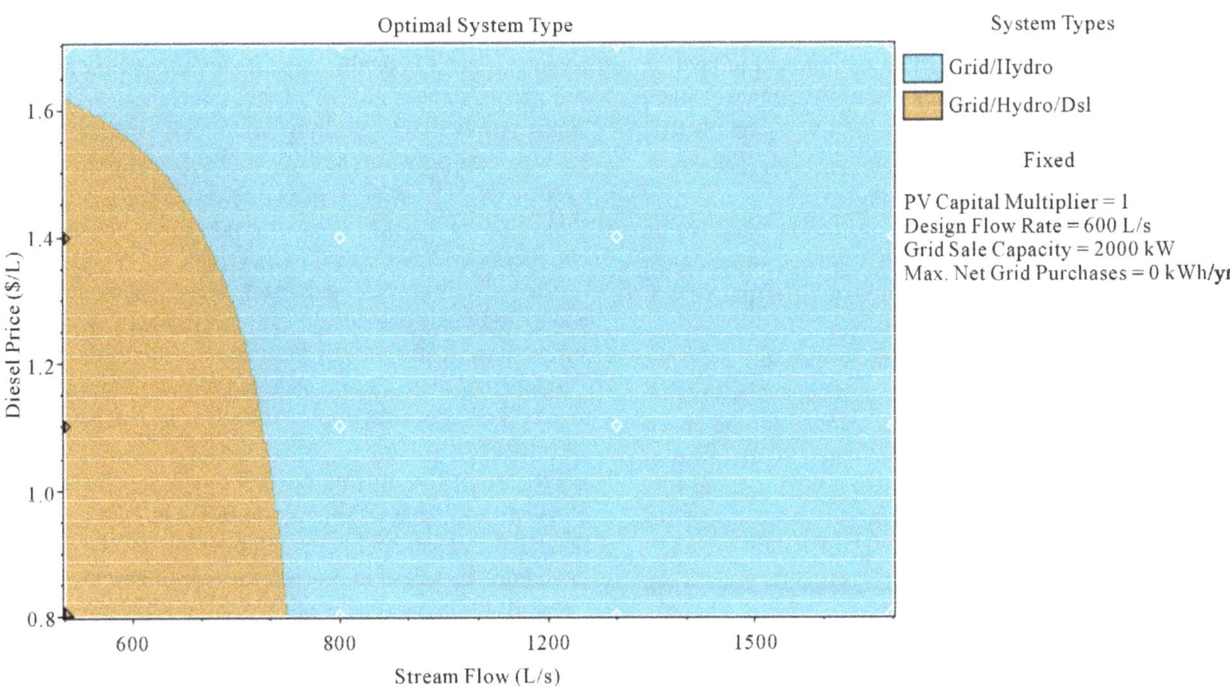

Figure 9. Results for the optimization space of the system of Figure 1, with components of Alternative B.

information that may be undersized. Second, they allow viewing combinations of system components and the cost of energy to higher water availability.

Alternative A would lead to an optimum combination of components that include the micro hydro power plant with design flow rate of 740 l/s, a 150 kW generator with no network connection, with probable failure in meeting the demand of around 10%.

The possibility of selling surplus power to the system makes this combination of components a solution quite profitable. The demand will represent only 15% of the total energy produced per year.

The price of diesel currently prevailing in the region is $1.1 per liter. The diesel generator is required only in

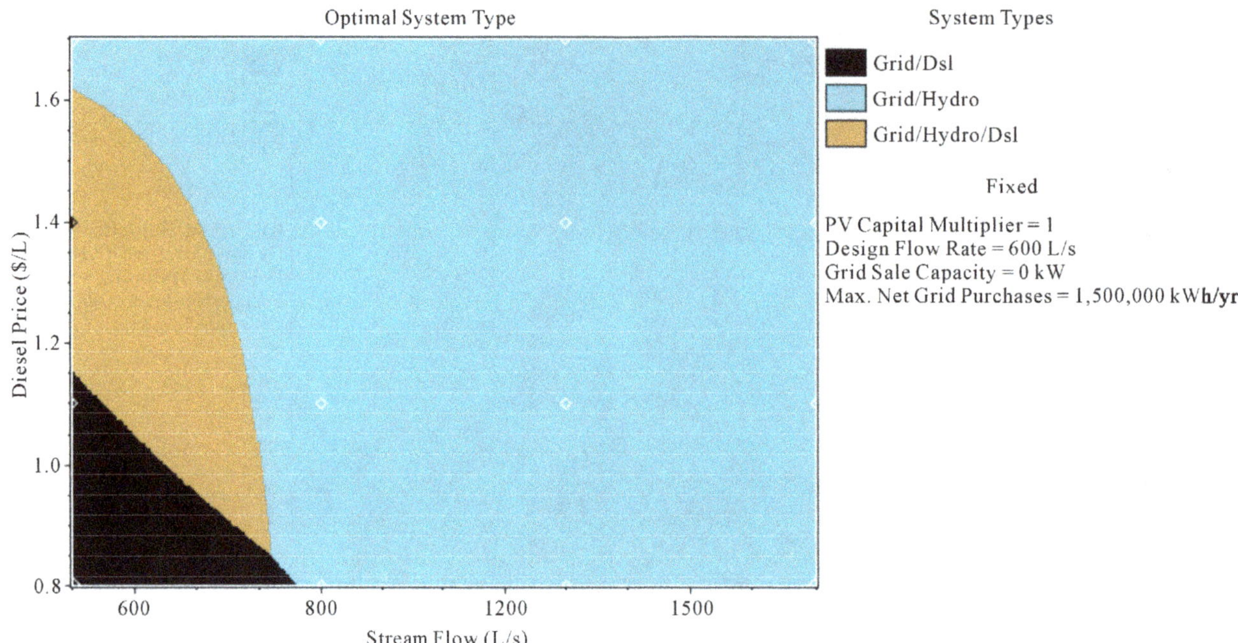

Figure 10. Results for the simulation of the complete system of Figure 1, Alternative B, without grid sale capacity.

periods of low water availability, when it would also be necessary to acquire power from the grid.

Alternative B would lead to an optimum combination of components that include the micro hydro power plant with design flow rate of 600 l/s, a 150 kW generator with no network connection, also with probable failure in meeting the demand of around 10%.

In Alternative A, the micro hydro power plant would have an installed power of 252 kW with a design flow rate of 740 l/s. In Alternative B, the installed power will be 427 kW, with a design flow rate of 600 l/s.

The results of these four figures indicate the presence of the connection to the grid as a component of optimal solutions. In part because the price charged for the provision of demand is relatively low, but also because at various times the cost of energy supplied by the hybrid system becomes greater than the cost of energy of the grid even at peak times.

The energy purchased from the system is always small, never exceeding 10% of all energy consumed during peak hours. If this supply is limited, the hybrid system would fail at most 10% in meeting demand. Considering that the company does not currently operate at peak times, go to operate for 90% of this period would be an increase in revenue that cannot be overlooked.

The modules do not appear in combinations of spaces optimization of simulations performed. Obviously, the high initial costs unfeasible solutions including solar energy. The difference between the hours of insolation and peak times requires an agreement for the sale of energy to the power system. The next section includes some results on this topic.

For Alternative A, an optimum combination of components, e.g., for 0.900 m^3/s, for diesel price of $1.10, and design flow rate of 0.740 m^3/s, include therefore the hydro power plant operating at 252 kW and diesel gen set with a capacity of 150 kW, with sale of surplus power to the grid. The energy price reaches $0.118 per kW·h sold, with 97% penetration of renewables and only 387 hours of operation of the diesel system.

The difference between the value of installed hydro power and value provided by Homer is due to the fact the software is identifying the flow that optimizes the system. The existing plant can be operated with a higher wattage or can be made a feasibility study for increasing the installed power.

For Alternative B, e.g., for 0.900 m^3/s, for diesel price of $1.10, and design flow rate of 0.720 m^3/s, the simulations indicate an optimum combination of components including the micro hydro power plant operating at 513 kW and selling surplus energy to the grid, without diesel gen set. The energy price reaches $0.388 per kW·h sold, with 97% penetration of renewables obviously equal to 100%. In this case it is easier to deal with the difference between the design flow and flow provided by Homer because the plant is still in the design phase.

These results and the above discussion indicate that the best choice is the Alternative B.

6. How Much Should Be the Cost of PV Modules to Increase Its Penetration?

PV modules do not appear among the optimal solutions

in the previous section, from **Figures 7** to **10**. A large collection area and the high cost per kW installed, are certainly the reasons for this absence.

It is reasonable to seek prices from which the PV modules present a higher penetration. These prices could be considered as a minimum price for achieving greater penetration. With these values in mind, it will be possible to devise strategies to promote the installation of modules in greater quantity. With these values in mind, negotiate better prices with suppliers, obtaining benefits by buying in large quantities will also be possible.

Figure 11 shows simulation results for the optimization space of the system of **Figure 1**, Alternative A, with sales capacity to the grid equal to 380 kW, with PV capital multiplier equal to 0.05, above, and 0.01, lower.

The PV capital cost multiplier is simply the price of

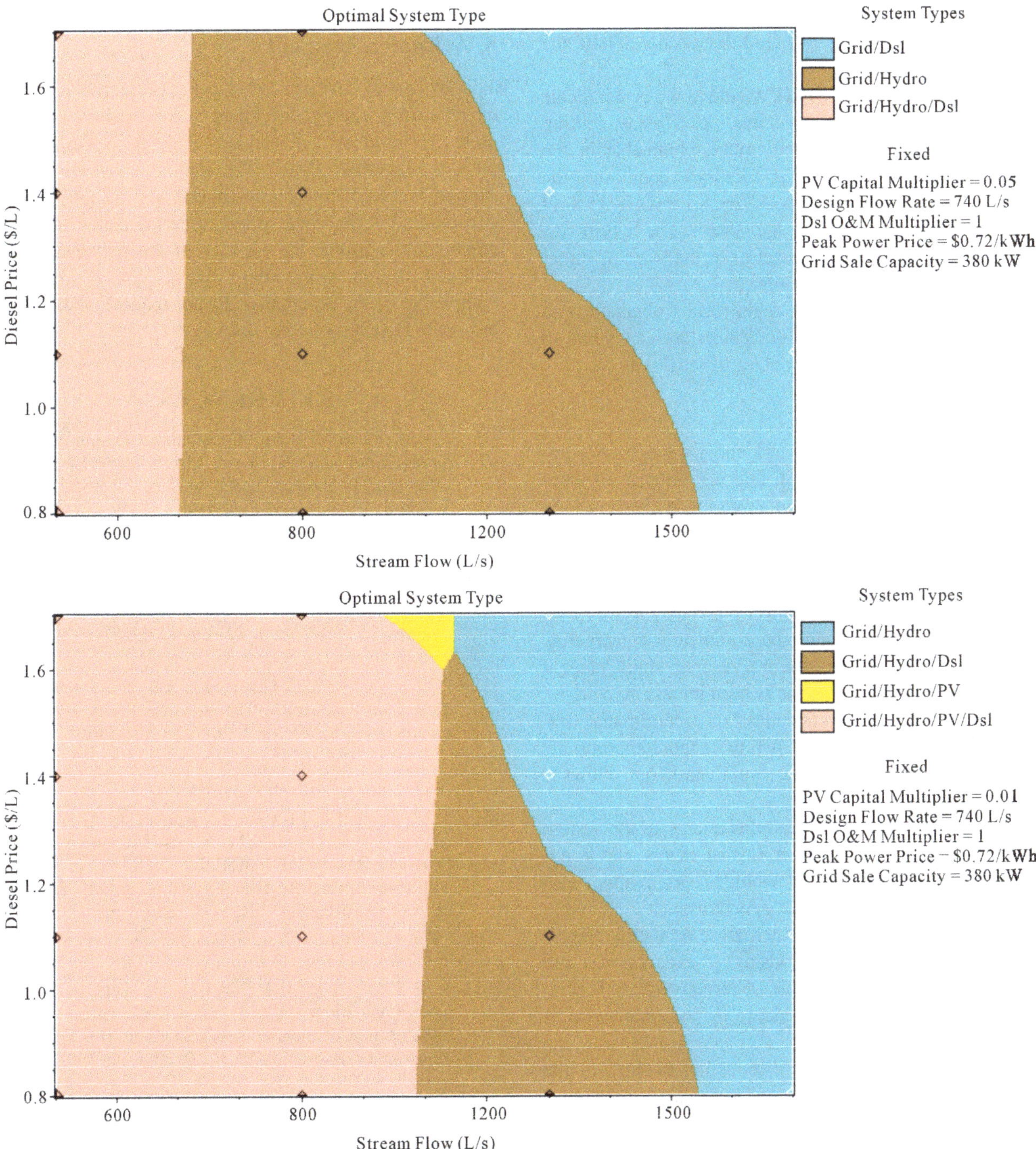

Figure 11. Results for the optimization space of the system of Figure 1, Alternative A, with sales capacity to the grid according to the installed PV power, with PV capital multiplier equal to 0.05 (above) and 0.01.

PV modules, multiplied by the indicated value. A multiplier 0.05 or 0.01 therefore corresponds to a hypothetical cost equal to 5% or 1% respectively of the current price. Simulations indicate the PV modules in optimization space only if the multiplier reaches 12%, corresponding to lower stream flow values. In such conditions, PV penetration becomes greater than zero.

In **Figure 11**, above, combinations that include PV modules appear only on the extreme left, where there is less water availability. Below, the area corresponding to combinations including PV modules occupies half the optimization space.

The simulations with 2000 kW sold to the grid indicate the operation of photovoltaic modules already starting with price at 40% of current prices, obtainable in the market for large acquisitions. Under these conditions, the company's demand during peak hours becomes 15% of the total energy produced by the system. The hybrid system becomes a supplier of energy to the system, rather than an alternative to peak hours.

Reference [17] shows the main results obtained with Homer, with input and with the spaces corresponding to the different values of the variables of sensitivity.

7. Final Remarks

This paper presented a study to identify the optimal solution for a recycled paper mill that also intends to work during the period from 19 to 22 hours without relying solely on the power supply provided by the grid.

Alternatives considered for power generation were two micro hydro power plants, a diesel generator set and PV modules. One of the hydro power plants requires a restoration while the other must be completely constructed. PV modules were considered in a separate study, since it did not appear in solutions for its high initial cost.

The alternative which includes this second plant, designated as B in the text, is shown to be more appropriate. This alternative has a higher initial cost, but allows a lower cost of energy.

The possibility of selling surplus energy to the system is the key to enabling a lower cost of energy and to enabling the installation of PV modules, occupying a vast area available for photovoltaic generation.

The results indicate that Alternative A, with a combination of hydro and diesel leads to sell excess power and a return $0.188 per kW·h sold. Alternative B, only with hydro, leads to the sale of energy by the final value of $0.388 per kW·h sold.

Simulations indicate that a reduction to 40% of the actual cost of photovoltaic modules makes possible a higher PV penetration. Larger reductions can obviously further increase this penetration.

The study indicated that the photovoltaic power generation was more interesting in this case as an investment rather than as an alternative generation at peak times. Its viability will depend on the bargaining agreement of sale of surplus power to the grid.

The photovoltaic power generation was included in this study by the availability of large area for installation of the modules. The study should be extended, considering different values of installed power and different amounts of energy sold to the grid, to determine the optimal investment to be made.

8. Acknowledgements

This work was developed as a part of research activities on renewable energy developed at the Instituto de Pesquisas Hidráulicas, Universidade Federal do Rio Grande do Sul, in southern Brazil. The authors acknowledge the support received from the institution and responsiveness encountered during data collection at the paper mill. For commercial reasons, the identity of the recycled paper mill and location of the micro hydroelectric power plant component of the Alternative B, per request of the owners of the factory, were not disclosed.

REFERENCES

[1] J. Vetterli and M. Benz, "Cost-Optimal Design of an Ice Storage Cooling System Using Mixed Integer Linear Programming Techniques under Various Electricity Tariff Schemes," *Energy and Buildings*, Vol. 49, No. 1, 2012, pp. 226-234.

[2] A. F. Orlando, M. P. Málaga and M. M. Huamani, "Methodology for Generating Electric Load Profiles for Sizing an Electric Energy Generation System," *Energy and Buildings*, Vol. 52, No. 1, 2012, pp. 161-167.

[3] P. E. S. Santos, R. C. Leme and L. Galvão, "On the Electrical Two Part Tariff—The Brazilian Perspective," *Energy Policy*, Vol. 40, No. 1, 2012, pp. 123-130.

[4] Google Maps, 2012. maps.google.com

[5] S. M. Shaahid, I. El-Amin, S. Rehman, A. Al-Shehri, F. Ahmad, J. Bakashwain and L. M. Al-Hadhrami, "Techno-Economic Potential of Retrofitting Diesel Power Systems with Hybrid Wind PV Diesel Systems for Off-Grid Electrification of Remote Villages of Saudi Arabia," *International Journal of Green Energy*, Vol. 7, No. 6, 2010, pp. 632-646.

[6] S. M. Shaahid and M. A. Elhadidy, "Prospects of Autonomous/Stand-Alone Hybrid (PV Diesel Battery) Power Systems in Commercial Applications in Hot Regions," *Renewable Energy*, Vol. 29, No. 2, 2004, pp. 165-177.

[7] S. M. Shaahid, I. El-Amin, A. Rehman A. Al-Shehri, J. Bakashwain and F. Ahmad, "Potential of Autonomous/Off-Grid Hybrid Wind Diesel Power System for Electrification of a Remote Settlement in Saudi Arabia," *Wind Engineering*, Vol. 28, No. 5, 2004, pp. 621-627.

[8] M. A. Elhadidy and S. M. Shaahid, "Decentralized/Stand-Alone Hybrid Wind-Diesel Power Systems to Meet Residential Loads of Hot Coastal Regions," *Energy Conversion and Management*, Vol. 46, No. 15-16, 2005, pp. 2501-2513.

[9] S. M. Shaahid and I. El-Amin, "Techno Economic Evaluation of Off-Grid Hybrid PV Diesel Battery Power Systems for Rural Electrification in Saudi Arabia—A Way forward for Sustainable Development," *Renewable and Sustainable Energy Reviews*, Vol. 13, No. 3, 2009, pp. 625-633.

[10] S. M. Shaahid and M. A. Elhadidy, "Economic Analysis of Hybrid PV Diesel Battery Power Systems for Residential Loads in Hot Regions—A Step to Clean Future," *International Renewable and Sustainable Energy Reviews Journal*, Vol. 12, No. 2, 2008, pp. 488-503.

[11] HOMER, "The Micropower Opyimization Model, Homer Energy," Version 2.68 Beta, 2009. www.homerenergy.com

[12] A. Beluco, P. K. Souza and A. Krenzinger, "A Dimensionless Índex Evaluating the Time Complementarity between Hydraulic and Solar Energies," *Renewable Energy*, Vol. 33, No. 10, 2008, pp. 2157-2165.

[13] A. Beluco, P. K. Souza and A. Krenzinger, "A Method to Evaluate the Effect of Complementarity in Time between Hydro and Solar Energy on the Performance of Hybrid Hydro PV Generating Plants," *Renewable Energy*, Vol. 45, No. 1, 2012, pp. 24-30.

[14] A. Beluco, P. K. Souza and A. Krenzinger, "Influence of Different Degrees of Complementarity between Hydro and Solar Energy on the Performance of Hybrid Hydro PV Generating Plants," Energy and Power Engineering, 2013.

[15] T. W. Lambert, P. Gilman and P. D. Lilienthal, "Micropower System Modeling with Homer," In: F. A. Farret and M. G. Simões, Eds., *Integration of Alternative Sources of Energy*, John Wiley & Sons, Hoboken, 2005, pp. 379-418.

[16] P. D. Lilienthal, T. W. Lambert and P. Gilman, "Computer Modeling of Renewable Power Systems," In: C. J. Cleveland, Ed., *Encyclopedia of Energy*, Elsevier, Oxford, 2004, pp. 633-647.

[17] A. Beluco, C. P. Colvara, L. E. Teixeira and A. Beluco "Simulation Results with Homer on Power Generation during Peak Hours with a Hydro PV Diesel Hybrid Energy System in a Recycled Paper Mill," Internal Report, UFRGS, IPH, 2012. galileu.iph.ufrgs.br/beluco/docs/homer-2570018.pdf

22

Are there Monthly Variations in Water Quality in the Amman, Zarqa and Balqa Regions, Jordan?

Khaled A. Alqadi[*], Lalit Kumar
Ecosystem Management, School of Environmental and Rural Science, Faculty of Arts and Sciences, University of New England, Armidale, NSW 2351, Australia

ABSTRACT

This study investigated the monthly variation of water quality in the Amman-Zarqa and Balqa regions in Jordan in terms of pH, ammonium, nitrate and conductivity. During 2004 there was no monthly variation in water quality for most of the tested parameters. All readings were above the accepted range except for pH, indicating that land use does have an impact on water quality irrespective of urban, industrial or agricultural usage. The water quality remained for the most part below the maximum levels for drinking standards in Jordan, but these standards are often below the WHO recommendations. The pH was found to fluctuate through the year. Nitrate levels were highly seasonal in irrigated lands but remained stable over basin covered by other land uses. Ammonium levels were high in areas of urbanisation and intensive animal husbandry as a consequence of effluent infiltration, peaking during the wet season due to increased infiltration. These results indicate that, over an annual cycle, the variation in water quality remains constant; however the continued drawdown of the aquifer system will inevitably lead to deterioration in the parameters investigated.

Keywords: Water Quality; pH; Ammonium; Conductivity; Nitrate

1. Introduction

The water crisis in Jordan is a major geopolitical issue that threatens the future and stability of the whole country. The Amman, Zarqa and Balqa regions supply irrigation water to an estimated 33,000 Ha, however, most of demand is for domestic and industrial water with the basin supplying over half of the population of Jordan with water [3]. Agricultural production in the basin is dominated by grazing, with only limited olive and fruit trees and the production of cereals near areas of permanent water.

Ground water quality has a major impact on human welfare and affects all human activity [7]. Understanding changes in ground water quality allows for effective management of water resources in the face of increasing pressures from urbanization, agricultural and industrial development [15]. However, Jordan is faced with a lack of funding to maintain monitoring which has resulted in fragmented data sets [13]. The utilisation of aquifers has led to severe lowering of the ground water table and this has changed the ground water chemistry [10]. The Amman, Zarqa and Balqa regions is renewable and draws water from areas of urban and industrial development and landfill sites and therefore is at significant risk of waste water infiltration pollution [1].

Data from 2002 indicates that there was an unsustainable drawdown of ground water from the Amman-Zarqa Basin. In 2002 a total of 84 mcm year^{-1} was withdrawn from the aquifer for municipal supply, while 54 mcm year^{-1} was taken to supply the agricultural sector. This represents an excess of 72 mcm year^{-1} of water above the estimated safe yield of 65 mcm year^{-1}. This overdrawing from the 772 officially registered wells has led to a fall in the ground water table and declines in water quality [13]. Salameh [19] argued that the current drawdown of the aquifers has led to permanent damage of the hydrological system, and Al-Mahamid [4] argued that at current extraction rates areas in the middle of the basin will be completely dry in the next few decades. These findings were supported by Dottrige and Jaber [8] who argued that at current levels of extraction many of the aquifers dependent upon for urban and industrial supply will be dry by the middle of the twenty first century. The aim of this paper is to examine the water quality over an annual cycle in the Amman, Zarqa and Balqa regions under differing land use.

2. Hydrogeology of the Amman-Zarqa Basin

The geology of the Amman-Zarqa Basin is primarily

[*]Corresponding author.

sedimentary with ages ranging from the lower Cretaceous to the present (**Table 1**). The major aquifers in the Amman-Zarqa Basin from which water is drawn are considered to be hydraulically connected, but maybe separated in regions by geological layers which act as aquicludes [9].

There are three aquifer systems in the Amman-Zarqa Basin. There is evidence that water moves between the aquifers along the Zarqa fault system [4]. The upper aquifer is contained within the linked Campanian Amman (B2) and Turonian Wadi Sir (A7) limestone formations and the neighbouring basalt strata (V). The middle aquifer is contained within the Cenomanian Shue mar (A4) limestone formation. The lower aquifer is contained within the Albian-Aptian Kurnub (K) sandstone formation. **Table 2** provides an overview of the hydrology and hydrochemistry of the three aquifers. The long term effects of drawdown has resulted in declines in the water table in all aquifers contained within the basin and this has led to an increase in the level of dissolved chemicals that are naturally occurring as a consequence of the surrounding geological formations of each aquifer [4]. As a consequence of the geological formations in which the aquifers are contained, calcium and magnesium are the dominant cations, while bicarbonate is the dominant anion [6].

The rainfall over the Amman-Zarqa Basin is highly seasonal. Peak rainfall occurs during late autumn to early spring with summer receiving negligible to no rainfall reflected in the annual runoff (**Figure 1**). The annual depth of rainfall over the basin is variable, ranging from 50 mm in the east to 1000 mm in the west; however the total annual rainfall for the region has declined over the last three decades by 25 - 33 mm [4]. This has an impact on the total runoff potential and the water available for recharging of the aquifers within the basin. During the period of rainfall the B2/A7 aquifer is subjected to infiltration which then enters the connected A4 aquifer through fracture zones as the water flows north easterly down the Amman-Zarqa sincline [6].

Table 1. The geology and hydrogeology classification of the Amman-Zarqa Basin [18].

Age Hydro	Formation	Members	Primary Rock types	Thickness (m)	Permeability (ms^{-1})	Code
Holocene	Wadi Fill		Soil, sand, gravel	10-40	$2.4x10^{-7}$	H
Pleistocene	Basalt		Basalt, clay	0-50	-	V
Maestrichtian	Muwaqqar		Chalk, marl, limestone	60-70	-	B3
Campanian B2a	Amman	Limestone unit	Chert, limestone with	80-120	$1x10^{-5}$-$3x10^{-4}$	
		Phosphate unit	Phosphate			
B2b						
Santonian	Wadi Ghudran		Chalk, marl, limestone	15-20	-	B1
Turonian	Wadi Sir		Hard crystalline limestone	90-110	$1x10^{-7}$-$1x10^{-4}$	A7
			dolomite and chert			
Cenomanian	Shue mar		hard dense limestone dolomite	40-60	$8.1x10^{-7}$-$7.6x10^{-4}$	A4
	Fuheis		Marl, limestone	60-80	$5.3x10^{-7}$-$1.7x10^{-5}$	A4
	Na'ur		Marl, limestone	150-220	$2.x10^{-8}$-$3.1x10^{-5}$	
A1-2						
Albian-Aptian	Kurnub		Silt, shale, sandstone	+300	$6.9.x10^{-3}$-$5.2x10^{-2}$	

Table 2. The average hydrological and hydrochemical data for the three major aquifers in the Amman-Zarqa Basin for the period 1995-2003 [4].

Parameters	Upper Aquifer	Middle Aquifer	Lower Aquifer
Hydro code	B2/A7, V	A4	K
Transmissivity (m^2d^{-1})	0.38 to 38	9.6 to 117	1 to 146
Number of springs	4	54	35
Spring Discharge (10^6m^3)	4	9	1.5
Confinement	Phraetic/ Confined	Confined	Phraetic/ Confined
Annual abstraction (10^6m^3)	113.1	12.6	7.5
% of Water supply	81.6	9.1	5.4
Number of wells	456	63	81
Est Annual Ave change in water table (m)	-1 to -1.5	-0.5 to -9	-0.5 to -0.8
Ave Chemical Composition			
Ca (mgL^{-1}) (Mean/Max-Min)	92.7/24.3-192.5	72.3/51.7-94.6	66.8/41.7-83.4
Mg (mgL^{-1}) (Mean/Max-Min)	49.2/7.5-129	34.9/22.5-46.1	32.4/23.1-44.5
Na (mgL^{-1}) (Mean/Max-Min)	163.4/25.5-422.1	44.1/19.3-79.6	59.2/20.5-117.1
K (mgL^{-1}) (Mean/Max-Min)	8.2/2-15.6	2.8/0-7.4	4.7/1.6-9.4
HCO_3 (mgL^{-1}) (Mean/Max-Min)	211.5/66.5-366	290.3/248.9-372	257.1/133-321.5
SO_4 (mgL^{-1}) (Mean/Max-Min)	108.5/17.3-495.4	33.5/15.4-65.3	67.2/24.5-204.8
Cl (mgL^{-1}) (Mean/Max-Min)	338.4/43.3-958.5	89.5/27.3-163.3	92.4/42.6-204.8
NO_3 (mgL^{-1}) (Mean/Max-Min)	46.8/3.5-107.2	33.0/7.7-66.9	26.0/5.8-80
EC (μScm^{-1})(Mean/Max-Min)	1679/500-3680	830.5/536-1176	818.1/650-1211
pH (Mean/Max-Min)	7.4/6.2-8.5	7.7/7.3-7.9	7.5/6.4-8.0

3. Description of Study Area

The Amman, Zarqa and Balqa regions cover an area of 1939 km^2 and are located in north eastern Jordan uplands and are between 500 – 1000m in elevation with an annual precipitation of 150 to 600 mm $year^{-1}$ [13,17]. **Figure 2** illustrates the study area and location of the study area within Jordan. The basin contains significant population and industrial production centres. Agriculture is primarily restricted to plains of the water courses with rangelands dominating the remainder. The sporadic distribution of population and industry, as well as the restriction of agriculture to the water courses has been that the spatial quality of the ground water is highly variable with aquifer systems [5].

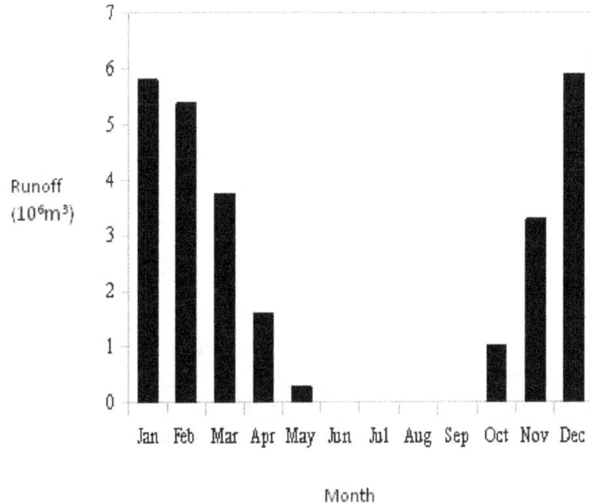

Figure 1. The monthly long-term rainfall average of the Amman-Zarqa Basin for the period 1970-2002.

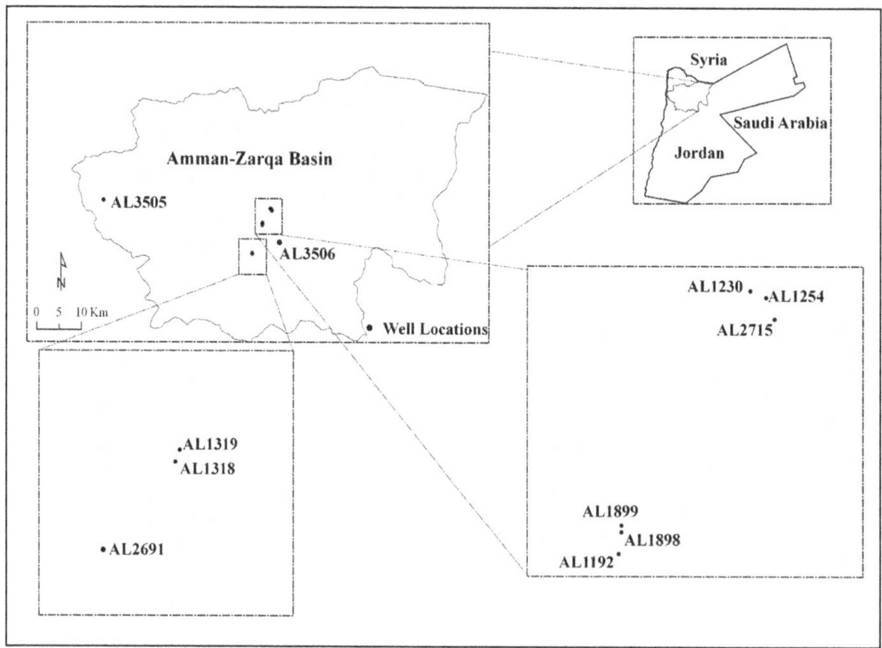

Figure 2. The location of the Amman, Zarqa and Balqa regions and the location of the wells investigated in this study.

4. Methods

The use of geographic information systems has proven to be an effective means of investigating spatial changes in water quality [7,15]. The spatial data support system, Arc View GIS, provides the tools to allow the seasonal mapping of temporal changes in water quality parameters [16]. The use of mapping systems in conjunction with aerial images allows for a greater understanding of the potential impacts of land use on water quality.

Historical data for wells in the Amman, Zarqa and Balqa regions were obtained from the Jordan Water Authority. From these historical records on the water quality, 11 wells from the three regions representing each of the three major aquifers, and with significantly detailed monthly data for 2004, were selected for investigation (**Table 3**). From the records of these 11 wells the monthly recorded conductivity, ammonium, nitrate and pH were graphed to determine if seasonal variation in water quality parameters could be determined. A map of the Amman-Zarqa Basin was digitised using Arc Map 9.3 and the location of the 11 wells plotted. Rainfall data for the area surrounding the wells in 2004 was also mapped. The depth of the wells was also investigated to determine the geological formation that contained and surrounded the wells, and this enabled an understanding of the influence of the surrounding strata on the quality of the water contained in the aquifer.

5. Results

The wells represent areas of five primary land uses. The well AL1230 is located in an area of heavy industrial use on the edge of a water course. The three wells AL1192, AL1898 and AL1899 are surrounded by residential areas; however there is a chicken farm nearby. The wells AL1254 and AL2715 are located on a small irrigation plain with a central water course surrounded by arid lands, while AL3505 is located in the nearby arid land. The three wells AL1318, AL1319 and AL2691 are all in light industrial with neighboring residential area, and are located near a water course, with AL1318 and AL1319 located in and around waste water treatment works. **Figure 3** illustrates the aerial view of the wells and their surrounds.

This investigation also indicates that in areas of low development represented by AL 3505 and AL3506, there was a low level of nitrate contamination. These two wells draw water from the middle and lower aquifer systems

Table 3. Identification, location and land use of the wells and surrounding areas used in this study.

Well ID	Well Name	Coordinates Palestine North(km)	East(km)	Land Use	Aquifer depth (m)	Altitude (masl)	Hydro Code
AL1192	New Municipality	166.535	254.830	Residential	281	568	A4
AL1230	Hashiniya No. 3	170.130	256.690	Heavy Industrial	102	540	B2/A7
AL1254	Hashiniya No. 5	170.040	256.910	Irr. Agricultural	106	540	ALL
AL1318	Awajan 22	160.340	252.468	Med. Industrial	-	587	B2/A7
AL1319	Awajan 21	160.410	252.494	Med. Industrial	148	583	B2/A7
AL1898	Zerqa Well No. 3	166.830	254.870	Residential	246	587	A4
AL1899	Zerqa Well No. 3	166.820	254.870	Residential	113	586	B2/A7
AL2691	Awajan 23	160.000	251.980	Med. Industrial	151	600	B2/A7
AL2715	Hashiniya No. 2	169.744	257.027	Irr. Agricultural	128	540	?B2/A7
AL3505	Dafali No. 3	172.011	218.905	Grazing	332	615	A4
AL3506	Supply No.4	162.750	256.900	Irr. Agricultural	500+	491	K

indicating that there is a natural tendency for low nitrate levels from this supply in underdeveloped areas. One well, AL1254 had a high nitrate indicting infiltration and may represent the use of nitrate fertilisers on the irrigated plains on which the well was located. This infiltration diffused through the aquifers to which it is connected and the well nitrate level returned to that of all the study area for the upper and middle systems. The results also indicate that, while there is a peak in the pH of the wells in late spring and summer, however, there is little change throughout the year, with all wells having a pH between 7 to 8. The ammonium level was only significantly raised in the residential areas, with all industrial and agricultural wells demonstrating an annual low ammonium level. The ammonium level in the residential areas peaked during the onset of the rains, indicating that there is a possible flushing of sewerage down into the water table from the individual dwellings septic systems and the nearby chicken farm. The peak in conductivity during the period of peak rainfall indicates that the infiltration of water through the geological strata is carrying dissolved salts into the aquifers.

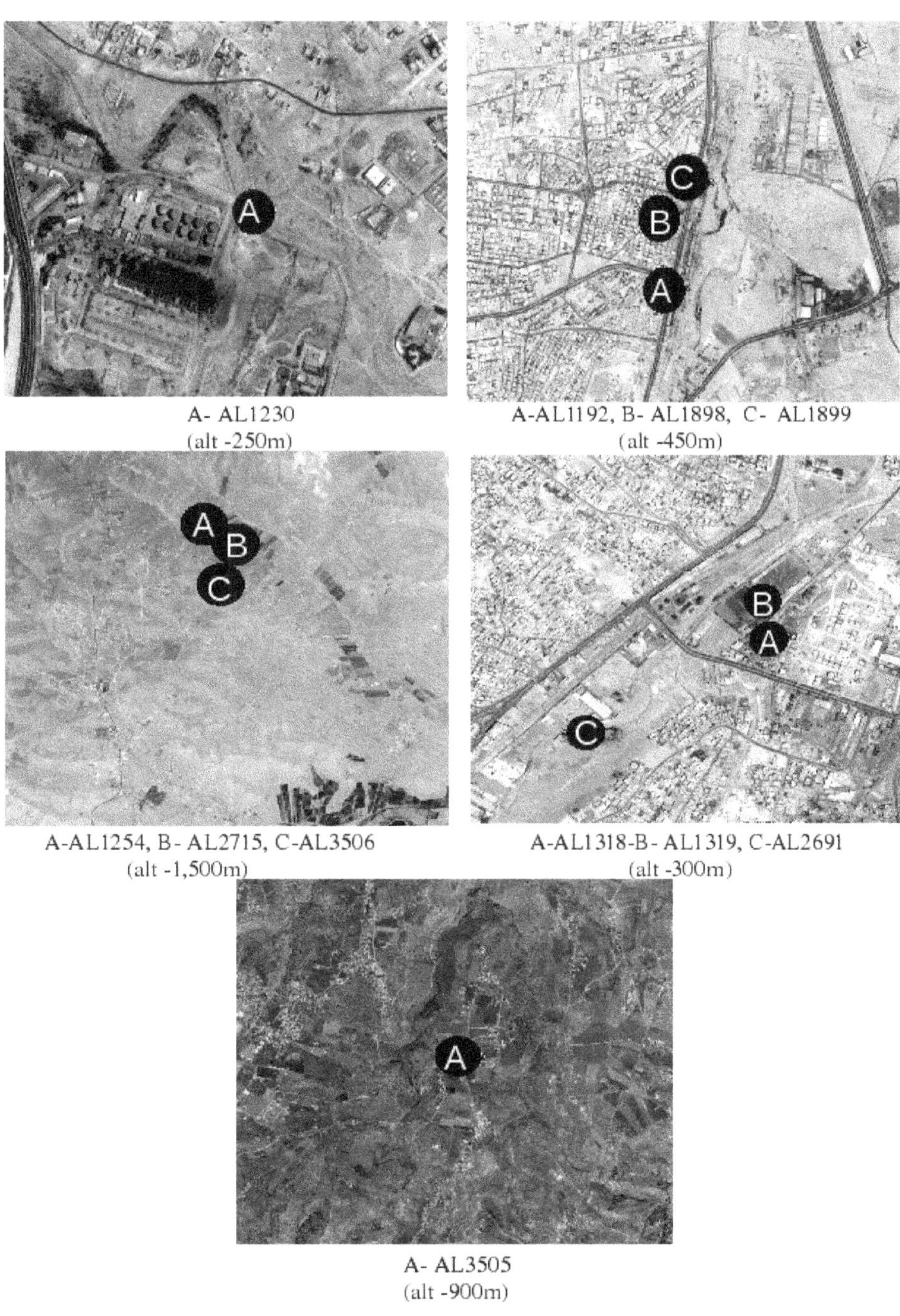

Figure 3. Land use images showing surrounding areas of each well (from Google Earth, 2004).

5.1. pH

At present there is no considered risk to consumers from water pH levels, and therefore, no safe health guidelines [15]. However, the pH has significant impact on the operational water quality parameters with the World Health Organisation (WHO, 2004) recommending an optimum range of 6.5 - 9.5. The Jordan Water Standard indicates that the permissible pH range is 6.5 - 85 for drinking water [4,11]. The maximum and minimum pH for the 11 wells in the Amman, Zarqa and Balqa regions investigated ranged from 6.84 to 8.06, respectively. This study determined that no well fell outside the operation water parameters of the WHO. **Figure 4** illustrates the pH on a monthly basis for each of the wells investigated. The results indicate that there was a variation in the pH of the wells throughout the year of ~1.2 to ~0.63. **Table 4** illustrates the maximum and minimum pH for each well and the month in which that level was reached. There is clear indication that water pH reaches a maximum during late spring to summer in all wells and with minimums reached in autumn, winter and early spring depending on the well.

Table 4. The pH for the maximal and minimal months and the percentage change over the period for the 11 wells.

Well ID	Maximum		Minimum		Variation
	pH	Month	pH	Month	pH
AL1192	8.01	July	6.84	April	1.17
AL1230	7.71	July	6.89	October	0.82
AL1254	8.06	July	7.19	March	0.87
AL1318	7.85	July	7.08	April	0.77
AL1319	7.82	July	6.96	January	0.86
AL1898	7.84	August	7.11	March	0.73
AL1899	7.82	July	6.97	March	0.85
AL2691	7.76	August	7.07	February	0.69
AL2715	7.78	May	7.10	October	0.68
AL3505	7.79	June	7.16	April	0.63
AL3506	7.99	May	7.33	April	0.66

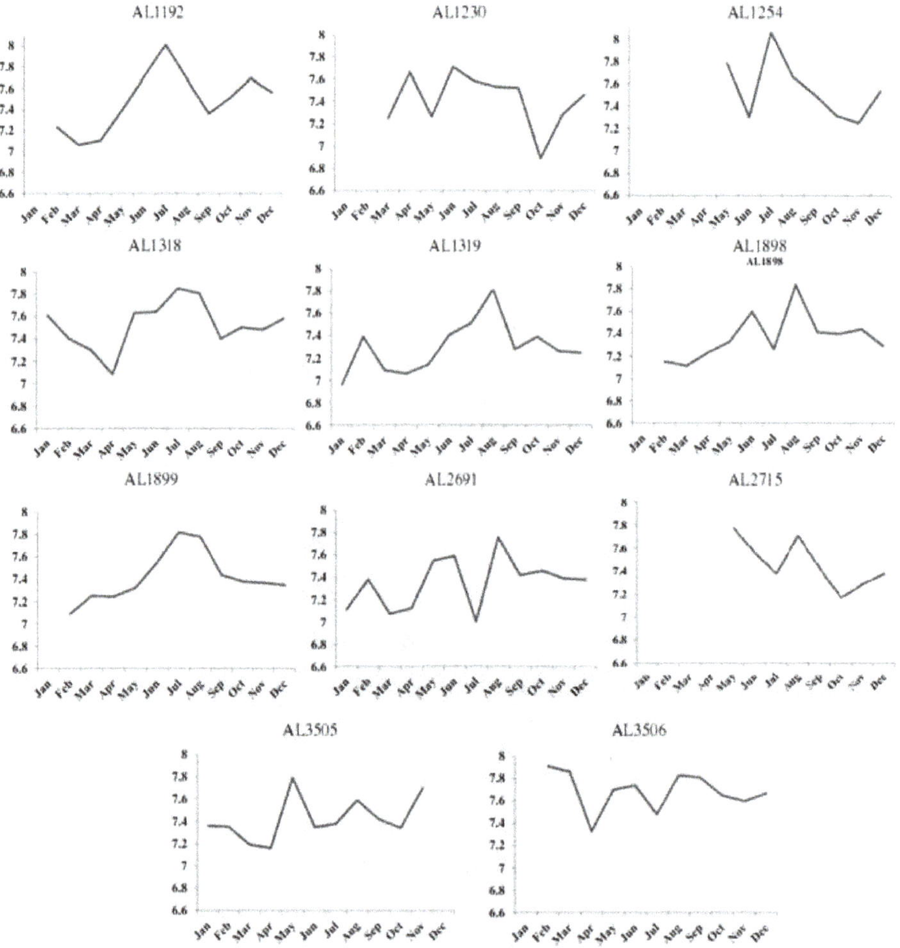

Figure 4. The pH records for each well in 2004.

5.2. Nitrate

High nitrate levels can have significant negative health consequences [11]. The WHO guidelines indicate a maximum nitrate level of 50 mgL^{-1} for drinking water; however while the authorities in Jordan recognize a permissible target level of 50 mgL^{-1}, the national standards allow for a concentration of up to 70 mgL^{-1}, [4]. The maximum and minimum nitrate levels for the 11 wells in the Amman, Zarqa and Balqa regions investigated ranged from <0.16 mgL^{-1} to 75.88 mgL^{-1}, respectively. The results of the investigation into the 11 wells found that only two of the wells (AL3505, AL3506) did not have year round dangerous nitrate levels making the water fit for consumption in WHO terms, with one other well (AL1318) being safe for only one month. However, the Jordan maximum permissible level for nitrate was only exceeded significantly in areas of irrigation (AL1254). Notwithstanding, the urban areas also contain significantly high levels of nitrates in the immediate wells. **Figure 5** illustrates the nitrate level on a monthly basis for each of the wells investigated. The results indicate that there was a variation in the nitrate within each well throughout the year of ~ 6 mgL^{-1} to ~ 50 mgL^{-1}. **Table 5** highlights the nitrate variation throughout the year for each of the wells investigated. These results indicate significant variation in nitrate concentration in each well on a monthly basis with no specific seasonal variation evidenced.

Table 5. The Nitrate concentration (mgL^{-1}) for the maximal and minimal months and the percentage change over the period for each of the 11 wells.

Well ID	Maximum		Minimum		Variation
	mgL^{-1}	Month	mgL^{-1}	Month	mgL^{-1}
AL1192	74.60	July	62.00	January	12.60
AL1230	61.33	July	52.30	October	9.03
AL1254	104.65	September	55.93	July	48.72
AL1318	74.50	January	38.94	August	34.56
AL1319	68.00	January	52.75	February	15.25
AL1898	75.88	April	66.75	January	9.13
AL1899	73.05	April	54.43	August	18.62
AL2691	71.05	November	50.58	February	20.47
AL2715	63.32	May(4th)	53.01	May(31th)	10.31
AL3505	25.30	January	0.66	February	24.64
AL3506	6.53	May	<0.16	Feb., Mar., Sept.	6.37

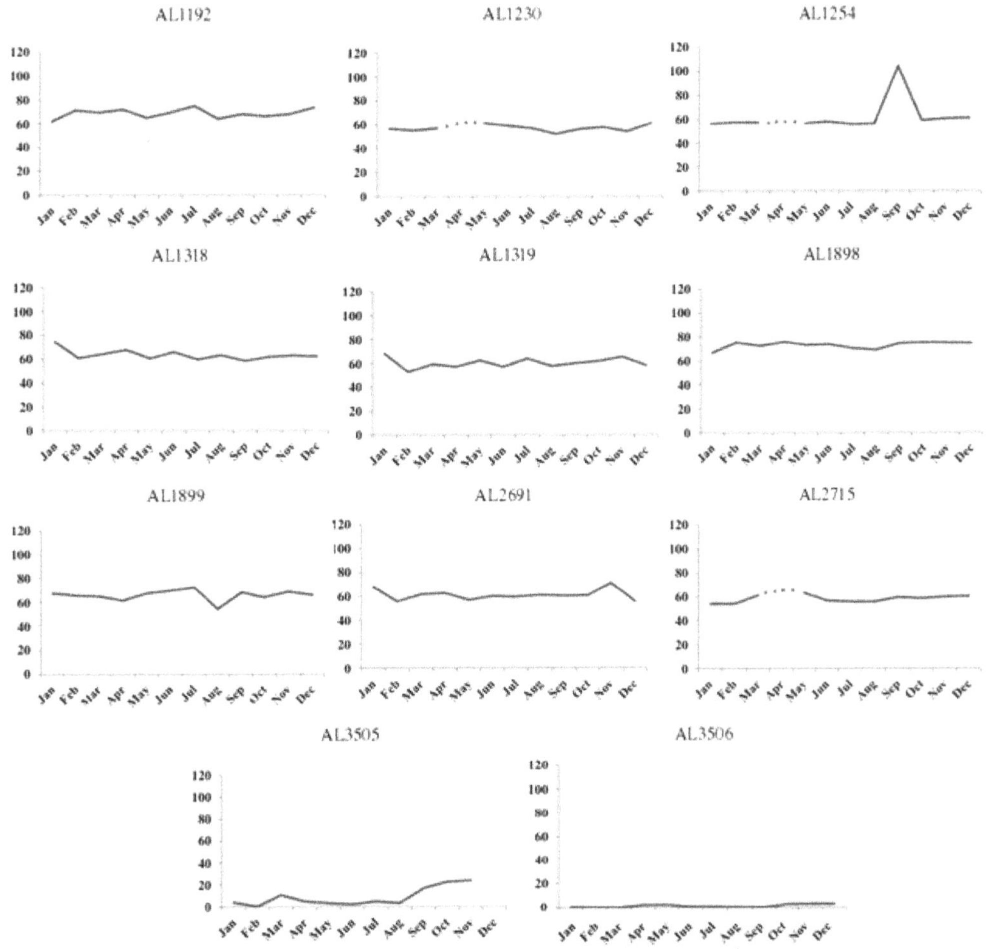

Figure 5. The Nitrate (mgL^{-1}) records for each well in 2004.

5.3. Ammonium

The maximum and minimum ammonium (NH_4) levels for the 11 wells in the Amman, Zarqa and Balqa regions investigated ranged from < 0.05 mgL^{-1} to 4.2 mgL^{-1} respectively. **Figure 6** illustrates the ammonium levels on a monthly basis for each of the wells investigated. The results indicate that there was a variation in the ammonium level within any one well throughout the year of ~0.05 mgL^{-1} to 3.41 mgL^{-1}. **Table 6** highlights the ammonium variation throughout the year for each of the wells investigated. These results indicate that only three wells (AL1192, AL1898, and AL1899) showed significant variation in nitrate concentration with all the other wells having an ammonium level of below ~0.1 mgL^{-1} throughout the year. Ammonium levels tended to rise towards the end of each year with the first months showing the lowest concentrations; however, the three wells with significantly higher ammonium levels (AL1192, AL1898, AL1899) showed the reverse with higher concentration during the first three months of the year.

Table 6. The Ammonium (NH_4 mgL^{-1}) for the maximal and minimal months and the percentage change over the period.

Well ID	Maximum*		Minimum*		Variatio
	Conc.	Month	Conc.	Month	Conc.
AL1192	3.62	January	1.55	December	2.07
AL1230	<0.10	Oct., Nov., Dec.	<0.05	Jan., Mar., Apr.	~0.05
AL1254	<0.10	Oct., Nov., Dec.	<0.05	Jan.-May	~0.05
AL1318	<0.10	Oct., Nov., Dec.	<0.05	Feb.- May	~0.05
AL1319	<0.10	Oct., Nov., Dec.	<0.05	Mar., Apr., May	~0.05
AL1898	4.16	February	0.75	January	3.41
AL1899	2.45	May	1.01	April	1.44
AL2691	<0.10	Sept.- Dec	<0.05	Feb.- May	~0.05
AL2715	<0.10	Oct, Nov., Dec.	<0.05	Jan., Mar., Apr.	~0.05
AL3505	<0.10	Oct, Nov., Dec.	<0.05	Jan. -May	~0.05
AL3506	<0.10	Oct, Nov., Dec.	<0.05	Jan.,-May	~0.05

*Partial result with data from some months unavailable.

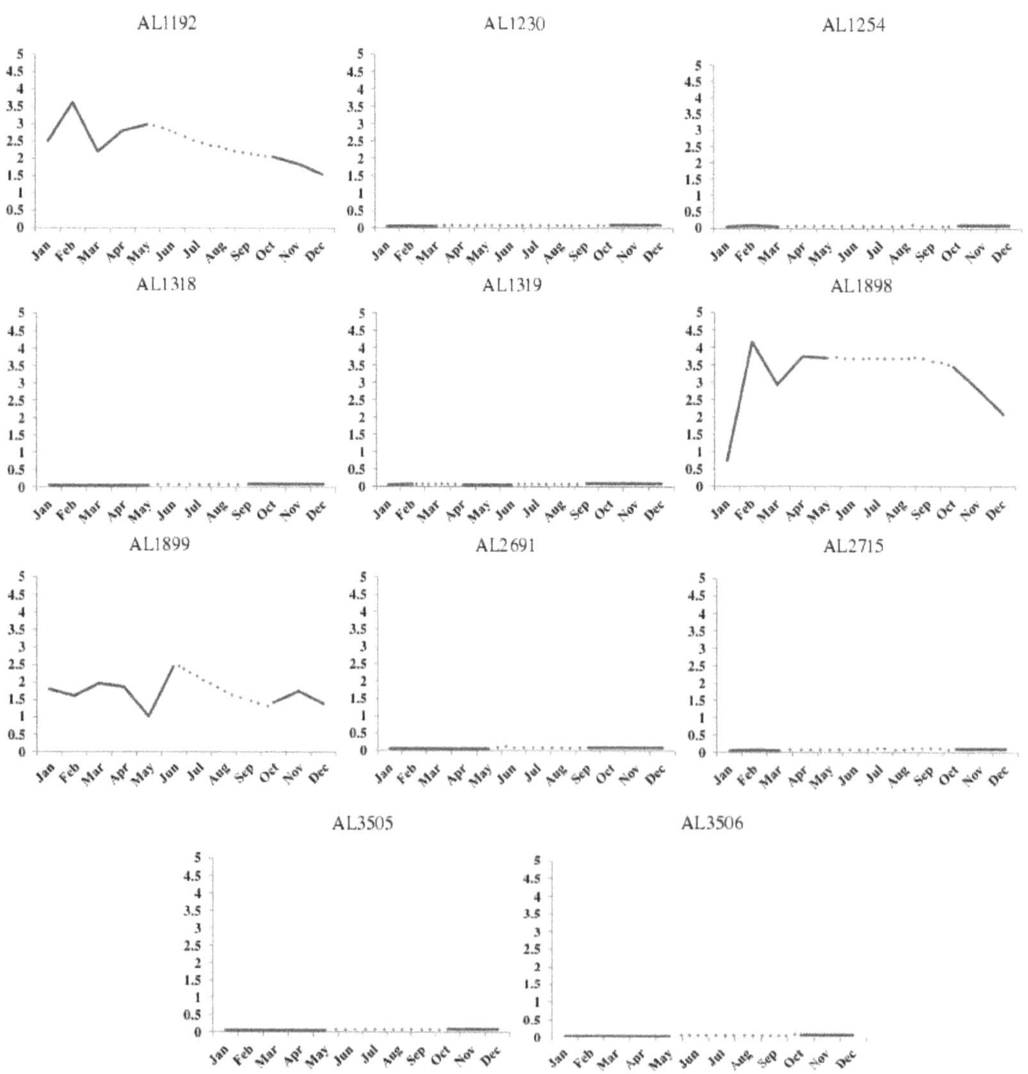

Figure 6. The ammonium (NH_4 mgL^{-1}) records for each well in 2004 with inferred data shown as a dotted line.

5.4. Conductivity

The maximum and minimum conductivity levels for the 11 wells in the Amman, Zarqa and Balqa regions investigated ranged from 790 μScm^{-1} to 3640 μScm^{-1}, respectively. **Figure 7** illustrates the conductivity level on a monthly basis for each of the wells investigated. The results indicate that there was a variation in the conductivity within any one well throughout the year of 276 μScm^{-1} to 1935 μScm^{-1}. **Table 7** highlights the monthly conductivity throughout the year for each of the wells investigated. These results indicate the early months of the year have the lowest conductivity; however this rises rapidly during May through June. These results also demonstrate that the conductivity within a well can change significant within any month with AL1192 showing a 145% rise in conductivity between the 4th April and 27th May.

Table 7. The conductivity for the maximal and minimal months and the percentage change over the period for each of the 11 wells.

Well ID	Maximum		Minimum		Variation
	μScm^{-1}	Month	μScm^{-1}	Month	μScm^{-1}
AL1192	3110	May	1175	April	1935
AL1230	3510	June	2800	Jan.-Feb.	710
AL1254	3630	June	2890	January	740
AL1318*	2290	March	1731	February	559
AL1319*	2140	March	1656	February	484
AL1898	3310	May	2960	November	350
AL1899	3370	May	2868	July	502
AL2691*	1886	March	1408	Feb. -Mar.	478
AL2715	3640	June	2910	January	730
AL3505*	1845	March	967	January	878
AL3506	1066	March	790	January	276

*Well subject to chlorination during some months.

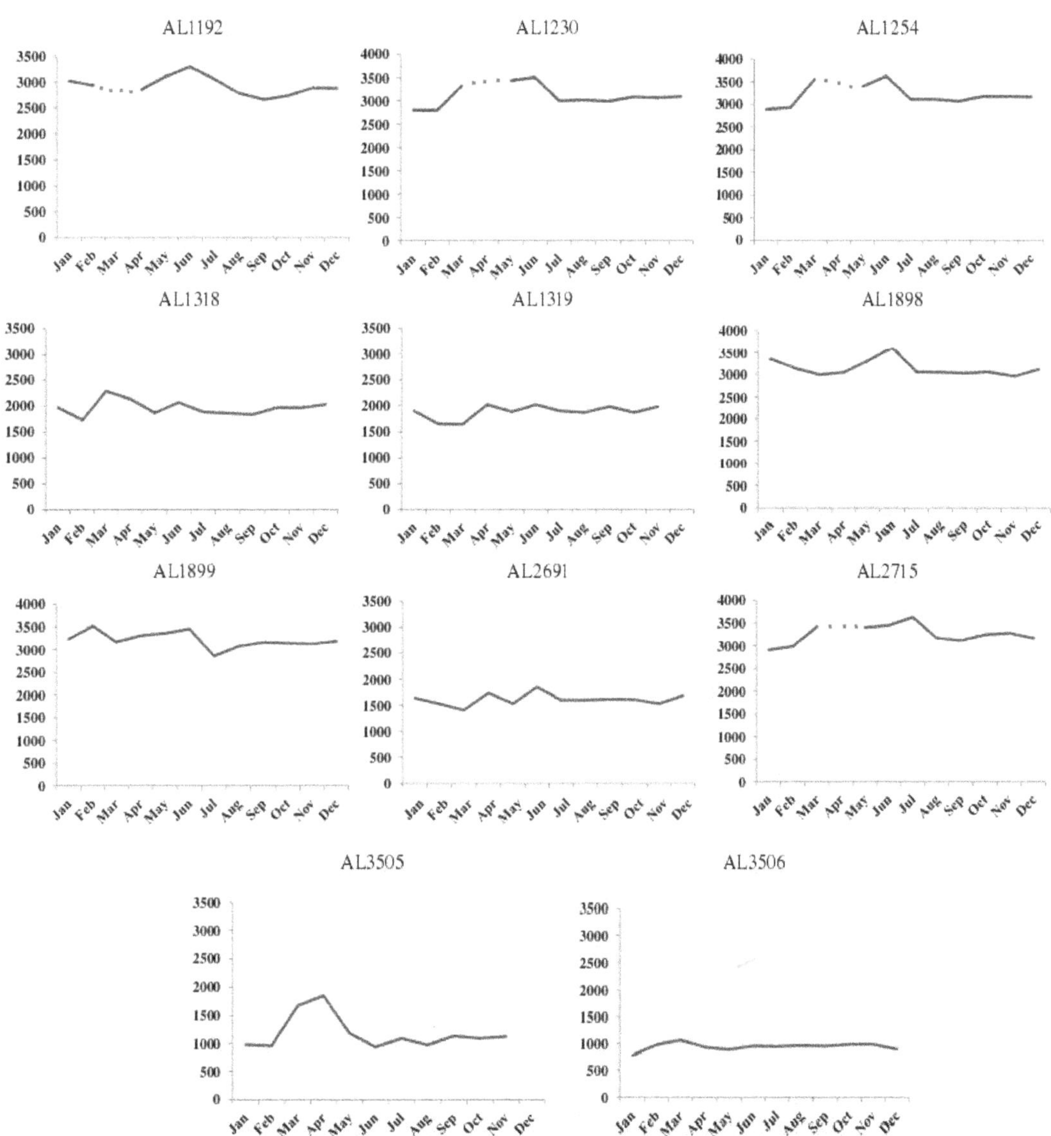

Figure 7. The conductivity (μScm^{-1}) records for each well in 2004 with inferred data shown as a dotted line.

6. Discussion

This study confirmed waste effluent can have a significant negative impact on the water quality of the aquifers in the basin. However, this study indicates that the problem may come from urban areas with low sanitation infrastructure. Also, areas under irrigation can have a marked, localised and temporary effect on water quality. Infiltration is areas of underdevelopment have significant increase in the conductivity, particularly as the first flush from the rains moves through the profile. Notwithstanding all aquifers were determined to comply with the guidelines for maximum permissible standards for drinking water, except for the A2/B7 aquifer from which water is drawn has regions of conductivity which is outside these standards. It must be noted that the maximum permissible level for most water contaminants in Jordan often exceed the WHO recommendations [4].

During this study the pH of all wells remained within the permissible standards of the Jordanian water authority [4]. This study concluded that irrigation can have a significant impact on the nitrate level of the aquifer with nitrate levels shown to significantly exceed the maximal permissible levels [4]. Areas of high urban density also had higher nitrate levels which seasonally peaked at or just above the Jordan Governmental standards for the maximum permissible level, while lands which are grazed have the lowest nitrates [4]. The high ammonium content of AL1192, AL1898, and AL1899 is postulated to come from infiltration of the effluent from the chicken farm, indicating that ammonium contamination is highly dependent on land use. The conductivity was found to be lowest in the rangelands with irrigated, urban, and industrial lands showing the highest readings. This reflects the level of water use in each of these areas, indicating that drawdown has a negative impact on the conductivity of the water. This study confirms the results obtained in neighboring aquifer systems in Jordan that indicate that urbanization and industrialization, coupled with intensive agriculture, have a negative impact on water quality when compared to natural rangeland systems [2,10].

Many of the aquifer systems in Jordan are under threat from overexploitation and this unsustainable drawdown will lead to many systems becoming dry by 2030 to 2040 [8].There is a need for increased monitoring throughout the Amman, Zarqa and Balqa regions over a long term in order to capture a more reliable image of the declines in water quality and regulate the water extraction or face the loss of the aquifer system [4,13,18].

7. Conclusions

The aquifers within the Amman, Zarqa and Balqa regions are affected by irrigation, industrialisation and urbansiation. The increased drawdown of the aquifers has led to declines in the water quality over the long term. The monthly variations in water quality parameters are signifcantly affected by the rainfall which leads to infiltration of water that carries pollutants from anthropogenic sources and dissolved salts from the geological strata that the water moves through. The drawdown of the aqufers has led to a concentration of salts and other contaminants that would have previously been dispersed in a historical larger water volume.

REFERENCES

[1] M. Al-Farajat, I. Hamdan, K. Jaber, and S. H. Mohammed, "GIS Mapping of Ground Water Vulnerability against Pollution in Amman Using DRASTIC dex, "*Hydrogeologie und Umwelt,* Vol. 33, No. 9, 2005, pp. 1-19.

[2] A. Al-Hanbali and A. Kondoh, "Ground Water Vulnerability Assessment and Evaluation of Human Activity Impact (HAI) within the Dead Sea Groundwater Basin, Jordan," *Hydrogeology Journal,* Vol.16, 2008, pp. 499-510.

[3] E. K. Al-Karablieh, A. S. Jabarin and M. A. Tabieh, "Jordan Horticultural Export Competitiveness from Water Perspective," *Journal of Agricultural Science and Technology,* Vol. B1, 2011, pp. 964-974.

[4] J. Al-Mahamid, "Integration of Water Resources of the Upper Aquifer in Amman-Zarqa Basin Based on Mathematical Modeling and GIS, Jordan," *Freiberg Online Geology,* Vol. 12, 2005, pp. 7-223.

[5] Al-Mashagbah, R. Al-Adamat, and E. Salmeh, "The Use of Kriging Techniques with in GIS Environment to Investigate Groundwater Quality in the Amman-Zarqa Basin/Jordan," *Research Journal of Environmental and Earth Sciences,* Vol. 4, No 2, 2011, pp. 177-185.

[6] E. Al-Tarazi, J. A. Rajab, A. Al-Naqa and M. El-Waheidi, "Detecting Leachate Plumes and Groundwater Pollution at Ruseifa Municipal Landfill Utalising VLF-EM Method," *Journal of Applied Geophysics,* Vol. 65, 2008, pp. 121-131.

[7] P. Balakrishnan, A. Saleem and N. D, "Mallikarjan, Groundwater Quality Mapping Using Geographic Information System (GIS): A Case Study of Gulbarga City, Karnataka, India," *African Journal of Environmental Science and Technology,* Vol.5, No. 12, 2011, pp. 1069-1084.

[8] J. Dottridge and N. A. Jaber, "Groundwater Resources and Quality in Northeastern Jordan: Safe Yield and Sustainability," *Applied Geography,* Vol. 19, pp. 313-323.

[9] A. El-Naqa, N. Hammouri and M. Kuisi, "GIS- Based Evaluation of Groundwater Vulnerability in the Russeifa Area Jordan," *Revista Mexicana de Ciencias Geológicas,* Vol. 23, 1999, pp. 277-287.

[10] N. Hammouri and A. El-Naqa, "GIS Based HydroGeological Vulnerability Mapping of Ground Water Resources in Jerash Area- Jordan," *Geofisica Internacional,*

Vol. 47, No. 2, 2008, pp. 85-97.

[11] P. F. Hudak, "Regional Trends in Nitrate Content of Texas Groundwater," *Journal of Hydrology*, Vol. 228, 2000, pp. 37-47.

[12] Jordanian Institute of Standards and Metrology (JISM) *Drinking water standards No. (286/2001)*, 2001, Government of Jordan, Amman.

[13] Jordan Ministry of Water and Irrigation (JMWI). *Disi-Mudawarra to Amman Water Conveyance System: Environmental and Social Management Plan Part 2*. 2009, Government of Jordan, Amman.

[14] G. Jousma, *Guideline on: Groundwater Monitoring for General Reference Purposes.* International Working Group I, International Groundwater Resources Assessment Centre, Utracht, GP 2008-1, 2008.

[15] R. T. Mehrjardi, M. Z. Jahromi, S. Hahmodi and A. Heidari, "Spatial Distribution of Groundwater Quaity and Geostatistics (Case Study: Yazd-Ardakan Plain)," *World Applied Sciences Journal,* Vol. 4, No. 1, 2008, pp. 9-17.

[16] B. Nas and A. Berktay, "Groundwater Quality Mapping in Urban Groundwater Using GIS," *Environmental Monitoring and Assessment,* Vol. 160, 2010, pp. 215-227.

[17] O. Rimawi, "*Hydrochemistry and isotope hydrology of groundwater and surface water in the north-east of Mafraq, Dhuleil, Hallabat, Azraq basin,* PhD. Thesis," Techn. University, Muenchen, 1985, p. 240.

[18] E. Salameh, "Over-Exploitation of Groundwater Resources and Their Environmental and Socio-Economic Implications: The Case of Jordan," *Water International* Vol. 33, No. 1, 2008, pp. 55-68.

Variations in the Water Quality of an Urban River in Nigeria

F. A. Oginni
Department of Civil Engineering, College of Science, Engineering and Technology, Osun State University

ABSTRACT

Sango-Ota is the industrial nerve centre of Ogun State in Nigeria. River Atuara is an urbanized river in this town. The aim of this study is to assess the quality of water in the river along its 13 km urbanized stretch within Owode – Ota and Gbenga quarters of Sango – Ota in Ogun State, Nigeria. A study of some physical and chemical analysis was carried out to determine the level of pollution in the river. Total Dissolved Solids, TDS, pH, Colour and Temperature measurements were obtained for nine locations on the 21 km river stretch. Laboratory analyses were carried out at 4 locations along the water course for the following parameters: pH, Conductivity, Turbidity, DO, BOD, COD, TDS, TSS. Others include Phosphate, Chloride, Nitrate, Sulphate, Cadmium, Lead, Iron, Copper, Zinc, and Nickel. Results indicate that the water quality reduces downstream of the urbanized stretch. Some of the level of heavy metals in the river calls for concern. At Owode, the lead content of 0.11 mg/L is too high compared to a maximum of 0.01 mg/l permissible, which can cause cancer. This can interfere with Vitamin D metabolism, and can affect mental development in infants. It is toxic to the central and peripheral nervous systems. Cadmium is below 0.002 which is just below the 0.003 mg/l permitted in Nigeria. Nickel content was 0.046 mg/l between Owode and Ewupe and this is above the maximum permissible level of 0.02 for Nigeria. This has the possibility of carcinogenic health impact. Owode and Ewupe have greater industrial impacts than the other two locations, Igboloye and Gbenga. The trends of each of the 21 parameters from the urbanized stretch of the river have been observed to follow a pattern that can be categorized as similar, mirrored, somersault and composite of mirrored and somersault. More studies were recommended in this direction as well as in determining the locations of factories and industries contributing to the pollution level around Ewupe and their effluent disposal programs will need to be ascertained.

Keywords: Urbanized Water; Physical Chemical; Analysis; Health; Impact

1. Preamble

Sango-Otta urban community is the industrial nerve centre of Ogun State. It can be regarded as an extension of Lagos. River Atuwara is the major river that drains the Sango-Otta community. The level of pollution in this urbanized stretch of the river is of concern to the community because of the level of industrialization within this community. Ekweozor and Agbozu, 2001 [1] had reported how environment degradation, environmental deterioration and underdevelopment issues in Nigeria had been of concern at both national and international levels. Nasrullah et al., [2] 2006 indicated that rapid industrialization is having direct and indirect adverse effects on our environment. Nwidu et al.,2008 [3] showed that quality of water and prevalent water related diseases in hospitals were casually related to contamination of the river within the community. Adekunle, 2009 [4] deduced that the qualities of wells were affected by proximity to river used as disposal of industrial effluent. There is therefore the need to carryout water quality studies on urbanized rivers within our communities. Studies on various aspects of water quality studies had been carried out on Nigerian waters. Workers in this respect include: Izonfuo and Bariweni, 2001 [5] on the Epie creek of in the Niger Delta; Wakawa et al., 2008 [6] on surface water of River Challawa Kano, Nigeria; Fakayode, 2005 [7] on Alaro River in Ibadan, Nigeria; Rim-Rukeh et al., 2006 [8] on effects of Agricultural Activities on Orogodo River, Agbor Nigeria; Ahmed and Tanko, 2000 [9] on water quality changes for irrigation within River Hadejia Catchment; and Arimoro, et al., 2007 [10] on impact of Sawmill Wood Wastes on the Water Quality and Fish Communities of Benin River, Niger Delta Area, Nigeria. Other workers in this field include Gaballah et al., 2005 [11]; Neal and Robson, 2000 [12]; FEPA,1991 [13]; USEPA,2000 [14]; WHO,2000 [15]; APHA, 2005 [16]; Robson and Neal, 1997 [17]; Inoue and Ebise, 1999 [18]; Walter, 1987 [19]; Isiorho and Oginni, 2008 [20]; Akpan, et al., 2008 [21]; and Fawell, 2007 [22]. The aim of this study is to determine the variations in the physical-chemical parameters that can be used to determine the level of pollution along the urbanized stretch of River Atuwara. This will further be analyzed in line with the land use within the basin.

River Atuwara is located in the Ifo – Ota district of Ogun state Nigeria. It takes its source from the Adenrele area of Ifo district in Ogun state having a relief of 200m above mean sea level. The river meanders from its source to cross the old Lagos-Abeokuta express road at Owode, passing through Ewupe, Benja and Igbolye after which it crosses another major road, Idiiroko road at Iju. The flood plain of the river is observed to be swampy indicating abundance of loamy and sandy soils within. The natural vegetation type in the catchment area is the tropical or lowland rain forest. The river together with another river, R. Iju have a confluence point at Iju located after Canaan land on Idiroko road, Ota where it is now named river Iju.

2. Materials and Methods

2.1. Reconnaissance Survey

A reconnaissance survey of the catchment area of the Atuwara basin from the source to beyond the urbanized stretch of the river was undertaken to be able to plan and collect necessary data and information on the land and water use of the basin. A 1:50,000 topographical map of Ifo/Ota was obtained from the Ministry of Lands and Survey located in Okelewo, Abeokuta. Materials employed for the reconnaissance survey include the followings:

Reconnaissance survey kits
Global positioning system (GPS)
TDS meter Stop watch
Measuring tape (50M) pH meter and
Thermometer

At areas where the river had to be traversed and or crossed in some other instances, canoes were hired. Services of some native Site assistants were also used so as to be able to create soft landing pads in possible hostile communities as well as making navigation less demanding.

Data collected during the reconnaissance survey include the following:

Communities that use the river
Economic activities
Religious practices/rites in the area
Usage of the river along its stretch
Sources of pollution to the river
Nature of pollution

2.2. Materials and Methods for the Physical Analysis

In assessing the quality of water in river channels, several physical, chemical, and biological analyses are required. The most common physical analyses are pH, temperature and dissolved solids. Survival of fishes and aquatic invertebrates depend on the right pH and temperatures. Akpan et al., 2008 [23] used pH as criteria for evaluation in their modeling and simulation of the effect of effluent from Kaduna Refinery and Petrochemical Company, KRPC, on the quality of R. Kaduna. The physical parameters considered in this study are Temperature, pH, and Total Dissolved Solids (TDS). These parameters were measured in-situ for nine locations and are considered as preliminary assessment of the quality of water in the channel. The locations are:

Adenrele Onihale Owode
Ewupe Mosafejo Benja
Igboloye Gbenga1 Gbenga

Mosafejo and Benja are located, respectively on rivers Iju and Atuwara, at same distance from the confluence point, Igboloye. Mosafejo, though not urbanized, is considered separately because of its proximity to both Igboloye and Benja. The urbanized stretch is from Onihale to Gbenga. Other parameters determined on the field include colour, odour, turbidity, and the velocity of flow.

Temperature was measured with the aid of a digital thermometer. Measurements of both the water temperature and the ambient temperature were measured and values recorded in degrees Celsius (℃).

The pH and total dissolved solids (TDS) or electrical conductivity (EC) of the raw water samples were measured with HANNA HI 9810 pH-TDS meter. The meter was standardized with a buffer solution (i.e. buffer 7 and 9). The buffer tablet was distilled in a 100ml of water in a beaker. The probe of the meter is then placed in the solution and adjusted to read 000 to standardize it. The electrode response was checked by measuring the pH of the test sample, first with distilled water and then with the sample. The system was allowed to stabilize before the final reading was made.

The velocity of the river flow was measured using the float method. The floats used were pieces of polystyrenes placed on river surface made to travel known distances along the river. The time taken to travel specific distances were measured. The velocity of flow was the computed.

2.3. Raw Water Sample Collection and Preservation

The materials required for raw water sample collection and preservation are as follows:

2l-bottles	De-ionized distilled water
Masking tape	Cooling facility
Canoe	Digital Camera
GPS and	Services of Local Assistants

Water samples were collected in clean containers during analysis and the quantity collected at any given time depends on the number of parameters wanted. In most cases 2-liter samples may be sufficient. It is essential to take the sample to the laboratory for testing immediately after collection. This is such that the parameters required are not altered due to variance in temperature or ambient condition. This could render the results from the laboratory invalid.

The raw water samples that could not be taken to the laboratory the same day were preserved. Since our water samples were for bacteriological analysis the collection bottles were always sterilised. The samples were left in a refrigerator set at 4℃, a temperature where bacteria are inactivated. The parameters, preservation required and maximum periods are indicated in **Table 1**.

Raw water samples were collected from four strategic locations along the urbanized stretch of the river. These are Owode, Ewupe, Igboloye and. Gbenga. The two liters capacity bottles were washed normally using detergent and then rinsed properly with de-ionized distilled water were used for collection and storage of samples. During collection, the sample bottles were rinsed again with the water samples twice. Each bottle was tagged with different labels on masking tape for identification purposes. The samples were put in cooling facility for adequate preservation after collection in readiness for the laboratory located some 15 km to the site. Laboratory services were provided by Messrs. Triple "E" Systems Associates Ltd. (RC No. 108,343), and with Laboratory Services DPR Permit No. RC 0733/2008.

Generally, the parameters considered are as follows:
Ph; Conductivity (μS/cm); Turbidity (FTU);
Colour; DO(mg/L); BOD (mg/L); COD (mg/L)
Total Dissolved Solids (mg/L); Total Suspended Solids(mg/L); Phosphate(mg/L);
Chloride (mg/L); Nitrate (NO3-) (mg/L);
Sulphate (SO_4^{-2}) (mg/L); Cadmium (mg/L);
Lead (mg/L); Iron (mg/L); Copper (mg/L);
Zinc(mg/L); Nickel (mg/L)

3. River Atuwara Drainage Basin and Land Use Survey Report

The catchment area of R. Atuwara at Igboloye as carved out is shown in **Figure 1**. This area also covered about 40% of Sango-Ota municipality. Within the basin, the right hand side of the main trunk of the river is more developed than the area on the left side of the basin.

The urbanized stretch of the river is from Owode to Igboloye. Social and economic activities around the river are indicated in **Table 2** below. Occupation of the people of Onihale, Ewupe and Igboloye is majorly agricultural activities which includes cultivation of crops, fish farming, piggery and snailry. The river crosses the Lagos – Abeokuta express road at Owode which serves as a market and has a large abattoir very close to the river. Here the river is used to dispose solid wastes from the market and organic matter from decomposed animal flesh from the abattoir. The river is also used for domestic purposes such as cooking, bathing, washing of clothes, meat buckets etc. Some areas of the river that may used to dispose industrial, chemical, and institutional wastes include Ewupe, Igboloye, Gbenga and Iju.

Table 1. Samples preservation and maximum periods.

PARAMETERS	PRESERVATIVES	MAXIMUM HOLDING HRS
Acidity – Alkalinity	Refrigeration at 4^0 C	24 – hours
Biochemical oxygen demand	Refrigeration at 4^0 C	6 – hours
Chemical oxygen demand	2ml conc. H_2SO_4 per litre	7 days
Colour	Refrigeration at 4^0C	24 – hours
Dissolved oxygen	Determine on site	No holding
Metals Total	Conc. HNO_3 to $_P$H 2- 3	6 months
Metals Dissolved	Filtrate 3ml 1 : 1 HNO_3 perL	6 months
Nitrogen Total	40mg Hg^{2+} or 2ml conc. H_2SO_4 at 4^0C	7 DAYS
Organic carbon Total	2ml conc. H_2SO_4 per litre (Ph2)	24 hours
pH	Determine on site	No holding
Phosphorus	40mg Hg^{2+} per litre at 4^0C	7 days
Pathogens		
Other organic substances		
Turbidity		

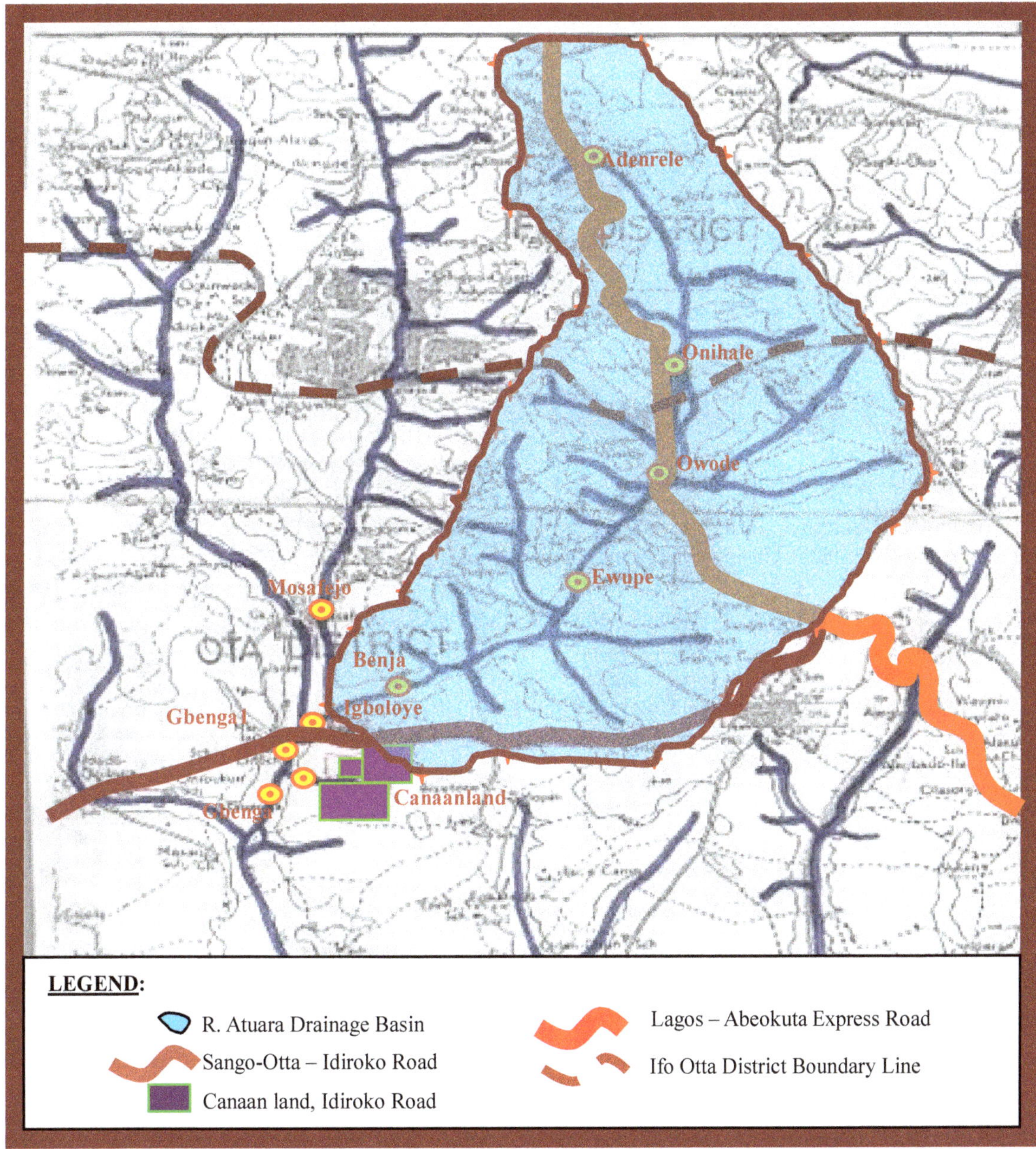

Figure 1. R. Atuara drainage network and catchment area at igboloye.

Table 2. Location of Settlements along R. Atuwara and Land use.

S/No	Sample Location	Distance From Source (Km)	Land / Water use
1	Adenrele	0	Fishing and domestic purposes
2	Onihale	6	Farming and waste disposal
3	Owode	8	Market, Abattoir, and Cloth washing
4	Ewupe	12	Domestic use but smelly water
5	Benja	16.8	Livestock and fishing. Bad odour
6	Igboloye	18.6	Domestic use and Shrine
7	Gbenga1	20.0	Domestic and Farming
8	Gbenga	21.0	Farming and Fishing

Typical uses of the urbanized river at different locations are indicated in Plates 1 - 4 below.

4. Analysis and Discussion of Results

Results of preliminary and physical assessment of nine locations around the water in the channel are presented in **Table 3**.

Occupation of the people around the river is majorly agricultural, including fish farming. The river is used for disposal of solid wastes both organic and inorganic. It is also used for domestic purposes such as bathing, cooking, and cloth washing, (laundering),

The variations of the ambient temperature and the water temperature along the river are shown in **Figure 2**.

4.1. Chemical Analysis

Chemical analysis was carried out on samples from four sites because these particular sites have peculiar events occurring around them such as the presence of pharmaceutical industries, contact with wastewater from institution, mixed agricultural activities e.t.c. The analysis carried out on them includes Conductivity(μS/cm), Turbidity (FTU), DO (mg/L), BOD (mg/L), COD (mg/L), Total Dissolved Solids (mg/L), Total Suspended Solids(mg/L), Phosphate (mg/L), Chloride (mg/L), Nitrate(NO_3^-)(mg/L), Sulphate(SO_4^{2-})(mg/L), Cadmium(mg/L), Lead(mg/L), Iron(mg/L), Copper(mg/L), Zinc (mg/L), Nickel (mg/L). Results of the analysis and expected standards by Nigeria and World Health Organization, WHO, are presented in **Table 4**.

Table 3. Temperature, Color, pH and Total Dissolved Solids in the Channel.

S/N	Location	Temperature (^{0}C)		Colour	pH	Total Dissolved Solids (ppm)	Remarks
		Ambient	Water				
1	Adenrele	31.5	26.7	Slightly Clear	5.5	0.49	Source
2	Onihale	32.4	26.3	Mudish brown	6.5	0.53	
3	Owode	31.4	26.5	Light brown	5.8	0.52	Urbanized
4	Ewupe	30.1	26.3	Greenish but slightly clear	5.3	0.57	Urbanized
5	Mosafejo	31.5	26.7	Clear	5.4	0.28	
6	Benja	28.0	26.4	Light brown	5.5	0.56	Urbanized
7	Igboloye	27.2	26.3	Light brown	5.5	0.53	Urbanized. Serves as shrine
8	Gbenga1	26.8	26.4	Light brown	5.5	0.44	
9	Gbenga2	28.5	26.2	Mudish brown	6.7	0.59	Lots of dirt

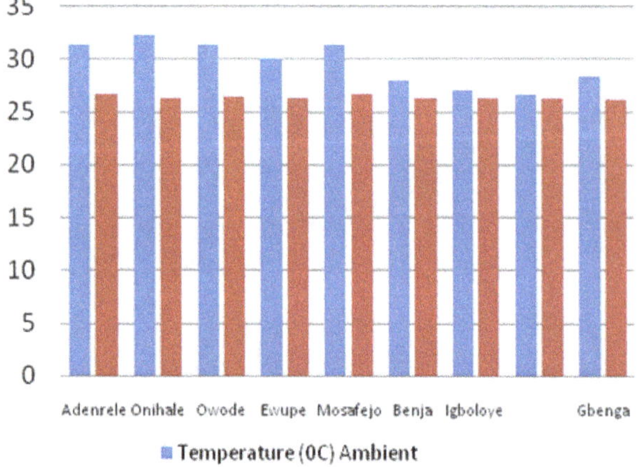

Figure 2. Ambient and water temperature (°C).

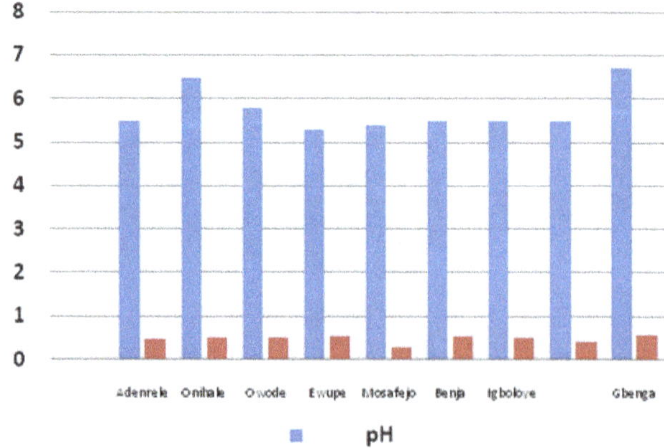

Figure 3. pH and total dissolved solids (TDS) at specified

locations.

Table 4. Results of water analysis and expected standards by nigeria and WHO.

S/No	Parameter	Owode	Ewupe	Igboloye	Gbenga	STANDARD	
						NIGERIA	WHO
	Distance from source (km)	8.0	12.0	18.6	21.0		
1	pH	7.69	7.16	7.38	7.39	6.5-8.5	6.5-7.5
2	Conductivity µS/cm	141	136	152	178	1,000	500
3	Turbidity	ND	2.00	26.00	85.00	-	5.0 NTU*
4	Color	Faint Yellow	Faint Yellow	Faint Yellow	Brownish Yellow		3.0 TCU**
5	DO mg/l	4.60	0.80	7.40	3.80	-	-
6	BOD mg/l	1.40	0.80	2.80	3.60	-	-
7	COD mg/l	40.00	60.00	28.00	20.00	-	-
8	TDS mg/l	82.00	70.00	79.00	92.00	500	1,000
9	TSS mg/l	35.00	40.00	45.00	108.00		
10	Phosphate mg/l	1.133	0.981	1.140	0.812	-	-
11	Chloride mg/l	3.80	4.70	6.80	3.10	250	250
12	Nitrate mg/l	0.003	0.181	0.001	0.012	50	50
13	Sulphate mg/l	10.00	13.00	17.00	14.00	100	250
14	Cadmium mg/l	<0.002	<0.002	<0.002	<0.002	0.003	Not done
15	Lead mg/l	0.11	<0.01	<0.01	<0.01	0.01	Not considered
16	Iron mg/l	1.922	1.161	1.850	1.927	0.3	"
17	Copper mg/l	0.091	0.067	0.080	0.115	1	"
18	Zinc mg/l	0.0654	0.0778	<0.001	0.0240	3	"
19	Nickel mg/l	0.046	0.046	0.012	0.116	0.02	"

*Nephelometric turbidity unit; **True color unit.

The results for the 4 sites are presented graphically from **Figure 4** to **Figure 21**.

The following are deductions from the analysis:

The areas that are of concern are the concentrations of the following parameters:

(i) Conductivity (ii) Turbidity
(iii) Lead at Owode (iv) Iron
(v) Nickel

The concentrations of these chemical parameters are above the WHO and Nigerian standards.

All the four sites except Owode, which has highest pH value of 7.69, (**Figure 4**), fell within the ranges given by Nigeria and WHO standards. The pH at Owode was above the range specified by WHO.

Generally conductivity in the river channel at all the locations were below both WHO and Nigerian standards. However, conductivity fell slightly below the Owode value at Ewupe to continue rising along the urbanized river stretch. This is shown in **Figure 5**.

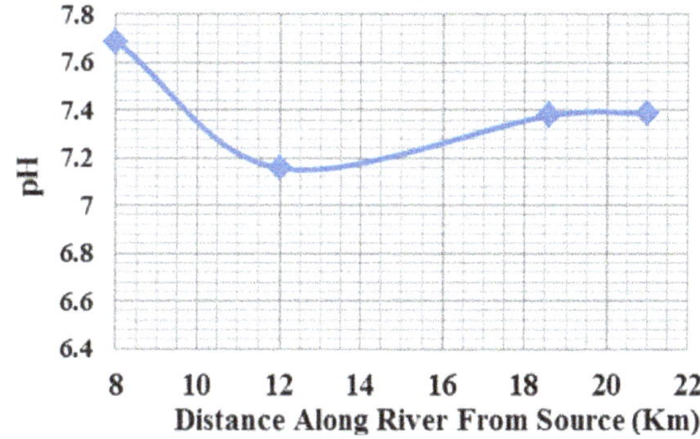

Figure 4. pH trend urbanized stretch.

The river is more turbid at both Igboloye and Gbenga at kilometers 18.6 and 21 respectively as shown in **Figure 6**. This can be accounted for by the use of the river for domestic purposes, like washing and farming at both locations. The first 4 km of the 13 km of the urbanized river stretch downstream (Owode and Ewupe) are more urbanized than the last 2.6 km of the 13 km stretch, (Igboloye – Gbenga). This explains the higher turbidity obtained in the less urbanized stretch along the river. It can be concluded that turbidity, T, is inversely proportional to the level of urbanization, U_L and indicated as equation (1).

$$T \; 1/\!\propto U_L \qquad (1)$$

If other factors that affect turbidity within the urbanized river stretch such as channel shape, slope, lining and bank slope stability are constant, then

$$T = k(1/U_L) \qquad (2)$$

where k is the constant coefficient to be determined.

The color of the water in the channel remained at faint yellow from commencement of the urbanized stretch at Owode, for 10.6 km downstream to Igboloye. The color at 2.4 km further downstream became brownish yellow at Gbenga. This indicates effect of erosion due to domestic use of the river within this section of the river.

The trend recorded for Dissolved Oxygen, DO, level along the river is shown in **Figure 7**. DO can be an indicator of how polluted the water is and how well the water can support aquatic plant and animal life. At Igboloye, 10.6 km from the commencement of urbanized stretch, the highest DO, level of 7.4 mg/L was recorded. The highest Biological Oxygen Demand, BOD, of 3.6 mg/L was recorded at Gbenga, 2.4 km downstream of Igboloye.

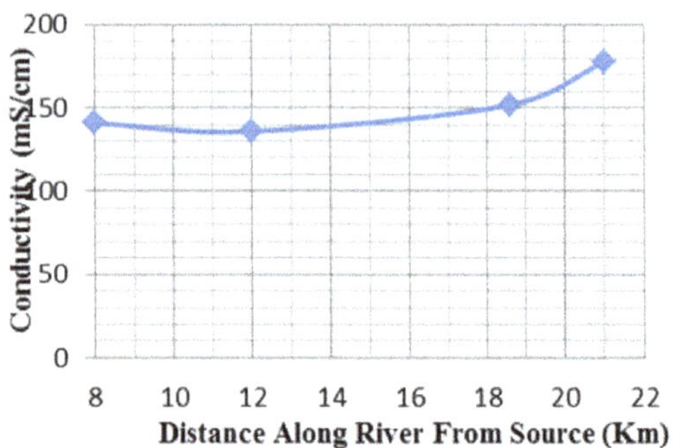

Figure 5. Conductivity (mS/cm).

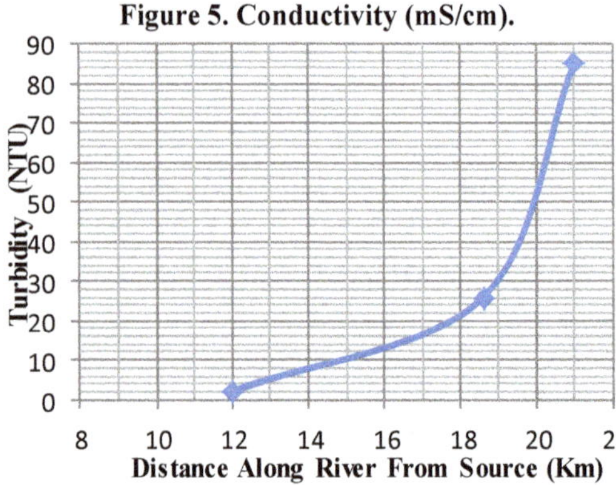

Figure 6. Turbidity (NTU).

Ewupe has the highest Chemical Oxygen Demand, COD, of 60.0 mg/L as well as lowest Biological Oxygen Demand, BOD and Dissolved Oxygen, DO of 0.80 mg/L in each case. These are shown in **Figure 8** and **Figure 9**.

The trends presented by the Total Dissolved Solids, TDS, and Total Suspended Solids, TSS, are shown in **Figure 10** and **Figure 11**. TDS values within the river stretch considered ranged between 70 and 92 mg/L which are far below the Nigerian and WHO standards, with the lowest recorded at 4 km away from downstream urbanized stretch at Ewupe. TSS rises gently from Owode, kilometer 8 from source to some 10.6 km downstream at Igboloye. The trend then rises sharply through the 2.4 km downstream to Gbenga as shown in **Figure 11**.

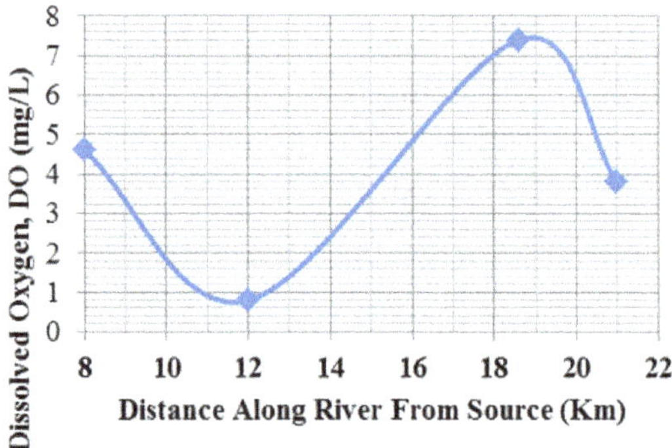

Figure 7. Dissolved oxygen, Do,(mg/L).

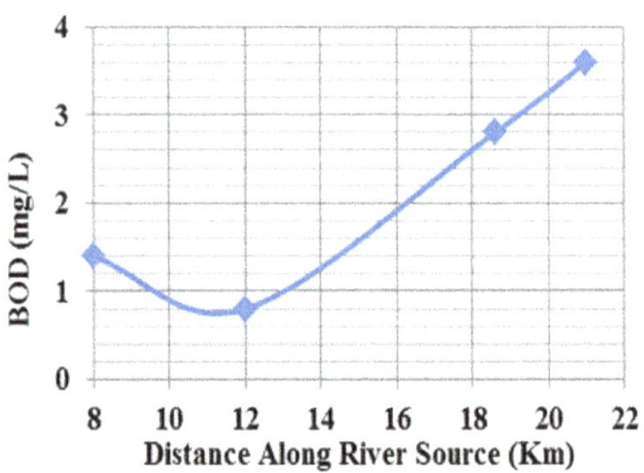

Figure 8. BOD (mg/L).

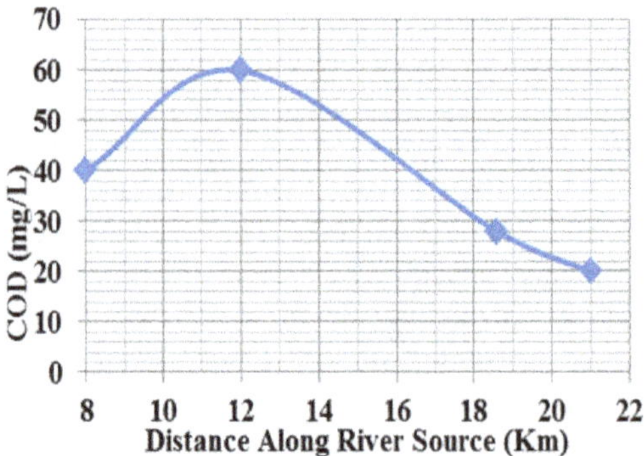

Figure 9. COD (mg/L).

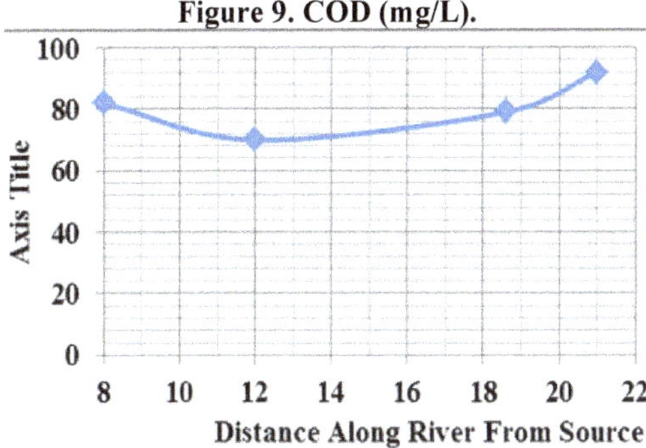

Figure 10. Total dissolved solids (mg/L).

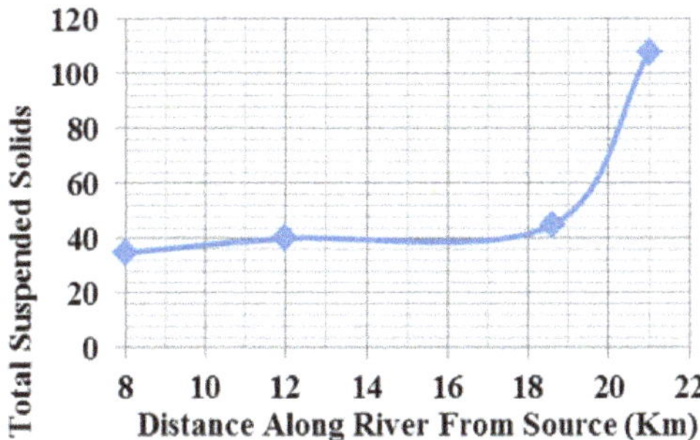

Figure 11. Total suspended solids (mg/L).

The trends presented by Phosphate and Chloride are indicated in **Figure 12** and **Figure 13**. The trends are similar from a distance of 12 km through downstream of the urbanized stretch. The trend presented by Nitrate, shown in **Figure 14** indicates a somersault trend of that of Phosphate. At kilometer 12 from source, Ewupe has the lowest Phosphate level as well as the highest Nitrate level. Also at 18.6 km from the source, the highest phosphate level was recorded with the lowest level for Nitrate.

The trend for Sulphate level along the urbanized stretch is shown in **Figure 15**. This trend is somehow similar to that of Chloride asshown in **Figure 13**. Both indicate highest levels at Igboloye, 18 km from the source, although the sulphate level did not fall below the Owode level as was the case for Chloride levels.

From results, the water in the channel is free from problems that may arise through higher levels of Cadmium, (**Figure 16**). At all the investigated stations, Cadmium levels were lower than 0.002 mg/L which in turn are lower than 0.003 mg/L limit stipulated by Nigerian standard.

Figure 17 indicates that the lead level in the river at the beginning of urbanization, is of great concern at Owode, 8 km from the river source. The expected Nigerian standard for lead is 0.01 mg/L and the levels at other locations are lower, that at Owode is 10 times greater at 0.11 mg/L.

The iron content in the river at all the locations ranged between 1.161 mg/L and 1.927 mg/L which are far above the 0.3 mg/L standard for Nigeria. However it records the lowest level at Ewupe (**Figure 18**).

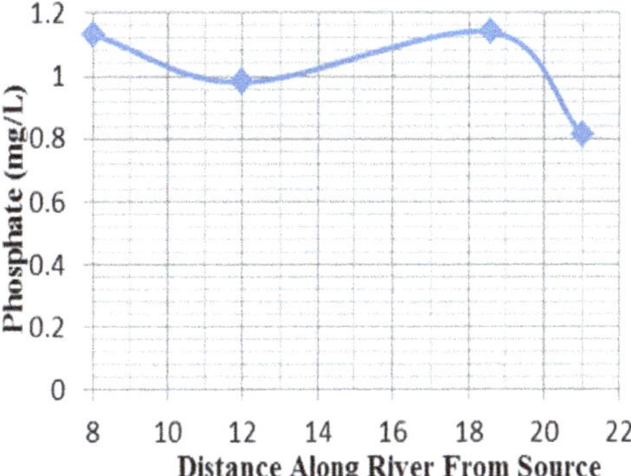

Figure 12. Phosphate (mg/L).

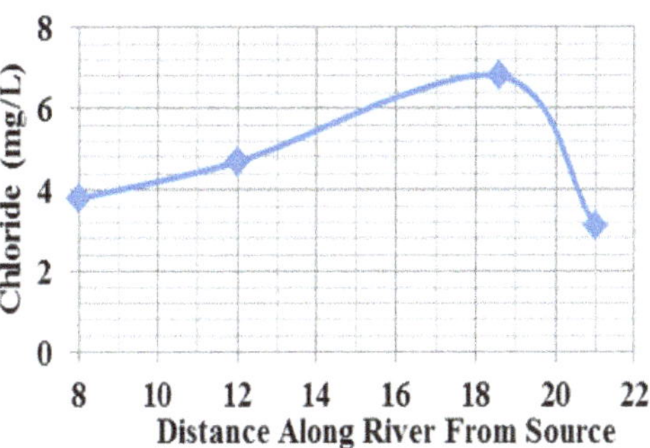

Figure 13. Chloride (mg/L).

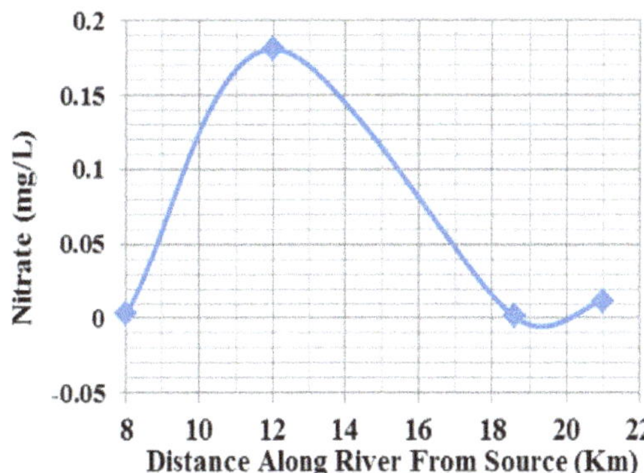

Figure 14. Nitrate (NO-3) (mg/L).

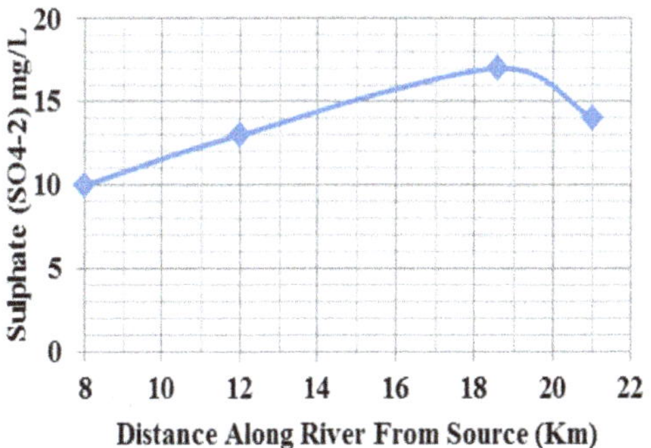

Figure 15. Sulphate (SO4-2)(mg/L).

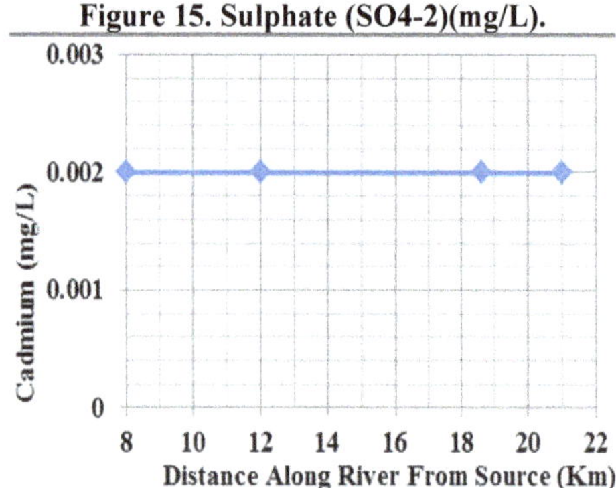

Figure 16. Cadmium (mg/L).

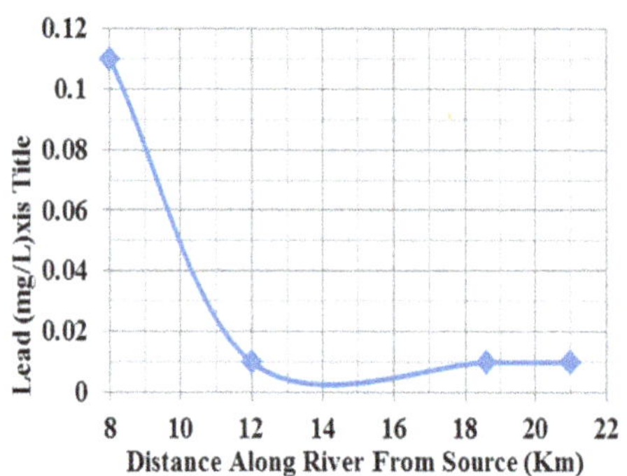

Figure 17. Lead (mg/L).

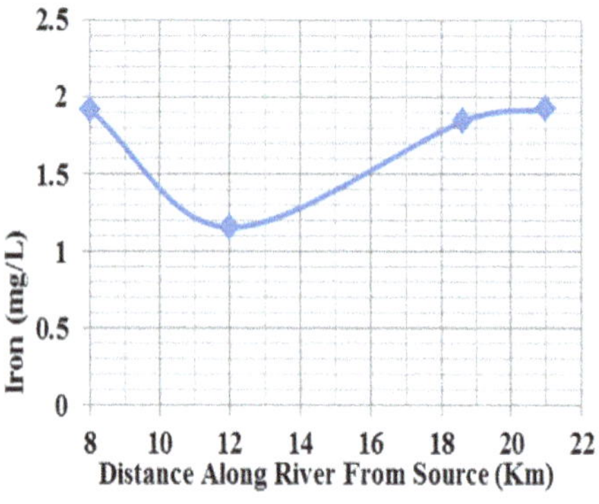

Figure 18. Iron (mg/L).

Copper contents in the river more or less followed the same trend as that of iron. This can be deduced from **Figure 19**. They are also below the Nigerian standard.

Zinc contents at all considered river locations fell below standards specified by Nigeria. Igboloye has the lowest Zinc level, as indicated by **Figure 20**.

The obtained result for Nickel content in the river (**Figure 21**) indicates that only Igboloye, 18.6 km from the source is below the standards set by Nigeria. The others are above thereby suggesting that it can possibly be carcinogenic.

Generally the quality of the water in the channel deteriorated greatest at Ewupe, 12 km downstream of the source and 4 km downstream of the commencement of

urbanized river stretch.

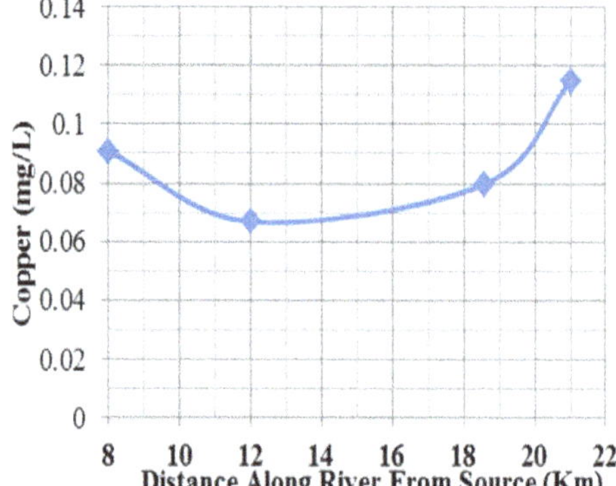

Figure 19. Copper (mg/L).

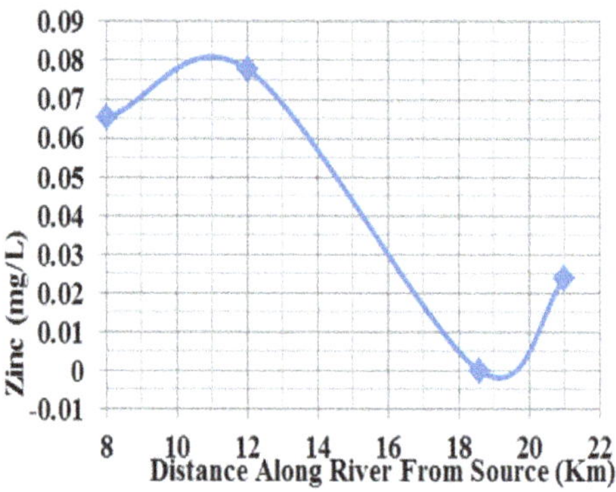

Figure 20. Zinc (mg/L).

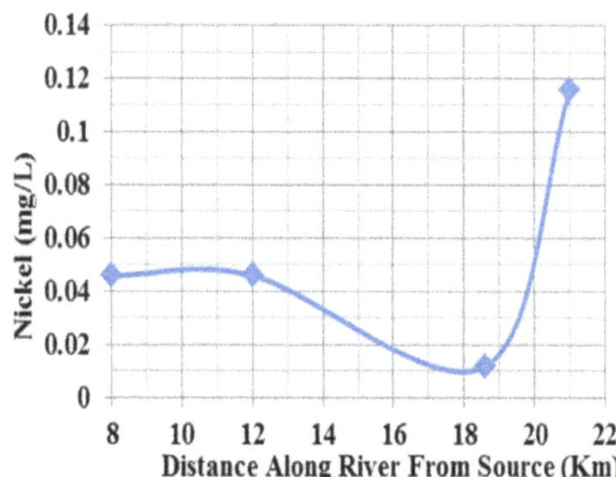

Figure 21. Nickel (mg/L).

4.2. Physical-Chemical Parameter Trend Analysis

It is observed that the results of many of the parameters along the urbanized stretch as presented by **Figure 4** through **Figure 21** have followed some interesting trends. The trends can be categorized as

(i) Similar trends (ii) Mirrored Trends

(iii)Somersault trends (iv) Composite trends

The parameters that are grouped under each category are indicated in **Table 5**.

The trends can be used for modeling water quality parameters.

Table 5. Grouped parameters and trend categories.

GROUP	CATEGORY				REMARKS
	SIMILAR	MIRRORED	SOMERSAULT	COMPOSITE	
1	pH and Iron	DO	DO	BOD	
	Iron	Zinc	Zinc	COD	
				Chloride	
				Iron	
				Sulphate*	*Apologetic
2	Conductivity	COD	BOD		
	TDS	Chloride	COD		
	Copper				
3	Chloride	TSS			
	Sulphate	Lead			

5. Conclusions and Recommendations

5.1. Conclusions

The urbanized stretch of R. Atuwara right from upstream at Owode to downstram at Gbenga had experienced different levels of pollution at different points as a result of varied activities along the river. Generally the quality deteriorated downstream.

Results of the physical and chemical analyses carried out pointed to some areas of concern. These are discussed below.

(i) The pH level at Owode, the beginning of the urbanized stretch is greater than the WHO standard, though within Nigeria's standard.

(ii) Conductivity levels are lower than both WHO and Nigerian standards. It is lowest at Ewupe, 4 km downstream of beginning of the urbanized river stretch. The sharpness of its rise within this stretch along the stream may be of concern since sudden increases in conductivity indicate that there is a source of dissolved ions in the vicinity. It can be used as an indication of potential water quality problems.

(iii) The DO level at Ewupe shows that the water quality at this point is the poorest, (ciese.org). This indicates that some fish and macro invertebrate populations will begin to decline (ciese.org). Most warm water fish need DO in excess of 2 mg/L to survive.

(iv) Lead level at Owode.

(v) Iron content levels through the river stretch is greater than recommended level.

(vi) The concentration of iron rose from Ewupe downstream through to Gbenga.

(vii) Nickel content in the river along the stretch except at Igboloye is greater than the Nigerian standard. Nickel above this shows possible carcinogenic condition. Nickel level at Gbenga is worst.

(viii) From observations of the **Figure 4** through **Figure 21**, the trends presented by some parameters may suggest categorization of the trends into 3 or 4. Some trends are similar to themselves, some seem to be mirror images of others, while some other ones seem to follow somersault trends of some other ones. A fourth trend can be a composite of the somersault and mirrored trends. This exercise may lead to modeling of behaviors of some parameters from known parameters.

5.2. Recommendations

There is need to carry out further studies on the locations of factories and industries around Ewupe and their effluent disposal programs will need to be ascertained.

2 Further studies is required in the areas of concerns highlighted in the conclusions above and necessary measures put in place to combat the problems that may result.

3 Further studies is also required to investigate the possibilities of modeling an unknown water quality from some known ones initiated by observed similarity or otherwise of the trends along the urbanized river.

4 The river should not be used as a source of disposal for in-habitants along its course. Rather wastes may be properly disposed using other disposal approaches to improve the river quality.

5 Government and stakeholders should put into place more stable laws and regulations with proper enforcement so as to guide river use and maintenance program that will incorporate users from the villages around the river course.

6 The different organizations using the river as a source of wastewater disposal should be well monitored by the concerned authorities.

7 Communities along the river course should be

enlightened on the possible consequences of pollution by introducing public awareness program.

REFERENCES

[1] I. K. E. Ekweozor and I. E. Agbozu, "Surface Water Pollution Studies of Etelebou Oil Field in Bayelsa State, Nigeria," *African Journal Science*, Vol. 2, 2001, pp. 246-254.

[2] R. N. Nasrullah, B. Hamida, I. Mudassar and M. I. Durrani, "Pollution Load in Industrial Effluent and Ground Water of Gadoon Amazai Industrial Estate (GAIE) SSWABI, NWFP," *Journal of Agricultural and Biological Science*, Vol. 1, No.3, 2006, pp. 18-24.

[3] L. L. Nwidu, B. Oveh, T. Okoriye and N. A. Vaikosen, "Assessment of the Water Quality and Prevalence of Water Borne Diseases in Amassoma, Niger Delta, Nigeria," *African Journal of Biotechnology,* Vol. 7, No.17, 2008, pp. 2993-2997.

[4] A. S. Adekunle, "Effects Of Industrial Effluent On Quality Of Well Water Within Asa Dam Industrial Estate, Ilorin Nigeria," *Nature and Science,* Vol. 7, No. 1, 2009, pp. 39-43.

[5] L. W. A. Izonfuo and P. A. Bariweni, "The Effect of Urban Run-off Water and Human Activities on some Physico-chemical Parameters of the Epie Creek of in the Niger Delta," *J. Appl. Sci. Environ. Manage.* Vol.5, No.1, 2001, pp. 47-55.

[6] J. I. Wakawa, A. Uzairu, J. A. Kagbu, and M. L. Balarabe, "Impact Assessment of Effluent Discharge on Physico-Chemical Parameters and some Heavy Metal Concentrations in Surface Water of River Challawa Kano, Nigeria," *African Journal of Pure and Applied Chemistry*, Vol. 2 No.10, 2008, pp. 100-106.

[7] S. O. Fakayode, "Impact Assessment of Industrial Effluent on Water Quality of the Receiving Alaro River in Ibadan, Nigeria," *AJEAMRAGEE,* Vol.10, 2005, pp. 1-13.

[8] A. Rim-Rukeh, O. G. khifa, and A. P. Okokoyo, "Effects of Agricultural Activities on the Water Quality of Orogodo River, Agbor Nigeria," *Journal of Applied Sciences Research*, Vol. 2, No. 5, 2006, pp. 256-259.

[9] Ahmed K and A. I. Tanko, "Assessment of Water Quality Changes for Irrigation in the River Hadejia Catchment," *J. Arid Agriculture,* Vol.10, 2000, pp. 89-94.

[10] F. O. Arimoro, R. B. Ikomi, and E. C. Osalor, "The Impact of Sawmill Wood Wastes on the Water Quality and Fish Communities of Benin River, Niger Delta Area, Nigeria," *International Journal of Science & Technology,* Vol. 2, No. 1, 2007, pp. 1-12.

[11] M. S. Gaballah, K. Khalaf, A. Beck, and J. Lopez, "Water Pollution in Relation to Agricultural Activity Impact in Egypt," *Journal of Applied Sciences Research,* Vol.1, 2005, pp. 9-17.

[12] C. Neal and A. J. Robson, "A Summary of River Water Quality Data Collection Within the Land-Ocean Interaction Study: Core Data for Eastern UK Rivers Draining to the North Sea," *Science of the Total Environment,* Vol. 251/252,2000,pp.585-665.

[13] Federal Environmental Protection Agency, (FEPA), "Guideline and Standards for Environmental Pollution in Nigeria. FEPA, Nigeria, 1991.

[14] United States Environmental Protection Agency, (USEPA), "2000 National Water Quality Inventory," http://www.epa.gov/305b/2000report/retrieved August 6, 2003.

[15] World Health Organisation, (WHO), "WHO Guidelines for Drinking Water Quality, Training Pack, Rome: WHO, 2000.

[16] American Public Health Association, (APHA), "Standard Methods for the Examination of Water and Wastewater," 21st Edition American Public Health Association, American Water Works Association, and Water Pollution Control Federation. Washington, DC., 2005.

[17] A. J. Robson and C. Neal, "A Summary of Regional Water for Eastern UK Rivers," *Science of the Total Environment*, Vol. 194-195, 1997, pp.15-39.

[18] T. Inoue and S. Ebise, "Run off Characteristics of COD, BOD, C. N and P Loading from Rivers to Enclosed Coastal Seas," *Marine Pollution Bulletin*, Vol. 23, 1991, pp. 11-14.

[19] L. Walter, "Handbook of Water Purification," *Ellis Horwood, New York*, 1987, pp. 85-90.

[20] S. A. Isiorho and F. A. Oginni, "Assessment of Wastewater Treatment in Canaan land, Ogun State, Nigeria," *1st Postgraduate Researchers' Conference on Meeting Environmental Challenges in the Coastal Region of Nigeria*, Sept.29-30, 2008. University of Abertay, Dundee, UK.

[21] U. G. Akpan, E. A. Afolabi and K. Okemini, "Modeling and Simulation of the Effect of Effluent from Kaduna Refinery and Petrochemical Company on River Kaduna," *AU Journal of Technology, AU J. T.*, Vol.12, No. 2, Oct. 2008, pp. 98-106.

[22] J. Fawell, "Drinking Water Standards and Guidelines. Foundation for Water Research," Allen House, The Listons, Liston Road, Marlow, Bucks SL 7 1FD, UK. February, 2007.

24

Evaluation of the Impact of Government Policy on the Overuse of Groundwater in the Minqin Basin in China

Lihua Yang
School of Public Administration & Workshop for Environmental Governance and Sustainability Science, Beihang University, Beijing, China

ABSTRACT

The existing literature simply concludes that the irrational behaviors of local people and natural factors are the major reasons for overuse of groundwater. Using the OLS and ARIMA (BJ) Statistical Methods and Trend Analyses, this article finds that government policy, as measured by four proxy variables, is a very important factor that strongly influences the overuse of groundwater at the collective level. This means the government is a very important actor in the game of groundwater usage. Although these findings cannot clearly separate government effects from local effects, using a Trend Analysis, they reveal that these significant effects are strictly consistent with variations in government policy. Moreover, they show that government policy effective at the county level strongly impacts the overuse of groundwater by influencing the behavior of the local people and that policy at the operative level impacts four policy domains: population, cultivated land, water assignments and peasant income.

Keywords: Groundwater; Government Policy; Institutional Analysis; Institutional Change; Water Resources

1. Introduction

Minqin County, also called Minqin Basin in some academic articles, is said to be typical of many of the desert-oasis counties of inland China with its combination of pastoral and agricultural production regimes [1]. It is located in the northeast of the Hexi Corridor (the most important part of the old Silk Trade Road) on the lower reaches of the Shiyang River on the Alashan Plateau, bounded by the Tengger and Badain Jaran deserts, two large deserts in northwest China, in the east, west and north, and covers an area of 16,000 square kilometers. This area falls under the jurisdiction of Wuwie city (some 70 or 80 km from Wuwei) in central-eastern Gansu province and was once a natural barrier in the path of encroaching sand (see **Figure 1**).

Groundwater has been extracted on a large scale, particularly since the 1970s, to irrigate the Minqin oasis. On average, the annual pumping amounts to 0.722 billion·m^3 [3]; large-scale exploitation of the groundwater has been lowering the groundwater table by 0.6 - 1.0 m annually, and many wells have been discarded and replaced more than 5 times [4]. Harris's study concludes "consequently, the water table has fallen from 2 meters in the 1950s to 8 meters during the 1980s" [5]. The existing literature about the overuse of groundwater in Minqin has stressed seven factors: 1) population, 2) land reclamation, 3) decrease of water from the upper reaches (or the overuse of water in the upper reaches), 4) some economic development factors, including agricultural structure, increased livestock numbers, and overgrazing, 5) temperature, 6) precipitation, and 7) evaporation [1,6,7]. Here, we evaluate whether government policies also influence the overuse of groundwater.

2. Modeling the Impact of Government Policy on the Overuse of Groundwater

2.1. The Dependent Variable and Data Description

The overuse of groundwater is the dependent variable in this research, denoted W. Theoretically, the accurate measure of the overuse of groundwater should be the water table or the remaining amount of groundwater. However, because of a lack of scientific methodology and tech-

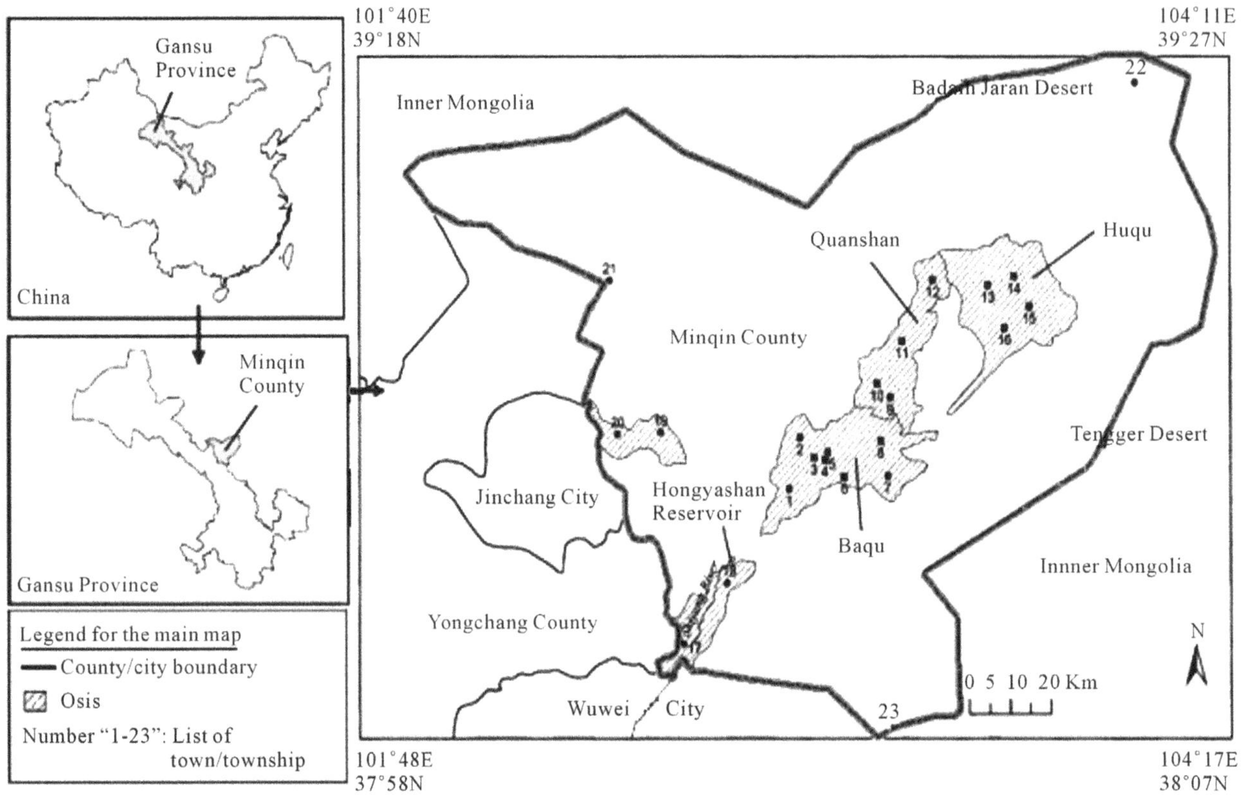

Figure 1. Map of Minqin County. Source: Adapted from the study by Lee and Zhang [2].

nological tools, these data have not been continually recorded in Minqin County. There is only information available for a few years. Therefore, we choose the average depth of wells (the unit is "m") as a proxy to measure the overuse of groundwater. Because the depth of the well depends on the water table, this proxy is a credible measure of overuse of groundwater. In addition, because the people in Minqin County mainly began to draw groundwater on a large scale starting in the 1960s, the annual data of the average depth of the wells from 1960 to 2004 was used in the analysis.

2.2. Maintaining the Integrity of the Specifications

Given the existing theories and studies mentioned above, we know that the major policies that strongly influence the use of groundwater mainly include the following: the population policy, the land policy, the water right policy, and the economic policy. We choose population (denoted as *P*, the unit is 10 thousand), the area of cultivated land (denoted as *L*, the unit is 10 thousand "mu"), the amount of water that flowed from the upper reaches (denoted as *F*, the unit is 0.1 billion·m^3) and the average peasant income (denoted as *I*, the unit is Yuan) as proxies to measure government impact. We deem them independent variables in the model. The annual data for these variables from 1960 to 2004 are all available from the government gazette of Minqin County. Furthermore, the reason why we choose land and population as the major policy proxies is based on the fact that agricultural water use, domestic water use and water used by humans and animals account for the vast majority of the water use.

To control for other factors that may also impact the overuse of ground water, depending on the existing theories and studies mentioned above, we add the average temperature (denoted as *T*, the unit is "c"), the average precipitation (denoted as *R*, the unit is "mm") and the average evaporation (denoted as *E*, the unit is "mm") as the control variables. The annual data for these variables from 1960 to 2004 can also be obtained from the government gazette of Minqin County. Finally, because the groundwater system of Minqin basin is a relatively closed system in terms of the geological characteristics [8], we can ignore the effects of other factors, such as the movement of the groundwater under the surface.

2.3. The Basic Research Model

Using the denotations given above, our basic research model can be specified as follows:

$$W_t = f\left(P_t, L_t, F_t, I_t, T_t, R_t, E_t\right) \tag{1}$$

where α is the intercept, β_i is the difference coefficient, and ε_t is the stochastic error term[1]. None of the data that will be used in this model is non-stationary. In a pre-study, we did a Dickey-Fuller test of unit roots and found the MacKinnon approximate P value for $Z(t)$ = 0.9950, indicating the unit root is problematic. Therefore, to technically eliminate some threats to the validity of our model caused by this attribute of the data, we develop the above variables as differenced forms. This is also reasonable because we suppose the differenced values of population, land, water, income and others are related to the differenced values of the overuse of groundwater. Thus, we now study the changes of independent values on the changes of dependent values. Then, a differenced model is created:

$$\begin{aligned} W_t - W_{t-1} &= \alpha + \beta_1(P_t - P_{t-1}) + \beta_2(L_t - L_{t-1}) \\ &+ \beta_3(F_t - F_{t-1}) + \beta_4(I_t - I_{t-1}) \\ &+ \beta_5(T_t - T_{t-1}) + \beta_6(R_t - R_{t-1}) \\ &+ \beta_7(E_t - E_{t-1}) + \varepsilon_t \end{aligned} \quad (2)$$

That is:

$$\begin{aligned} \Delta W_t &= \alpha + \beta_1 \Delta P_t + \beta_2 \Delta L_t + \beta_3 \Delta F_t + \beta_4 \Delta I_t \\ &+ \beta_5 \Delta T_t + \beta_6 \Delta R_t + \beta_7 \Delta E_t + \varepsilon_t \end{aligned} \quad (3)$$

where $\Delta W_t = W_t - W_{t-1}$, $\Delta P_t = P_t - P_{t-1}$, $\Delta L_t = L_t - L_{t-1}$, $\Delta F_t = F_t - F_{t-1}$, $\Delta I_t = I_t - I_{t-1}$, $\Delta T_t = T_t - T_{t-1}$, $\Delta R_t = R_t - R_{t-1}$, and $\Delta E_t = E_t - E_{t-1}$.

Then, after doing the differenced translation, the data are still available for 44 years from 1961 to 2004.

3. Empirical Results and Analysis

3.1. The OLS Results

Using the OLS method, the statistical results obtained from the STATA are shown in **Table 1**. The R^2 is 0.6951, and the adjusted R^2 is 0.6359. Because one of the important assumptions of the classical linear regression model is that disturbances ε_t are homoscedastic, we were initially concerned about heteroscedasticity. However, when a white test was done, the chi-squared value was 39.93708 and its P value was 0.26, indicating there is no strong evidence for the assumption that homoscedasticity is violated. To test whether there is a first order autocorrelation, a Durbin-Watson test was also done. The Durbin-Watson d-statistic is 1.847752, which means there is no significant evidence of first order autocorrelation. Generally, the coefficients of population (P), the area of cultivated land (L), the water amount that flowed from the upper reaches (F) and the average peasant income (I) are all significant. The P values of population and peasant income are less than or about 0.001, the P value of the water that flowed from the upper reaches is less than 0.005, and even the P value of the cultivated land is also less than 0.05. This means there is significant evidence that the population, the cultivated land, the water that flowed from the upper reaches and the peasant income strongly influence the depth of wells (a proxy of the overuse of the groundwater). Concretely, the population coefficient is 2.888201, which means that if the population increases by 10 thousand, the depth of the wells will increase 2.888201 meters on average, assuming the other variables remain constant. The coefficient of the area of cultivated land is 0.8832096, which means that if the cultivated land increases by 10 thousand "mu"s, the depth of the wells will increase 0.8832096 meters on average, assuming the other variables remain constant. The coefficient of water that flowed from the upper reaches is –1.563346, which means the depth of the wells will increase 1.563346 meters on average when the water flowing from the upper reaches decreases by 0.1 billion·m^3, assuming the other variables remain constant. The coefficient of the peasant income is 0.0094834,

Table 1. Government policy impacts on the overuse of groundwater using the OLS method.

Independent variables (Differenced)	Regression coefficients	Standard errors
Intercept	0.8914002*	0.3255335
Population (P)	2.888201***	0.5514038
Area of cultivated land (L)	0.8832096*	0.3327417
Amount of water that flowed from the upper reaches	–1.563346**	0.4977644
Average peasant income (I)	0.0094834**	0.0025329
Average temperature (T)	0.1173247^{++}	0.4584166
Average precipitation (R)	–0.012006	0.0059962
Average evaporation (E)	–0.0036052*	0.0015649
F statistic (7, 36)	11.73***	
R-squared	0.6951	
Adjusted R-squared	0.6359	
Root MSE	1.6228	
White's general test statistic	39.93708^{+}	
Durbin-Watson d (8, 44)	1.847752	
N = 44		

Note: Significance levels are: $^{***}P < 0.001$, $^{**}P < 0.005$, $^{*}P < 0.05$; $^{+}P > 0.1$, $^{++}P > 0.7$ (two-tailed).

[1]Some readers here may worry about the problem of multicollinearity. However, based on pre-studies, we found there is no perfect or very high multicollinearity. Furthermore, our task is to find whether the factors mentioned above are really impacting the overuse of groundwater, and it is therefore unreasonable for us to use the "dropping variables" method. Thus, we follow the thought of the "Do Nothing" or "God's Will" schools and leave this in our model. Then, the above simple linear function form is still a useful model for us to study the targeted topic.

which means, if the other factors remain constant, the depth of the wells will increase 0.0094834 meters when the income increases by 1 Yuan. By the same token, the depth of the wells will increase 0.94834 meters on average when the income increases by 100 Yuan. The connections between the population, the area of cultivated land and the water that flowed from the upper reaches and the overuse of groundwater are quite clear. However, some readers may wonder why the peasant income is related to the overuse of groundwater. When peasants have more money, they may dig more wells or dig the wells deeper to get more water, organize other programs that may use more groundwater, use more technology to reduce the water evaporation and other kinds of water loss, and so on, all of which are the facts of life in the Minqin basin.

As for the coefficients of the control variables, the results are different. The P value of temperature is about 0.8, which is very large, meaning there is no statistically significant evidence that the temperature is an important factor in the overuse of groundwater. This finding contradicts common sense. **Figure 2(a)** shows that over last 45 years, the temperature had an obviously increasing trend, which was consistent with the trend of global warming, and many scholars concluded from this that temperature was an important factor in the overuse of groundwater. The P value of precipitation is about 0.053, or roughly 0.05. This suggests its impact on the overuse of groundwater is not significant at the level of 0.05, but it is still significant at the significance level about 0.055, and its coefficient is –0.012006. The negative number means that as the rate of precipitation increases, the well depth will decrease. The P value of evaporation is less than 0.05, and its coefficient is –0.0036052, another negative number. This is somewhat surprising, as many people generally expect that evaporation will have a positive effect on the groundwater use. Our model may give a negative value for several reasons: first, as shown in **Figures 2(b)** and **(c)**, we found that, in general, levels of precipitation and evaporation were relatively stable and changed little over the past 45 years. Second, although the evaporation had no obvious change over the past 45 years, the variation in the levels of evaporation exhibited a modest diminishing trend. Finally, at the beginning of the research period (in the early years of the 1960s), it had an extremely high value, but at the end of the research period (in 2001, 2002 and 2003), it had an extremely low value. These characteristics of the data may make the estimate of this particular coefficient inaccurate. Taken together as a total trend, the impacts of natural factors on the overuse of groundwater are not very significant, especially when we compare these effects to the effects of the four government policy proxy variables.

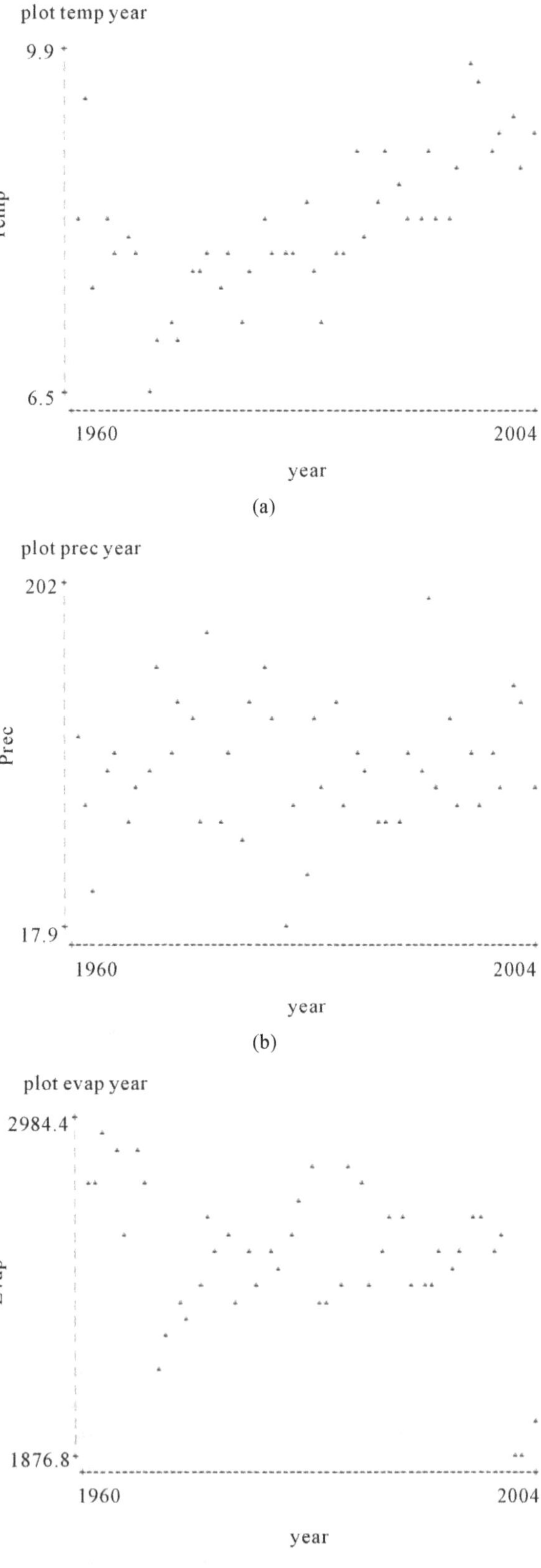

Figure 2. The change in temperature, precipitation and evaporation from 1960 to 2004. (a) Temperature; (b) Precipitation; (c) Evaporation.

3.2. The BJ Results

As mentioned above, our original time series are non-stationary, that is, they are integrated. We know the emphasis of the Box-Jenkins (BJ) methodology, technically known as the ARIMA (Autoregressive Integrated Moving Average) methodology, "is not on constructing single-equation or simultaneous-equation models but on analyzing the probabilistic, or stochastic, properties of economic time series on their own under the philosophy let the data speak for themselves" [9], and after doing some pre-study, we found we have a time period which is stationary except for one difference. Therefore, to get a more accurate result from our model, the BJ methodology was used to identify and estimate our statistical model. The results obtained from the STATA are shown in **Table 2**.

From **Table 2**, we know the value of Wald Chi2 is 87.04, and its P value is less than 0.001. All the coefficients of our independent variables remain the same; only the standard errors and P values are changed. The P values of the intercept, the area of cultivated land, and the amount of water that flowed from the upper reaches are all less than 0.05. The P values of population and the average peasant income are both less than 0.001. This represents a slight increase in the P value of the amount of water that flowed from the upper reaches from less than 0.005 to 0.05, but the P value of the average peasant income decreases a little bit from less than 0.005 to less than 0.001. They all remain significant, however. By contrast, for our control variables, the P value of temperature is larger than 0.8, and the P values of precipitation and evaporation are both larger than 0.1, indicating that by the BJ methodology, none of the coefficients of our control variables are significant. These results also tell us that the proxies of the government policy are the important factors in the overuse of groundwater, whereas the effects of the three natural factors on the overuse of groundwater are not significant.

Table 2. Government policy impacts on the overuse of groundwater using the BJ methodology.

Independent variables (Differenced)	Regression coefficients	Standard errors
Intercept	0.8914002*	0.4094415
Population (P)	2.888201***	0.5367814
Area of cultivated land (L)	0.8832096*	0.4331895
Amount of water that flowed from the upper reaches	−1.563346**	0.7772646
Average peasant income (I)	0.0094834**	0.0026848
Average temperature (T)	0.1173247^{++}	0.567676
Average precipitation (R)	−0.012006	0.0085309
Average evaporation (E)	−0.0036052*	0.0035872
Wald Chi2 (7)	87.04**	
Log likelihood	−79.32203	
N = 44		

Note: 1) Standard errors are given in parentheses beneath coefficients. 2) Significance levels are: $^{***}P < 0.001$, $^{**}P < 0.005$, $^{*}P < 0.05$; $^{+}P > 0.1$, $^{++}P > 0.8$ (two-tailed).

4. Does Government Policy Really Impact the Overuse of Groundwater?

4.1. Trend Analyses and Government Policy Matching

From the above analysis, we conclude that population, the cultivated land, the amount of water that flowed from the upper reaches and the peasant income all strongly influences the overuse of groundwater. But does this mean that government policies play an important role in the overuse of groundwater in Minqin? Some people may say these effects can also be attributed to the irrational behavior of the local people, on which most of the existing literature is premised, although this is often based solely on intuition. It may also be argued that although the government has some effect on these factors, the local people also influence these factors, and it may be impossible to separate the impact of government policy from local influence on groundwater overuse. Thus, it is important to determine which of these effects are caused by the government policy and to what degree. Essentially, however, the effects are the joint product of government policies and local behaviors because in general, government policies can only affect the overuse of groundwater by influencing local behaviors. Nevertheless, we believe these effects strongly reflect government policy based on two facts: first, as mentioned above and discussed in the next section, in a non-democratic political system, the behavior of the local people is always strongly managed and influenced by government policies. Second, there is a strict consistency between the variations of the government policies and the variations of our four proxy variables. This suggests the variations in our four proxy variables are really caused by and relatively strictly controlled or at least strongly influenced by the government and its polices. Certainly, the effects of local behavior on the overuse of groundwater can be deemed to be government policy effects on the overuse of groundwater. In this section, we assess the validity of these statements. To study this problem, we evaluate whether the variables used in our model are suitable proxies for government policy.

In terms of population, we know that the Chinese population was strongly influenced by political factors during Mao's era from 1949 to 1976, and China has had a one-child policy since 1974. Thus, the population is strictly controlled by government, except for some natural factors. This can be seen in **Figure 3(a)**. From this figure, we see from 1960 to 1964, the population dramatically decreased in Minqin County. This is due not to natural factors but to the failure of Mao's Great Leap, which left numerous peasants dead. Then, the government began to strongly encourage the peasants to bear more children to compensate for the population gap caused by the tragedy of Mao's Great Leap policy. Accordingly, after 1964, the population in Minqin began to increase again until 1976. During this period, one strange phenomenon should be considered: after a high rate of increase in 1968, the population suddenly decreased in 1969. This can also be explained by government policy because in this year many people died or moved to other places due to the political movement of the Expansion of the Anti-right Movement during the Great Cultural Revolution. After 1973, and even more so after 1975, the sharp increase in population ceased until 1983. This was a result of the one child policy in effect in Minqin County beginning in 1973 and in full force from1975. From 1984 to 1989, the population did not increase very fast and had two stages—1984 to 1986 and 1987 to 1989—which reflected the implementation effects of the one child policy under different governmental leaders. In the 1990s, the population also increased relatively quickly because young men born between 1965 and 1973 began to have their own children. At this point, the government encouraged the people to bear more children according to Mao's philosophy "the more people, the better". After 1999, this increase was stopped due to the one child policy and government-mandated emigration of peasants because of the serious desertification in Minqin County.

With regard to the amount of water that flowed from the upper reaches, as is clear from **Figure 3(c)**, the trend was continually decreasing. This trend is largely due to government policies. We can see that from 1967, the amount of water was dramatically decreased until the early years of the 1980s, and the speed was very fast. This is the result of many reservoirs being built after the Hongyashan reservoir was built in line with the government policy encouraging peasants to build more reservoirs to develop the economy. Much water was stored in these upper reach reservoirs, and the water use of the upper reaches also dramatically increased. **Table 3** shows some information about the exploitation of the water resources in the Shiyang River watershed. This table tells us the increases in population, irrigation area, number of reservoirs, amount of water in the reservoirs, length of the canals, surface water flowing in the canals and the water efficiency of the canal system from 1949 to 1995 are all consistent with the decrease in the amount of water that flowed from the upper reaches in Minqin.

The area of cultivated land is also strongly affected by government policy. **Figure 3(b)** shows that from 1964 to 1975, due to the Great Leap and the Great Cultural Revolution, many peasants died and more people took part in the political movement, leaving cultivated land fallow. This trend stopped in 1976 with the death of Mao Zedong, and the influence of the political movement on the peasant life diminished. Then, from 1976 to 1979, the area of cultivated land began to increase. However, beginning in 1980, due to the reform and opening up policy after Deng Xiaoping became the new leader of China in 1978, some cultivated land was used for city construction, and the total cultivated area decreased until 1984. It then increased again until 1988 because of new economic policies conducted by the central and local governments. At that point, the agricultural structure of Minqin became stable and governments also began to devote attention to industrial development in cities, so peasants could not cultivate more land. After 1992, due to the overheated economy effect from 1992 to 1995 especially in 1993 and 1995 after Deng Xiaoping's peak in 1992 in Southern China, more land was cultivated, particularly in 1993 and 1996, and this trend continued until 2000. After 2000, the policy of converting the land for forestry and pasture had to be conducted by the government according to Zhu Rongji, the new leader of the State Council of China, so the area of cultivated land continually decreased. This policy was continued when Hu Jingtao and Wen Jiabao became the new leaders of China in 2002.

Peasant income is also strongly influenced by government policy. From **Figure 3(d)**, we see there was almost no change in the average peasant income from 1960 to 1979 before the reform and opening up policy was implemented, due to the policy failures of Mao's era and too many political movements. From 1980 to 1993, although the overall trend of the average peasant income increased after Deng Xiaoping became the real leader of China, its pace was not fast. The increase had three different levels: 1980 to 1985, 1986 to 1989, and 1990 to 1993, all of which are completely consistent with the government economic patterns and reform steps from 1980 to 1993 in China, including different economic reforms, the overheated economic conditions, and important political events. From 1994 to 1997, peasant income rapidly increased, but this was due to the overheated economy effect after Deng's peak in 1992 as mentioned previously. From 1998 to 2001, the increase was less rapid, as the government focused on the problems of state-owned firms when Jiang Zeming and Zhu Rongji were the real leaders of China. From 2002 to 2004, pace picked up again and incomes increased quickly because the new peasant policy had been implemented to increase peasant income after Hu Jingtao and Wen Jiabao became

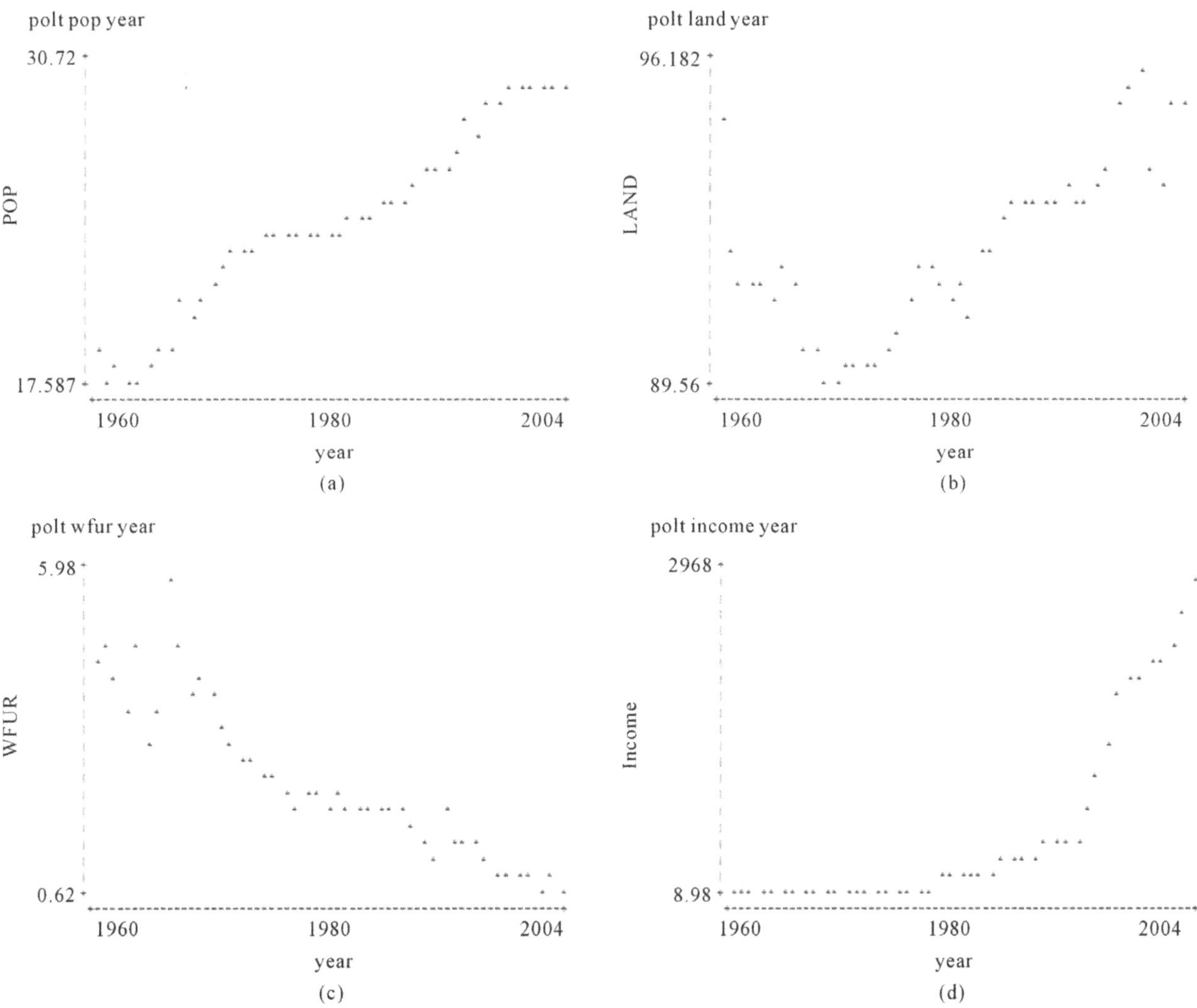

Figure 3. Change in population, cultivated land, amount of water that flowed from the upper reaches and peasant income from 1960 to 2004. (a) Change in population; (b) Change in cultivated land; (c) Change in the amount of water that flowed from the upper reaches; (d) Change in peasant income.

Table 3. Exploitation of water resources in the Shiyang River watershed.

Time	1949	1959	1972	1979	1985	1995
Population/($\times 10^6$)	0.9	1.08	1.1	1.43	1.55	2.2
Immigration area/($\times 10^5$ km^2)	1.34	1.55	2.07	2.44	2.72	3.03
Number of reservoirs	0	4	22	22	22	23
Amount of water in reservoirs/($\times 10^8$ m^3)	0	0.54	2.34	2.71	3.09	4.5
Length of canals/km	30	82	200	2046	2849.2	6716
Surface water flowing in canals/($\times 10^8$ m^3)		1	2	11.4	11.5	14.1
Water efficiency of canal system/%	20	35	45	50	58	65
Number of wells	0	0	582	11,700	121,000	142,000
Amount of groundwater exploitation/($\times 10^8$ m^3)	0	0	4.76	8.95	10.3	11.6

Source: Zhao *et al.* [10].

the new leaders of China in 2002.

In sum, from the analyses above, we found the variations of our four proxy variables are consistent with the government policy variation, particularly the central government policy variations. Thus, we can safely conclude that the four proxy variables in our model are indeed suitable proxies for government policy. It could be argued that these kinds of rough matching analyses are not reliable, but our answer is, if only one or two periods of the trends of different proxy variables match the government policy variation, this challenge would be reasonable. However, from the above analyses, we have seen that these matches are systematic and strict, indicating the reliability of the proxies. Furthermore, in the absence of more accurate data or more concrete information, and because of the difficulty of a continued statistical analysis in this domain, although there may be some caveats that will reduce the credibility and the causal validity of these analyses, this method is still a good choice to find the causal relationship between the proxy variables and the related government policies, and statistically and theoretically, these analyses are valuable.

4.2. The Government Itself Is the Problem

From the above analyses, we can conclude that, although we cannot clearly separate local effects from the effects showed by the four policy proxy variables, we do find the variations of these four policy proxies are highly consistent with the variation in government policy. This means that the four variables we chose to study in our basic model are credible proxies and indexes for government policy. Because the coefficients of these proxies are all significant, which means these factors strongly influence the overuse of groundwater, we can conclude that government policy really impacts the overuse of groundwater, which means the government itself is actually the problem despite its real and original desire to resolve the problem. This finding is consistent with E. Ostrom's arguments [11] that there is nothing special about the state, which is deemed as "the political science equivalent of Isaac Newton's recognition that the moon was falling" [12].

The effects of government policy on the overuse of groundwater can be divided into three basic dimensions. First, the government polices themselves are the institutions at the constitutional choice level or the collective choice level. The constitutional level institutions comprise the related government policies made by the central government in Beijing. The collective level institutions comprise the related government policies directly created by the provincial or county governments themselves. Second, the government policies create the new collective level institutions or change the old ones, like the implementation rules or polices made, chosen or changed by the provincial or county level governments under the requirement or directions of the central government policies, or the implementation rules, strategies, and concrete policies made, chosen or changed by the county level governments following the directions of the provincial rules created by the provincial policies. Third, the above two kinds of institutions create or change the operative level institutions or rules used by the locals. For example, the local peasants often should decide concrete rules, strategies under the large rule framework created by the constitution and collective level rules based on related government policies. Certainly, these rules and institutions, as a language-based phenomenon, cannot impinge directly on the real world [13]. However, by affecting the shared understandings of individuals making choices within decision situations, they impact the actors of decisions, the information structure, actions and their sequences, etc. [14]. As to the process, like Kiser and E. Ostrom [13] argued, these effects also have three steps: "First, the individuals affected by a change in rules must be cognizant of and abide by the change. Second, institutional change has to affect the strategies they adopt. Third, the aggregation of changed individual strategies must lead to different results" (1982: 56-57).

Although three levels of institutions always interact each other, within a non-democratic political system, the locals have little rights and power, and because there is little or asymmetric information to impact the other institutional creators like the bureaucrats in central, provincial and county governments, the operational level rules have little or no effect on the collective and constitutional institutions. This means that within a hierarchical and politicized system, the interaction directions of different levels of institutions are limited. Typically, only the strongly top-down effect exists, and the bottom-up effects do not. In such a system, "those who are assigned prerogatives to govern are assigned prerogatives that radically unequal to those who are subject to rules... Attributes of justice ascribed to equality under law say nothing of the inequalities between those who exercise governmental prerogatives and those who are subject to governmental prerogatives". Similarly, "the radical inequality between rulers and ruled is reinforced by the use of sanctions as instruments for enforcing rules" (V. Ostrom 1980: 384) [14]. This is the very reason why the behaviors of the local people and operative rules for using groundwater are strongly influenced by the two higher level rules, but they seldom have any effects on the higher level rules. Certainly, this is also the reason why we say local behavior and operative rules for using groundwater are strongly impacted by government policies. In this situation, all the locals can do is develop some new rules and strategies to try to adopt or comply with the collective and constitutional rules. If these kinds of behaviors lead them to pay too high a cost and they cannot or do not want to bear this cost, they will try to shirk or avoid the limitations of the exogenous rules, although sometimes these behaviors also have high costs and risks. If the collective and constitutional level rules are developed to try to limit the overuse of groundwater, these sequential strategies may lead to more serious problems of overuse of groundwater. In particular, if the collective level and constitutional level institutions are changed to try to eliminate these locals following operative rules, the locals have to find other new strategies and form new potential operative rules to avoid the new negative effects on them by these exogenous institutions. Thus, a paradox of an institutional competition game appears. However, in this competition game, the actors often pay a much higher cost, and it will often deteriorate the conditions of the CPR (common pool resources), which has been proven by the overuse of groundwater in the Minqin basin.

5. Conclusion

Chinese scholars and officials often overestimate the capability and knowledge of the locals and other entities, such as the enterprise, to resolve environmental problems after 1949. Therefore, they often assume the behaviors of the local people are to blame and deem them responsible for environmental problems, such as the overuse of groundwater [15]. However, the findings in this article show that the overuse of groundwater has been strongly affected by government policy in the Minqin basin after 1960. Because of the lack of the relevant, accurate data, this article uses proxy variables to estimate the effects of policy; however, the findings are still relatively credible, reliable and important, as shown by the above analyses. Thus, we conclude that the notion that all the overuse of groundwater in Minqin is due to the irrational behaviors of the local people or natural factors like temperature, precipitation and evaporation may be wrong and should be studied more carefully. We find there is strong statistical evidence that government policy has positive influences on the overuse of groundwater, but contrary to what might be expected, the effects on the overuse of the groundwater of all three natural factors—temperature, precipitation and evaporation—are not very significant in our model. Although these findings cannot distinguish local effects from government policy effects and clearly tell us the degree to which government policy impacts the overuse of groundwater through influencing the behaviors of the local people and their operative level institutions or rules, using the Trend Analysis, we find these significant effects are strictly consistent with variations in the government policies effective at the county level of the Minqin basin. If other scholars still argue that local behaviors really impact the overuse of the groundwater in the Minqin basin, we can reply that that such behaviors have been strongly influenced by government policy, based on our findings. Indeed, this statement is also supported by the record of the county annals of Minqin, which claimed that before 1949, the peasants in Minqin had developed a very complex self-governance system to resolve the problems of the oases and it worked very well for over several hundred years [15]. After 1949, this system was abolished on the grounds that it was ignorant, uneducated and based on backward traditions [16]. This phenomenon was described by E. Ostrom [11] in the West and Central basin in California and bears repeating in the Minqin case: "After several decades of institutional change, the resulting institutional infrastructure that had been created represented a major investment that dramatically changed the incentives and behaviors of participants and the resulting outcomes. Each institutional change became the foundation for the next change" (1990: 141). The main difference in the Minqin basin is that these kinds of institutional changes were caused by the exogenous government actors, rather than the local people themselves, and the institutional diversity was diminished.

6. Acknowledgements

The study was supported by the National Natural Science Foundation of China (71073008). The author wishes to thank Professor Vincent Ostrom and Elinor Ostrom for their kind help. I benefited tremendously from their thoughtful comments and suggestions.

REFERENCES

[1] S. G. Reynolds (Food and Agriculture of United Nations), "Sustainable Development of Grassland Ecosystems: Two Case Studies from China," *Proceedings of the International Symposium on Sustainable Development of Grassland Ecosystems*, Inner Mongolia, 27-30 August 2001. http://www.fao.org/ag/AGP/AGPC/doc/Present/China/china1.htm

[2] H. F. Lee and D. D. Zhang, "Perceive Desertification from the Lay Perspective in Northern China," *Land Degradation & Development*, Vol. 36, No. 5, 2004, pp. 529-542.

[3] L. Chen, Y. Qu, *et al*., "Water and Land Resources and Their Rational Development and Utilization in the Hexi Region," Science Press, Beijing, 1992.

[4] J. Shi, "Effect of Rainwater Reuse on Ecological Environment in Shiyanghe Catchment," *Proceedings of 7th International Rainwater Catchment System Conference*, Vol. 1, No. 3, 1995, pp. 1-7.

[5] P. Harris, "Sixth and Final Report on Range Management," Gansu Integrated Desert Control and Sustainable

Agriculture, CPR/91/111, 1997.

[6] Joey, "Will the Desert Claim Minqin?" *China Daily*, 2004.

[7] X. Ma, B. Li, C. Wu, H. Peng and Y. Guo, "Predicting of Temporal-Spatial of Ground Water Table Resulted from Current Land Use in Minqin Oasis," *Advances in Water Science*, Vol. 14, No. 1, 2003, pp. 85-90.

[8] Y. Li and F. Chen, "Water Resources Sustainable Utilization Countermeasures in Minqin Basin of Gansu Province," *Journal of Mountain Science*, Vol. 19, No. 5, 2001, pp. 465-469.

[9] D. N. Gujarati, "Basic Econometrics," 4th Edition, Tata McGraw-Hill Publishing Company Limited, New York, 2003.

[10] H. Zhao, J. Ma, G. Zhu and X. Li, "The Study on the Change of the Groundwater Environment and Its Causes in the Minqin Basin, Gansu Province," *Arid Zone Research*, Vol. 21, No. 3, 2004, p. 212.

[11] E. Ostrom, "Governing the Commons the Evolution of Institutions for Collective Action," Cambridge University Press, Cambridge, 1990.

[12] P. A. Schrodt, "Meso-Level Regimes and Robust Plans," Comment 1 on Workshop in Political Institutions, Workshop in Political Theory and Policy Analysis, Indiana University, Bloomington, 1992.

[13] V. Ostrom, "Artisanship and Artifact," In: McGinnis D. Michael, Ed., *Polycentric Governance and Development*, The University of Michigan Press, Ann Arbor, 1980, pp. 377-393.

[14] L. L. Kiser and E. Ostrom, "The Three Words of Action: A Metatheoretical Synthsis of Institutional Approaches," In: M. D. McGinnis, Ed., *Polycentric Games and Institutions*, The University of Michigan Press, Ann Arbor, 1982, pp. 56-88.

[15] L. Yang, "Scholar Participated Governance: Combating Desertification and Other Dilemmas of Collective Action," Ph.D. Dissertation, Arizona State University, Phoenix, 2009.

[16] L. Yang, Z. Lan and J. Wu, "Roles of Scholars in the Practice of Combating-desertification: A Case Study in Northwest China," *Environmental Management*, Vol. 46, No. 2, 2010, pp. 154-166.

25

Voltage Stability Constrained Optimal Power Flow Using NSGA-II

Sandeep Panuganti[1], Preetha Roselyn John[1], Durairaj Devraj[2], Subhransu Sekhar Dash[1]
[1]Department of Electrical and Electronics Engineering, Sri RamaswamyMemorial, Chennai, India
[2]DEAN (Research), Kalasalingam University, Krishnankoil, India

ABSTRACT

Voltage stability has become an important issue in planning and operation of many power systems. This work includes multi-objective evolutionary algorithm techniques such as Genetic Algorithm (GA) and Non-dominated Sorting Genetic Algorithm II (NSGA-II) approach for solving Voltage Stability Constrained-Optimal Power Flow (VSC-OPF). Base case generator power output, voltage magnitude of generator buses are taken as the control variables and maximum L-index of load buses is used to specify the voltage stability level of the system. Multi-Objective OPF, formulated as a multi-objective mixed integer nonlinear optimization problem, minimizes fuel cost and minimizes emission of gases, as well as improvement of voltage profile in the system. NSGA-II based OPF—case 1—Two objective-Min Fuel cost and Voltage stability index; case 2—Three objective—Min Fuel cost, Min Emission cost and Voltage stability index. The above method is tested on standard IEEE 30-bus test system and simulation results are done for base case and the two severe contingency cases and also on loaded conditions.

Keywords: Voltage Stability; Optimal Power Flow; Multi Objective Evolutionary Algorithms

1. Introduction

GA, invented by Holland in the early 1970s, is a stochastic global search method that mimics the metaphor of natural biological evaluation.Genetic Algorithms (GA) [1] operates on a population of candidate solutions encoded to finite bit string called chromosome. In order to obtain optimality, each chromosome exchanges the information using operators borrowed from natural genetic to produce the better solution. The combined Economic-Emission multiobjective problem seeks to simultaneously minimize both fuel costand the emissions produced by power plants. Environmental concerns on the effect of SO_2 and NOX emissions producedby the fossil-fueled power plants led to the inclusion ofminimization of emissions as an objective in the OPF formulation.

1.1. Voltage Stability

Voltage instability stems from the attempt of load dynamics to restore power consumption beyond the capability of the combined transmission and generation. Voltage stability constrained OPF—Voltage stability indicator is incorporated in the OPF formulation through the L-index value. The voltage stability index is an appropriate measure of the closeness of the system to voltage collapse. NSGA-II is a popular non-domination based genetic algorithm for multi-objective optimization which has a better sorting algorithm and incorporates elitism and no sharing parameter needed to be chosen as compared to the original NSGA. Emission cost of generators also play a vital role and is thus formulated in the minimization OPF problem. Since OPF was introduced in 1968, several methods have been employed to solve this problem, e.g. Gradient base, Linear programming method and Quadratic programming. However all of these methods suffer from three main problems. Firstly, they may not be able to provide optimal solution and usually getting stuck at local optima [2]. Secondly, all these methods are based on assumption of continuity and differentiability of objective function which is not actually allowed in a practical system.

1.2. VSC-OPF

The Contingencies such as unexpected line outages in a stressed system may often result in voltage instability, which may lead to voltage collapse. After a voltage collapse, the system becomes dismantled owing to the widespread operation of protective devices. Studies have been performed to predict the voltage instability with both static and dynamic approaches.

In this paper three different cases along with the system loaded conditions are considered. In the first case base case OPF as a single objective optimization problem is solved using GA [3]. In the second case VSC-OPF problem is formulated in MOGA with minimization of fuel cost and L-index value. In the third case economic emission of gases along with VSC-OPF problem is considered as a multi-objective problem and L-index is solved using the NSGA-II approach in an IEEE 30 bus system. NSGA [4] is a popular non-domination based genetic algorithm for multi-objective optimization. It is a very effective algorithm but has been generally criticized for its computational complexity, lack of elitism and for choosing the optimal parameter value for sharing parameter *σ share*. A modified version, NSGA-II [5] was developed, which has a better sorting algorithm, incurporates elitism and no sharing parameter needs to be chosen *a priori*.

2. Voltage Stability Index

The voltage stability analysis involves determination of an index known as voltage collapse proximity indicator. This index is an approximate measure of the closeness of the system to voltage collapse. There are various methods of determining the voltage collapse proximity indicator. One such method is the L-index method proposed in Kessel and Glavitsch. It is based on load flow analysis. Its value ranges from 0 (no load condition) to 1 (voltage collapse). The bus with the highest L-index value will be the most vulnerable bus in the system. The technique is incorporated from [6]. The L-index calculation for a power system is briefly discussed below.

Consider an N-bus system in which there are N_g generators. The relationship between voltage and current can be expressed by the following expression:

$$I_{bus} = Y_{bus} \cdot V_{bus} \tag{1}$$

By segregating the load buses (PQ) from generator buses (PV), Equation (1) can write as

$$\begin{bmatrix} I_G \\ I_L \end{bmatrix} = \begin{bmatrix} Y_{GG} & Y_{GL} \\ Y_{LG} & Y_{LL} \end{bmatrix} \begin{bmatrix} V_G \\ V_L \end{bmatrix} \tag{2}$$

where I_G, I_L and V_G, V_L represent currents and voltages at the generator buses and load buses.

Rearranging the above equation we get:

$$\begin{bmatrix} V_L \\ I_G \end{bmatrix} = \begin{bmatrix} Z_{LL} & F_{LG} \\ K_{GL} & Y_{GG} \end{bmatrix} \begin{bmatrix} I_L \\ V_G \end{bmatrix} \tag{3}$$

where:

$$F_{LG} = -[Y_{LL}]^{-1}[Y_{LG}]$$

The L-index of the jth node is given by the expression:

$$L_j = \left| 1 - \sum_{i=1}^{N_g} F_{JI} \frac{V_i}{V_j} \angle(\theta_{ji} + \delta_i - \delta_j) \right| \tag{4}$$

where:

V_i	Voltage magnitude of *i*th generator
V_j	Voltage magnitude of jth generator
θ_{ji}	Phase angle of the term F_{ji}
δ_i	Voltage phase angle of ith generator unit
δ_j	Voltage phase angle of jth generator unit
N_g	Number of generating units.

V_L, I_L: Voltages and Currents for PQ buses; V_G, I_G: Voltages and Currents for PV buses; Where, Z_{LL}, F_{LG}, K_{GL}, Y_{GG}: sub matrices generated from Y_{bus} partial inversion.

L_j: L-index voltage stability indicator for bus *k*.

The values of F_{ji} are obtained from the matrix F_{LG}. The L-indices for a given load condition are computed for all the load buses and the maximum of the L-indices (L^{max}) gives the proximity of the system to voltage collapse. The L-index has the advantage of indicating voltage instability proximity of the current operating point without calculation of the information about the maximum loading point.

3. Problem Formulation

In general, the OPF problem is formulated as an optimisation problem in which a specific objective function is minimised while satisfying a number of equality and inequality constraints[7]. The objectives of the OPF problem considered here are minimisation of fuel cost in the normal state and the minimisation of the voltage stability index L^{max} in the emergency state. Power flow equations are the equality constraints of the problem, while the inequality constraints include the limits on real and reactive power generation and bus voltage magnitude as follows.

$$\text{Minimise } F_1 = \sum_{i=1}^{N_G} (a_i P_{Gi} + b_i P_{Gi} + c_i) \tag{5}$$

$$\text{Minimise } F_2 = \sum L^{max.} \tag{6}$$

$$\text{Minimise } F_3 = \sum_{i=1}^{N_G} (d_i P_{Gi} + e_i P_{Gi} + g_i) \tag{7}$$

$$\text{Inequality constraints } P_{Gi}^{min.}{}_{Gi} \le P_{Gi} \le P_{Gi}^{max.} \tag{8}$$

$$V_i^{min.} \le V_i \le V_i^{max.} \tag{9}$$

$$Q_{Gi}^{min.} \le Q_{Gi} \le Q_{Gi}^{max.} \tag{10}$$

$$\text{Equality Constraints } P_G = P_D \tag{11}$$

where:

N the number of total buses

N_G the number of generator buses

N_L the number of load buses

N_b the number of transmission lines

P_i,Q_i real and reactive power injected at bus *i*

$|V_i|$ Voltage magnitude at bus *i*

The equality constraints given by the above equations are satisfied by running the power flow program. The active power generation (P_{gi}) (except the generator at the slack bus) and generator terminal bus voltages (V_{gi}) are the optimization variables and they are self-restricted by the optimization algorithm.

4. Non-Dominated Sorting Genetic Algorithm II (NSGA-II)

NSGA introduced by Srinivas and Deb [8], implements the idea of a selection method based on classes of dominance of all solutions. This algorithm identifies non-dominated solutions in the population, at each generation, to form non-dominated fronts, based on the concept of non-dominance of Pareto. After this, the usual selection, crossover, and mutation operators are performed.

However, there are some disadvantages in NSGA. It has been generally criticized for its computational complexity, lack of elitism and for choosing the optimal parameter value for sharing parameter σ_{share}. A modified version, NSGA-II was developed, which has a better sorting algorithm, incorporates elitism and no sharing parameter needs to be chosen a priori [9]. In this algorithm, the population is initialized as random, and the number of population is N. Once the population in initialized the population is sorted based on non-domination into each front. The first front being completely non-dominant set in the current population and the second front being dominated by the individuals in the first front only and the front goes so on. Each Individual in the each front are assigned rank values or based on front in which they belong to. Then, crowding distance is calculated for each individual. The crowding distance is a measure of how close an individual is to its neighbours.

The NSGA-II procedure is also shown in **Figure 1**. Parents are selected from the population by using binary tournament selection based on the rank and crowding distance. The individual with lesser rank or greater crowding distance is selected. The selected population generates offspring from crossover and mutation operators. The population with the current population and current offspring is sorted again based on non-domination and only the best N individuals are selected. The selection is based on rank and on crowding distance on the last front. Then the new population will be selected as parents at the next round.

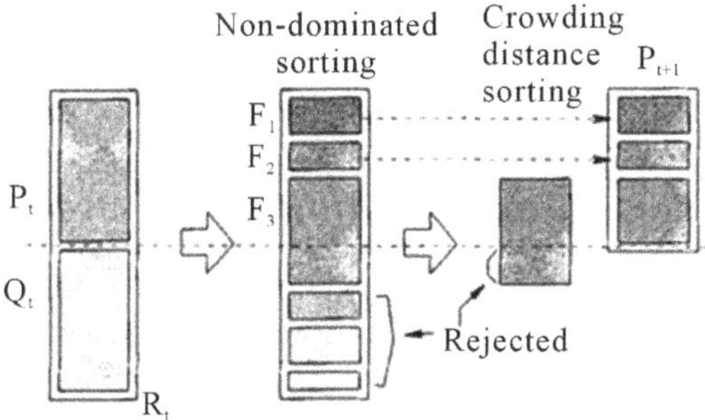

Figure 1. NSGA-II procedure.

4.1. Population Initialization

The population is initialized based on the problem range and constraints if any.

4.2. Non-Dominated Sort

The The initialized population is sorted based on non-domination.The fast sort algorithm is described as below.

- For each individual p in main population P do the following
  Initialize $Sp = \phi$. this set would contain all the individuals that are being dominated by p.
  Initialize $n_p = 0$. This would be the number of individuals that dominate p.
  For each individual q in P
  If p dominated q then
  Add q to the set S_p
  Else if q dominated p then
  Increment the dominated counter for p *i.e.* $n_p = n_p + 1$.
  If $n_p = 0$ *i.e.* no individual dominate p then p belongs to the first front, set rank of individual p to one *i.e.* $p_{rank} = 1$. Update the first front set by adding p to front one *i.e.* $F_1 = F_1 \cup \{p\}$
- This is carried out for all the individuals in main population P.
- Initialize the front counter to one $I = 1$.
- Following is carried out while the ith front is non-empty *i.e.* $Fi \neq \phi$.
  $Q = \phi$. The set for storing the individuals for $(i + 1)^{th}$ front.
  For each individual p in front F_i
  For each individual q in S_p (S_p is the set of individuals dominated by p)
  $n_q = n_q - 1$, decrement the domination count for individual q.
  If $n_q = 0$ then none of the individuals in the subsequent fronts would dominate q. hence set $q_{rank} = i + 1$.
  Update the set Q with individual q *i.e.* $Q = Q \cup q$.
  Increment the front counter by one.
  Now the set Q is the next front and hence $F_i = Q$.

This algorithm is better than NSGA [10] since it utilize the information about the set that an individual dominate (S_p) and number of individuals that dominate the individual (n_p).

4.3. Crowding Distance

Once the non-dominated sort is complete the crowding distance is assigned. Since the individuals are selected based on rank and crowding distance all the individuals in the population are assigned a crowding distance value. Crowding distance is assigned front wise and comparing the crowding distance between two individuals in different front is meaningless. The crowing distance is calculated as below

- For each front *Fi*, *n* is the number of individuals.
- Initialize the distance to be zero for all the individuals *i.e.* *Fi*(*dj*) = 0, where *j* corresponds to the jth individual in front *Fi*.
- For each objective function m.
- Sort the individuals in front *Fi* based on objective m *i.e.*, *i* = sort (*Fi*, *m*).
- Assign infinite distance to boundary values for each individual in F_i *i.e.* $I(d_1)=\infty$ and $I(Dn)=\infty$.
- For *k* = 2 to (*n* − 1)

$$I(d_k)=I(d_k)+\frac{I(k+1)\cdot m-I(k-1)\cdot m}{f_m^{\max}-f_m^{\min}}. \quad (12)$$

- $I(k)\cdot$m is the value of the mth objective function of the kth individual in *I*.

The basic idea behind the crowing distance is finding the euclidian distance between each individual in a front based on their m objectives in the m dimensional hyper space. The individuals in the boundary are always selected since they have infinite distance assignment.

4.4. Selection

Once the individuals are sorted based on non-domination and with crowding distance assigned, the selection is carried out using a ***crowded-comparison-operator*** (a_n). The comparison is carried out as below based on

1) Non-domination rank p_{rank} *i.e.* individuals in front F_i will have their rank as $p_{\text{rank}} = i$.

2) Crowding distance $F_i(d_j)$

- $p\ a_n\ q$ if
- $p_{\text{rank}} < q_{\text{rank}}$
- or if *p* and q belong to the same front F_i then $F_i(d_p) > F_i(d_q)$ *i.e.* the crowing distance should be more.

The individuals are selected by using a binarytournament selection with crowed-comparison-operator.

4.5. Genetic Operators

NSGA-II use **Simulated Binary Crossover (SBX)** [10,11] and **polynomial mutation** [10,12].

4.5.1 Simulated Binary Crossover

The Simulated binary crossover simulates the binary crossover observed in nature and is give as below.

$$\left.\begin{aligned} C_{1,k}&=\frac{1}{2}\left[(1-\beta_k)p_{1,k}+(1+\beta_k)p_{2,k}\right]\\ C_{2,k}&=\frac{1}{2}\left[(1+\beta_k)p_{1,k}+(1-\beta_k)p_{2,k}\right]\end{aligned}\right\} \quad (13)$$

where $C_{i,k}$ is the ith child with k^{th} component, $P_{i,k}$ is the-selected parent and β_k ($\geq$) is a sample from a random number generated having the density

$$\left.\begin{aligned} p(\beta)&=\frac{1}{2}(\eta_c+1)\beta^{\eta_c}, \text{if } 0\leq\beta\leq 1\\ p(\beta)&=\frac{1}{2}(\eta_c+1)\frac{1}{\beta^{\eta_c+2}}, \text{if } \beta>1\end{aligned}\right\} \quad (14)$$

This distribution can be obtained from a uniformly sampled random number *u* between (0, 1). η_c is the distribution index for crossover. That is

$$\left.\begin{aligned} \beta(u)&=(2u)^{\frac{1}{(\eta+1)}}\\ \beta(u)&=1\Big/\left[2(1-u)\right]^{\frac{1}{(\eta+1)}}\end{aligned}\right\} \quad (15)$$

4.5.2. Polynomial Mutation

$$c_k = p_k + \left(p_k^u - p_k^l\right)\delta_k \quad (16)$$

where c_k is the child and p_k is the parent with p_k^u being the upper boundon the parent component, p_k^l is the lower bound and δ_k is small variation which is calculated from a polynomial distribution by using

$$\left.\begin{aligned} \delta_k&=(2r_k)^{\frac{1}{\eta_m+1}}-1, \text{if } r_k<0.5\\ \delta_k&=1-\left[2(1-r_k)\right]^{\frac{1}{\eta_m+1}}, \text{if } r_k\geq 0.5\end{aligned}\right\} \quad (17)$$

r_k is an uniformly sampled random number between (0,1) and η_m is mutation distribution index.

4.6. Recombination and Selection

The offspring population is combined with the current generation population and selection is performed to set the individuals of the next generation. Since all the previous and current best individuals are added in the population, elitism is ensured. Population is now sorted based on non-domination. The new generation is filled by each front subsequently until the population size exceeds the current population size. If by adding all the individuals in front *Fj* the population exceeds *N* then individuals in front *Fj* are selected based on their crowding distance in the descending order until the population size is *N*. And hence the process repeats to generate the subsequent generations.

5. Best compromised Solution

Upon having the pareto-optimal set of non-dominated solution, the proposed approach [8] presents a best compromise solution tothe decision maker. Due to the imprecise nature of the decision maker's judgement, the ith objective function *Ji* is represented bya membership function defined as

$$\mu\left(J_i^-\right)=\begin{cases}1, & J_i \le J_i^{\min} \\ \dfrac{J_i^{\max}-J_i^-}{J_i^{\max}-J_i^{\min}}, & J_i^{\min} < J_i^- < J_i^{\max} \\ 0, & J_i \le J_i^{\max}\end{cases} \quad (18)$$

where $J_i^{\max}$ and $J_i^{\min}$ are the maximum and minimum values of the ith objective function among all non-dominated solutions.

For each non-dominated solution k, the normalized membership function μ_D^K is calculated as

$$\mu_D^K = \frac{\sum_{l=1}^{K}\mu\left(J_i^K\right)}{\sum_{k=1}^{K}\sum_{i=1}^{5}\mu\left(J_i^K\right)} \quad (19)$$

6. Simulation Results

The proposed NSGA-II approach has been applied to solve the VSC-OPF problem in an IEEE 30-bus test system. The system has six generator buses, 24 load buses and 41 transmission lines.The generator cost coefficients and the transmission line parameters are taken from [12]. Three different cases were considered for simulation, one without considering the voltage stability *i.e*, to solve the VSC-OPF problem using MOGA and the second one is solved having economic emission of gases including VSC-OPF in NSGA-II.These simulations were implemented using the MATLAB program. The results of these simulations are presented, **Figure 2**.

In this case the two objectives are minimization of fuel cost and minimization of L-index using multi-objective Genetic Algorithm**.** The results of VSC-OPF using MOGA is shown in **Table 1**.

6.1. (Case 1): VSC-OPF Using NSGA-II

The voltage stability index (L-index) was included as the second objective function of the OPF problem along with the base fuel cost. The NSGA-II based algorithm was applied to solve this VSC-OPF problem. The optimal control variable setting obtained in this case is presented in **Table 2** alongwith the L-index value. In **Figure 4** shows the pareto optimal front of generation cost and L-index is shown and the **Table 2** shows the line outage 27 - 28 along with line outage 27 - 30 is shown in **Table 3**.The solution is a set of non-dominated solutions. The comparison of the results obtained in NSGA-II and three objective is shown in **Table 5**. From this table it is clear that the performance of NSGA-II is better than MOGA in VSC-OPF problem.

Contingency analysis was conducted on the system with 125% loaded condition by simulating the single line outages and in each case the maximum L-index value was evaluated. From the contingency analysis it was found that line outage 28 - 27 is the most severe one from the voltage security point of view during this contingency state.

Table 6 gives the fuel cost, $L^{\max}$ and minimum voltage value of the contingency constrained VSC-OPF using NSGA-II. This reduction in $L^{\max}$ is obtained at the expense of increased fuel cost. **Figure 5** shows the pareto optimal front of contingency constrained VSC-OPF.

The line outage for 27 - 30 as shown in **Figure 4**, in **Tables 4**, **5** and is also performed along with the same loaded condition as in line outage 27 - 28 as shown in **Figure 3** and results are tabulated. The reduction in L-max is obtained at the extent of increased fuel cost.

Table 1. NSGA-II base case.

Control Variables	Min F1 (Fuel cost)	Min F2 (L-index)	Best compromised sol.
P1	169.6545	140.9243	168.6642
P2	50.0000	80.0000	50.4760
P5	23.9893	42.9811	24.1108
P8	22.0474	18.9395	22.1680
P11	13.2559	17.0000	13.4598
P13	14.8000	20.4000	14.8000
V1	1.0000	1.0000	1.0000
V2	1.0000	1.0040	1.0000
V5	0.9826	1.0000	0.9991
V8	0.9884	1.0000	0.9891
V11	0.9899	0.9903	0.9918
V13	0.9874	0.9888	0.9933
Fuel cost, F1	807.1765	872.7911	807.9227
L-index , F2	0.1101	0.1075	0.1095

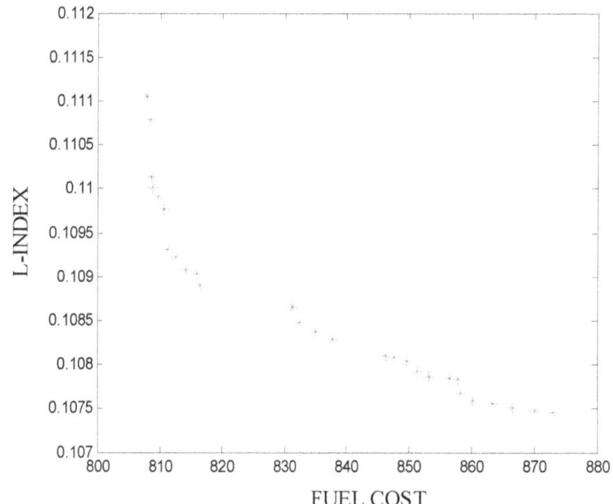

Figure 2. NSGA-II base case.

Table 2. NSGA II—Line outage 27 - 28.

Control Variables	Min F1 (Fuel cost)	Min F2 (L-index)	Best compromised sol.
P1	174.9058	112.3927	149.0853
P2	50.000	80.0000	61.6056
P5	22.0000	41.7412	27.7161
P8	22.500	22.4915	19.6752
P11	12.8957	16.1521	16.8888
P13	14.8000	20.3995	20.4000
V1	1.000	1.0000	1.0000
V2	1.000	0.9901	0.9988
V5	0.9874	0.9874	0.9856
V8	0.9917	1.0040	0.9911
V11	0.9903	0.9902	1.0000
V13	0.9866	0.9866	0.9859
Fuel cost, F1	814.0790	876.6776	814.2402
L-index, F2	0.2905	0.2877	0.2895

Table 3. NSGA-2—Line outage 27 - 30.

Control Variables	Min F1 (Fuel cost)	Min F2 (L-index)	Best compromised sol.
P1	173.2358	107.6728	147.3212
P2	50.0000	80.0000	55.1566
P5	22.2229	42.9728	30.5305
P8	19.2142	22.5000	22.3421
P11	15.1615	17.0000	17.0000
P13	14.8000	20.4000	20.2354
V1	1.0000	1.0000	1.0000
V2	1.0000	0.9990	0.9984
V5	0.9793	0.9876	0.9844
V8	0.9866	1.0020	1.0020
V11	0.9875	0.9930	0.9914
V13	0.9835	0.9902	0.9886
Fuel cost, F1	810.0262	876.3660	812.5044
L-index , F2	0.1989	0.1953	0.1964

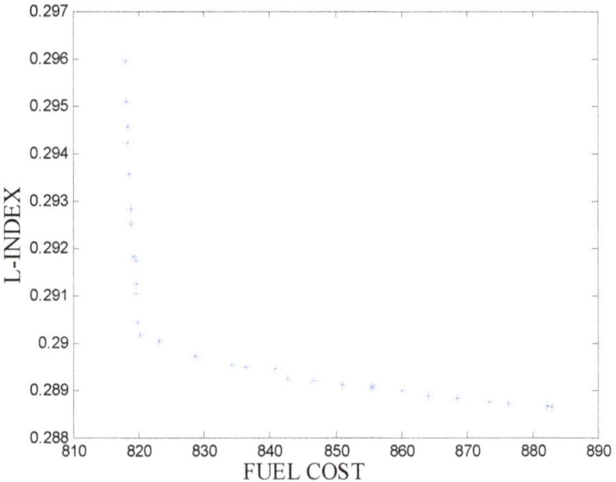

Figure 3. NSGA-II—Line outage 27 - 28.

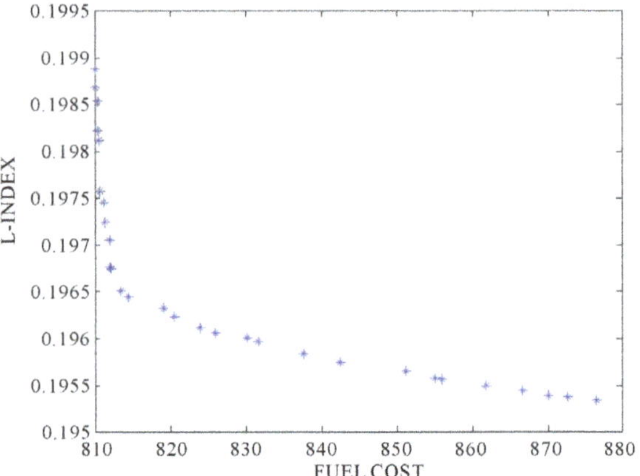

Figure 4. NSGA-2—Line outage 27 - 30.

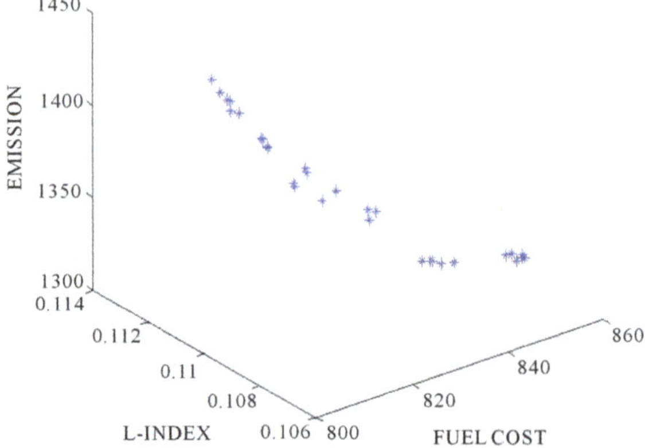

Figure 5. NSGA 2—3 Objective base case.

6.2. (Case 2): Economic Emission Based VSC-OPF Using NSGA-II

The economic emission of the gases are included as the third objective along with the voltage stability index and base fuel cost. The NSGA-II based algorithm was applied to solve this VSC-OPF problem. The optimal control variable settings are similar to that of the two objective case. In **Figure 5** shows the pareto optimal front of generation cost, L-index and economic emission dispatch of gases for base case and in **Figures 6, 7** the line outage 27 - 28 and line outage 27 - 30 are also included.

Table 4. NSGA-II—3 objective line outage 27 - 28.

Control Variables	Min F1 (Fuel cost)	Min F2 (L-index)	Min Emission, F3	Best compromised sol.
P1	160.3011	165.1097	169.5373	162.3011
P2	65.7505	60.3991	57.8956	66.7505
P5	23.3074	25.8083	22.000	24.3172
P8	16.6748	14.5257	15.4874	16.6738
P11	12.8745	13.0869	14.3629	12.6730
P13	14.8000	14.9005	14.8981	14.8230
V1	1.000	1.000	1.000	1.000
V2	1.000	1.000	1.000	1.000
V5	0.9999	1.000	0.9999	0.9999
V8	0.9999	1.000	0.9999	0.9899
V11	0.9999	1.000	1.000	0.9994
V13	0.9999	0.9999	1.000	0.9999
Fuel cost, F1	807.0019	853.6211	856.5133	809.0019
L-index, F2	0.1102	0.1077	0.1083	0.1099
Emission, F3	1424.5	1324.0	1316.2	1386.5

Table 5. NSGA-2—Line outage 27 - 30.

Control Variables	Min F1 (Fuel cost)	Min F2 (L-index)	Min Emission, F3	Best compromised sol.
P1	170.8424	167.0531	145.6723	168.5707
P2	50.000	57.9636	69.0360	57.4857
P5	23.1042	23.8583	23.2324	23.0908
P8	21.3731	16.0485	20.4102	15.9655
P11	14.2896	12.6677	15.5218	12.8070
P13	14.8000	16.8955	19.3528	16.6933
V1	1.000	1.000	1.000	1.000
V2	1.000	1.000	1.000	1.000
V5	0.9989	1.000	0.9999	0.9999
V8	0.9883	0.9991	0.9999	1.000
V11	0.9914	1.000	0.9999	0.9999
V13	0.9925	0.9999	1.000	0.9999
Fuel cost, F1	809.9571	874.2141	844.7509	810.5175
L-index, F2	0.1978	0.1944	0.1961	0.1954
Emission, F3	1413.7	1305.8	1325.7	1316.8

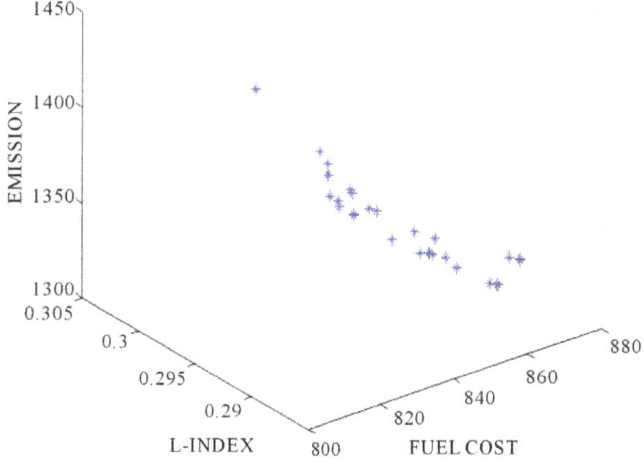

Figure 6. NSGA-2—Line outage 27 - 28.

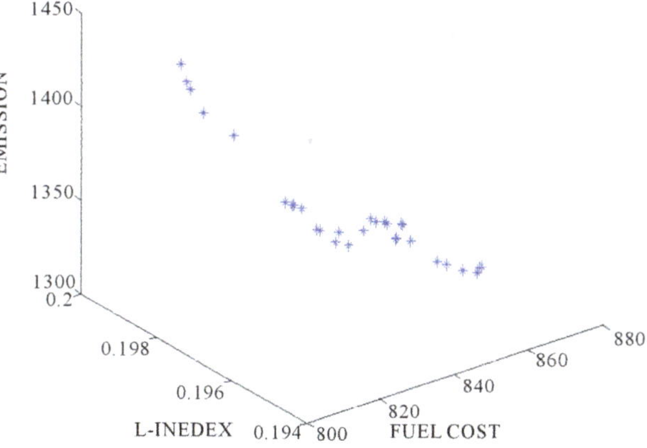

Figure 7. NSGA-2—Line outage 27 - 30.

7. Conclusion

In this paper, the various aspects of single-objective optimal power flow and multi-objective voltage stability constrained optimal power flow are studied. An efficient and diversified approach using NSGA-II algorithm is identified to solve the above multi-objective optimization problems. Several case studies have been employed separately for single & multi-objective optimization problem. Firstly, the results are obtained for single objective OPF and contingency constrained VSC-OPF using genetic algorithm for the optimization of Fuel cost which are then compared with the power flow results of other papers. The multi-objective VSC-OPF problem is formulated using NSGA-II algorithm. The proposed algorithm occupies less memory space and takes CPU time than conventional GA approach. Simulation results of the IEEE 30-bus system have been presented to illustrate the effectiveness of the proposed approach to solve the VSC-OPF problem. This simulation results were carried out using NSGA-II and are found that voltage stability is improved in NSGA-II than multi-objective GA of the proposed algorithm than the other approaches.

REFERENCES

[1] N. Srinivas and K. Deb, "Multi-Objective Optimization Using Nondominated Sorting Ingenetic Algorithms," Technical Report, Department of Mechanical Engineering, Indian Institute of Technology, Kanpur, 1993.

[2] N. Srinivas and K. Deb, "Multi-Objective Optimization Using Nondominated Sortingin Genetic Algorithms," *Evolutionary Computation*, Vol. 2, No. 3, 1994, pp. 221-248.

[3] A. J. Wood and B. F. Wollenberg, "Power Generation Operation and Control," John Wiley & Sons, Inc., New York, 1996.

[4] M. S. Kumari, "Enhanced Genetic Algorithm Based Computation Technique for Multi-Objective, Optimal Power Flow Solution," *Electrical Power and Energy Systems*, Vol. 32, No. 6, 2010, pp. 736-742.

[5] K. O. Alawode, A. M. Jubril and O. A. Komolafe, "Multi-Objective Optimal Power Flow Using Hybrid Evolutionary Algorithm," *International Journal of Electrical Power & Energy Systems Engin*, Vol. 3, No. 3, 2010, p. 196.

[6] D. Devaraj and J. P. Roselyn, "Improved Genetic Algorithm for Voltage Security Constrained Optimal Power Flow Problem," *International Journal of Energy Technology and Policy*, Vol. 5, No. 4, 2007, pp. 475-488.

[7] H. Sadat, "Power Systems Analysis," McGraw Hill Publication, New Delhi, 1997.

[8] O. Alsac, and B. Scott, "Optimal Load Flow with Steady State Security," *IEEE Transactions on Power Systems*, Vol. PAS-93, No. 3, 1974, pp.745-751.

[9] H. W. Dommel and W. F. Tinney, "Optimal Power Flow Solutions," *IEEE Transactions on Power Apparatusand Systems*, Vol. PAS-87, No. 10, 1968, pp. 1866-1876.

[10] S. Dhanalakshmi, S. Kannan, K. Mahadevan and S. Baskar, "Application of Modified NSGA-II Algorithm to Combined Economicand Emission Dispatch Problem," *International Journal of Electrical Power & Energy Systems*, Vol. 33, No. 4, 2011, pp. 992-1002.

[11] R. He, G. A. Taylor and Y. H. Song, "Multi-Objective Optimal Reactive Power Flow including Voltage Security and Demand Profile Classification," *International Journal of Electrical Power & Energy Systems*, Vol. 30, No. 5, 2008, pp. 327-336.

[12] K. O. Alawode, A. M. Jubril and O. A. Komolafe, "Multi-Objective Optimal Reactive Power Flow Using Elitist Nondominated Sorting Genetic Algorithm: Comparison and Improvement," *Journal of Electrical Engineering & Technology*, Vol. 5, No. 1, 2010, pp. 70-78.

26

Introspections and Suggestions on the Amount Fixing of Administrative Penalty for Environmental Pollution

Xiaohong Zheng
Law school Southwest University,Chongqing ,China,400715

ABSTRACT

Environmental pollution has seriously damaged the health of mankind and the development of future generations. Because pollution damages are irreversible, taking effective measures to prevent contamination accidents is the primary task for administrative penalty on environmental pollution. However, the specific amount of such penalty is a little bit on the low side, and the standard for fixing the penalty is not rational somewhat. As a result, the original functions and purpose of administrative penalty on environmental pollution cannot be fully achieved. By comparing and using for reference related systems at home and abroad, this paper advances the drawbacks and suggestions for the amount fixing of China's administrative penalty on environmental pollution in the hope of supplying theoretical basis for the improvement of China's administrative penalty mechanism on environmental pollution.

Keywords: Environment; Administrative Penalty; Amount Fixing

1. Introduction

Over the past twenty years, China has experienced the period of high environmental risks. According to statistics, the direct economic losses merely brought about by environmental pollution each year amount to RMB 120 billion Yuan. Besides, environmental pollution will result in potential and irreversible damages to human health. As an important measure for environment-oriented law enforcement, the administrative penalty mechanism on environmental pollution made some achievements in the past, but it is rather weak and seriously lags behind in its performance with the increasingly serious environmental pollutions. The most typical case was the ConocoPhillips's oil spillage accident in June 2011: the polluted sea area covered 6,200 square kilometers, which led to a direct economic loss of more than RMB 100 million Yuan, which seriously damaged the public health. However, the corresponding administrative penalty was only RMB 200,000 Yuan, which was only a minuscule amount for a company with an asset of tens of billions of USD like ConocoPhillips, and could not truly show its warning effect to the violator. Another case is the American's Mexico Gulf oil leakage accident in 2010: the responsible party was imposed a fine of 1.256 billion USD, and was required to establish a foundation of 20 billion USD as well as make a subsequent payment of 3.269 billion USD for solving the accident. The responsible party for an oil spillage accident in November 2011 in Brazil was promptly imposed a heavy fine of 50 million BRL (equivalent to 28 million USD). The greatly different results between China and foreign countries in handling environmental pollution accidents push us to re-examine the design of China's administrative penalty system on environmental pollution and to re-think of the fixing on the amounts of such penalties so as to find out the crux of the problem.

2. Related Legal Provisions on Administrative Penalty for Environmental Pollution in China

Presently, China has been implementing the *Law of the People's Republic of China on Environmental Protection* that is adopted in 1989. Article 35 of the law entitles the competent administrative department of environmental protection to give warning or impose a fine to "such behaviors as refusing the spot inspection by the competent administrative department of environmental protection or other departments that exercise administration and supervision on environment according to law, or resorting to deception during the inspection", that is, the competent administrative department of environmental protection was granted the power of imposing administrative penalty directly. The *Law of the People's Republic of China on Administrative Penalty* implemented on Oct. 1, 1996 applies the principle of "moderate punishment on fault," such as Article 4, which stipulates "the administrative penalty shall be fixed and implemented on the basis of fact and shall be moderate to the act, nature and circumstances of the offence as well as to the damage to the society," In Article 51, it is stipulated that the faulty party shall, if failing to pay the penalty over due time, be imposed a fine of 3% of the penalty for each overdue day; in the Article 73 of the Law of the People's Republic of China on Marine Environment Protection implemented on April 1, 2000, it is stipulated that the faulty party who discharges the pollutant that is forbidden to discharge to the sea area by this law shall be imposed a fine amounting to no less than RMB 30,000 Yuan but no more than RMB 200,000 Yuan," which is a provision on maximum punishment for the marine environmental pollution in China; the Article 75 of the new *Law of the People's Republic of China on Water Pollution Prevention and Control* (revision in 2008) stipulates that "the people's government at county level or above shall order the party who sets up sewage outlet within the protective area of the drinking water source to dismantle such outlet within the limited time and impose a fine no less than RMB 100,000 Yuan but no more than RMB 500,000 Yuan, provided the responsible party fails to dismantle such outlet in the due time, the outlet will be dismantled forcefully and the expenses arising from such forceful dismantling shall be borne by the responsible party who is also imposed a fine of no less than RMB 500,000 Yuan but no more than RMB 1 million Yuan or the responsible party is even ordered to stop production for rectification," which is a provision on maximum punishment for the environmental pollution in China and is strictly limited to apply within protective area of the "drinking water source," as shown to us, the amount of administrative penalty on environmental protection is on increase in China. Prior to September 2000, China adopts the principle of collecting charges for the sewage drainage exceeding permits and such drainage was deemed as the legal act after the party pay the drainage fee," but the new *Law of the People's Republic of China on Prevention and Control of Atmospheric Pollution* (revision in 2000) adopts the principle that "the party exceeds the sewage drainage permit is required to make harness within limited time and is imposed a fine," and that "the sewage drainage exceeding the permit is deemed as an illegal act;" In the aspect of laws on prevention and control of noise pollution, the measure of "collecting the drainage fee for the sewage drainage exceeding the permit" is still applied. The inconsistency between the laws and norms brings inconvenience in implementing the pollution penalty system in the practice; and the extents to which the penalty is determined are also different in particular situations as they are determined by different persons or in different areas. It is stipulated in the Article 9 in the new *Law of the People's Republic of China on Water Pollution Prevention and Control* (revision in 2008) that "the water pollutants discharged shall not exceed the national or local standards on water pollutant drainage and the limited index of the total amount of the key water pollutants discharged." China has moved a great step forward in the aspect of environmental protection by controlling the total amount of the pollutants.

In accordance with the *Constitution Law,* the *Law of the People's Republic of China on Environmental Protection* and the relevant provisions of laws on protection, prevention and control of natural resources, the objective of the environmental protection law and the ultimate goal of the administrative penalty on environmental pollution is to "safeguard the health of mankind and promote the sustainable development of economy and society." The 18th CPC national congress that was concluded not long before has blueprinted the future development of China, which will put the ecological civilization construction in priority in China's modernization. The annual meeting of China Council for International Cooperation on Environment and Development (CCICED) that was opened on December 12, 2012 has the theme of "regional balance and green development." Therefore, as one of the most common and most effective measures for environmental protection, the administrative penalty on environmental pollution is to safeguard the health of the mankind, protect the environment and achieve the sustainable economic and social development in the end.

3. Introduction & Analysis on Internationally Advanced experience of Administrative Penalty on Environment Pollution

3.1. US System of Administrative Penalty on Environmental Pollution

When speaking of the legislation on administrative penalty for the environmental pollution, the US is the first country to be introduced as its administrative legislation on environment is in the pioneering rank in the world. At present, the commonly used system of penalty by day in the world was first implemented in the US and it is the most characteristic of the US administrative penalty; what's more, the amount of the US's administrative penalty is usually enormous figure, it is natural that the violator will evaluate the serious consequences arising from the violation act through the cost-benefit analysis in such pressurized condition, and thus greatly reducing the intentions to violate the laws.

The US has worked out the *National Environmental Policy Act* in 1969, which is one of the earliest basic environmental laws in the world. The country has passed the *Clean Air Act* in 1970, the *Clean Water Act* in 1972, etc. As the chemical pollution accident in the Love Canal was revealed, the US Congress passed the *Comprehensive Environmental Response, Compensation, and Liability Act* (CERCLA) [1] in 1980 under the pressure of the public. The federate administrative authority in charge of environment is the United States Environmental Protection Agency (USEPA), and deterrent force of US's administrative penalty on environmental pollution mainly arises from the penalty type and fine amount. The USEPA has the power of enforcement on civil legislations in which the judicial organ participates, and its penalty is at high degree. In the aspect of fine amount, the US has stipulated the upper limit on the maximum fine amount and adopts the "penalty by day" system in practice, that is, the limit to the daily penalty is stipulated in dynamic state, i.e., the amount of daily penalty on a violation act is limited and varies with time, the longer the time passed, the more the penalty amount was.

It is indicated in the *Administrative Penalty Policy* formulated by EPA that "permitting the violators to earn benefit from the violation act will put the legalists in the unfavorable position, and thus making a penalty on them, therefore, the main economic benefits arising from the illegal act shall be collected via the administrative penalty." [2]The first step to determine the amount of the administrative penalty is to determine the cardinal number of the fine mainly on the basis of the benefits obtained from the illegal act. Unless otherwise stipulated, the penalty amount shall not be less than the benefits obtained from the illegal act [3]. For example, EPA requires the enforcement official to give the reason in the files provided the penalty amount is less than the illegal benefits; the second step is to adjust the coefficient, to ensure the violator reluctant to repeat the violation act, and the potential violator to cancel the intention of violation due to the huge penalty, it is necessary to make adjustment on the basis of the cardinal number. As is shown by the court in the case of complaint of SPIRG against SPIRG v. Monsanto Co. that "it is not sufficient to prevent the violator or potential violator from stopping its illegal act under the condition that the penalty amount is equivalent to the economic returns of the violator, it is necessary to add the extra fine to the penalty amount to achieve the punishment effect.

3.2. Legal System of Brazil on Environment

The content and system of the Brazilian environmental legislation have been relatively matured, and some scholars think it is one of the most advanced legal systems on environment in the world. However, like most of developing countries, the environmental law is in poor enforcement in Brazil. The Brazilian government has issued the *National Environmental Policy Act* on August 31, 1981, which is a milestone in the Brazilian legislation; the country has issued another epoch-making law of *Environmental Offence Act* in 1998, which features the definitive provisions on fines, it is stipulated in the Article 75 that the maximum fine imposed on the violator may reach 50 million BRL. The key that the Brazilian administrative authority can make a penalty in no time is that the Brazilian judicial system has granted the strong power to the supervision organs, which intervenes widely in the environmental protection and forms the unique background for the enforcement of supervision organs. In addition, according to the environmental penalty system of Brazil, the administrative penalty may be determined prior to the penalties of other kind in the case of the serious violation against the environment administration. The Brazilian environmental legislation, with the detailed provisions and complete system, forms the relatively complete and reciprocally complementary legal system on environmental protection with the *Constitution Law* at the core, which is a system that may rival with the system of the developed countries because Brazil selects a law enforcement road different from that of the other countries in Latin America, in particular, the Brazilian legal system on environment has introduced and referenced the German environment law and the US environment law, which are the most advanced laws in the world.[4]

The design of penalty system on environmental pollution in other countries may provide reference to China to complete our administrative penalty system against environmental pollution. The above-mentioned introduction tells us that the relatively large amount of administrative penalty on environmental pollution has been an international trend, which has the following in common: First, intensify the penalty force and enforcement; second, the penalty procedure is dominated by the administrative authority and participated by the judicial organ; the penalty amount is determined by the coefficient calculation based on the cardinal number. Compared with the practical experience abroad, the introspection on the drawbacks and the system design of penalty amount fixing has called our concerns.

4. Introspections on the Amount Fixing of Administrative Penalty on Environmental Pollution

4.1. Amount Fixing Mode of Administrative Penalty

"Numerical Interval" mode is mostly applied to the penalty amount fixing in China, and its upper limit is still low to achieve the ideal of justice that "making the illegal cost considerably higher than compliance", which may result in a higher compliance cost and a lower illegal cost. "Multiple Penalty" mode is adopted in Article 73 of *Law of the People's Republic of China on Prevention and Control of Water Pollution*, which stipulates that the faulty party who mishandled the treatment facilities for water pollutants shall be imposed a fine of more than doubled and not more than three time payable pollutant discharge fee. However, due to the severe low payment base, the penalty, even triple pollutant discharge fee, is still on the low side. Generally speaking, penalty amount shall be sharply increased and make sure an easy and accessible calculation mode anyway.

4.2. Limitation of the Principle of "Moderate Punishment on Fault"

Such principle mainly applies to those incurred fault or loss. However, many faults or losses needed to be handled are not incurred during the environmental administrative enforcement, for instance, the river water has been polluted but such pollution is not impaired anyone yet. Due to the lack of specific data of such fault or loss, the principle of "moderate punishment on fault" may not be helpful for related administrative department to determine the amount of penalty imposed on violator. Apart from the said principle of "moderate punishment on fault", the *Law of the People's Republic of China on Administrative Penalty* has not given any detailed specification on the amount of administrative penalty. The abovementioned laws and regulations specify quite different penalty amounts for different (but similar) illegal activities as well as fix a broader interval (more than RMB 10,000 Yuan but not more than RMB 100,000 Yuan) on the penalty amount for single illegal activity. As a result, penalty will be imposed at the discretion of the law enforcement officials, which may lead to arbitrary law enforcement.

4.3. Afterwards Indemnity Functions

The *Law of the People's Republic of China on Administrative Penalty* places an over-emphasis on maintaining legal sanctity by penalty and reflects the responsibility to the State from administrative counterpart, but neglects the indemnity for victims. In the United States, penalty system has functions of pre-restraint and post-punishment as well as indemnity functions, which is to compensate for the loss of victims and social public interests.

Study on the penalty system design of other countries may provide salutary reference and lesson for China to perfect our administrative penalty system. We shall not simply copy or imitate foreign administrative penalty system but make a thorough investigation on how to better improve it considering China's national conditions when we use it for reference. By this, it is expected that such reference will improve and optimize the administrative penalty system on the whole to produce the best possible value and effect. Therefore, through comparison and analysis on the common practice of current international administrative penalty system, the writer puts forward following suggestions:

5. Suggestions on Improving the Amount Fixing of Administrative Penalty on Environmental Pollution

5.1. Establish New Concept on Environmental Protection Based on Sociology and Ecology ——Prevention of Environmental Protection

Administrative penalty on environmental pollution shall have functions of prevention and restraint beforehand as well as post-punishment and deterrent. Such practice as pollution prior to control and damage prior to rehabilitation would inevitably duplicate the unsustainable development mode on economy and society. [5] Administrative penalty on environmental pollution has a theoretical prerequisite that the government shall represent the society and be commissioned to make an overall plan and to rationally eliminate and diminish violations of laws and regulations that impair the society. The ultimate solution is to prevent the problem from happening. Similarly, the most effective solution to eliminate pollution is to keep it from occurring. [6] Use "Society Standard" instead of "National Standard" to fully achieve the tenet of "Law enforcement for the people" into specific administrative penalty fixing on environmental pollution. Make an analysis on legal issues by using economic theory and method and adopt proper penalty amount fixing mode is to solve the problem with the minimum legal resource, i.e. to take economic efficiency as the goal and evaluation criterion of the laws and to make legislation, enforcement and justice advantageous for the social wealth increment.

5.2. Use "The Hand Formula" for Penalty Amount Fixing

"The Hand Formula" was formally put forward in 1947 by Learned Hand, the United States federal judge at the case of the United States' litigation of Carol Towing Company. [7] This Formula is of heuristic and connotative view of society. In a sense, it integrates the plaintiff and the defendant to make a "unified settlement", and then regards the society as a whole, which has a coincident goal with the administrative penalty on environmental pollution. Therefore, such administrative penalty fixed by this formula should be able to effectively impel parties to take adequate and rational preventive action to avoid related impairs.

5.3. Carry Out "Fined per Diem" Rule

Wang Canfa said that the stipulated maximum fine of RMB 3 million Yuan is not necessary when amending the *Law of the People's Republic of China on Prevention and Control of Water Pollution* and that the fine should be imposed by the day no matter how much the fine is. For example, Du Pont Company was imposed a fine of 25 thousand USD per day and ultimately 310 million USD for Teflon that may impair people's health from the day it got to know such risk until it halted sales, and Du Pont Company was deterred to break the law again. *Friends of Nature*, a non-governmental environmental organization has held a seminar in regard to amendment of environmental protection law and the experts participating in the seminar believed that "Fined per diem" rule should be regarded as core content in the published amendment of environmental protection law. Liang Xiaoyan, the executive director of *Friends of Nature* suggested supplementing the Law with "Fined per diem" rule that worked in the practice on environmental protection in many other countries around the world.

5.4. Implement Environmental Liability Insurance System and Set up Damage Compensation Fund

Environmental liability insurance system and damage compensation fund can be regarded as two compensation modes for post-punishment of administrative penalty on environmental pollution. Such modes can be put into trial implementation in accordance with the specific circumstance in China and both of them pertain to the category of environmental tort remedy socialization. This insurance and compensation system has been maturely used in the US, Germany and Sweden. Theoretical foundation for environmental tort remedy socialization mainly originates from social responsibility theory, balance of interests theory and economic externality theory. Speaking of Superfund Law, it indeed works in the abovementioned Gulf of Mexico oil spill. China may use for reference to share risks in a socialized way and to keep polluting enterprises from bankruptcy that may influence economic development. This is of importance for such developing countries as China.

REFERENCES

[1] J. Wang, "Science of Environmental Law," Peking University Press, Beijing, 2006.

[2] *Policy on Civil Penalties*, EPA General Enforcement Policy#GM — 21, recodified as PT. 1—1(Feb.16, 1984), P. 3.

[3] Atlantic States Legal Found. V. Tyson Foods, Ine., 897 F.2d 1 128, 1 141(1 1 Ih Cir.1990).

[4] C. Fan, "Analysis and Assessment on Brazilian Legal System on Environmental Protection," *Northern Legal Science*, Vol. 5, No. 2, 2011.

[5] D. P. Han, "Courses on Environmental Protection Law" Law Press China, 2003.

[6] J. R. Ye, "Environmental Policy and Law," China University of Political Science and Law Press, Beijing, 2003.

[7] Z. Y. Wei, "Civil Law," Peking University Press, Beijing, July.

Permissions

The contributors of this book come from diverse backgrounds, making this book a truly international effort. This book will bring forth new frontiers with its revolutionizing research information and detailed analysis of the nascent developments around the world.

We would like to thank all the contributing authors for lending their expertise to make the book truly unique. They have played a crucial role in the development of this book. Without their invaluable contributions this book wouldn't have been possible. They have made vital efforts to compile up to date information on the varied aspects of this subject to make this book a valuable addition to the collection of many professionals and students.

This book was conceptualized with the vision of imparting up-to-date information and advanced data in this field. To ensure the same, a matchless editorial board was set up. Every individual on the board went through rigorous rounds of assessment to prove their worth. After which they invested a large part of their time researching and compiling the most relevant data for our readers. Conferences and sessions were held from time to time between the editorial board and the contributing authors to present the data in the most comprehensible form. The editorial team has worked tirelessly to provide valuable and valid information to help people across the globe.

Every chapter published in this book has been scrutinized by our experts. Their significance has been extensively debated. The topics covered herein carry significant findings which will fuel the growth of the discipline. They may even be implemented as practical applications or may be referred to as a beginning point for another development. Chapters in this book were first published by Scientific Research Publishing Inc.; hereby published with permission under the Creative Commons Attribution License or equivalent.

The editorial board has been involved in producing this book since its inception. They have spent rigorous hours researching and exploring the diverse topics which have resulted in the successful publishing of this book. They have passed on their knowledge of decades through this book. To expedite this challenging task, the publisher supported the team at every step. A small team of assistant editors was also appointed to further simplify the editing procedure and attain best results for the readers.

Our editorial team has been hand-picked from every corner of the world. Their multi-ethnicity adds dynamic inputs to the discussions which result in innovative outcomes. These outcomes are then further discussed with the researchers and contributors who give their valuable feedback and opinion regarding the same. The feedback is then collaborated with the researches and they are edited in a comprehensive manner to aid the understanding of the subject.

Apart from the editorial board, the designing team has also invested a significant amount of their time in understanding the subject and creating the most relevant covers. They scrutinized every image to scout for the most suitable representation of the subject and create an appropriate cover for the book.

The publishing team has been involved in this book since its early stages. They were actively engaged in every process, be it collecting the data, connecting with the contributors or procuring relevant information. The team has been an ardent support to the editorial, designing and production team. Their endless efforts to recruit the best for this project, has resulted in the accomplishment of this book. They are a veteran in the field of academics and their pool of knowledge is as vast as their experience in printing. Their expertise and guidance has proved useful at every step. Their uncompromising quality standards have made this book an exceptional effort. Their encouragement from time to time has been an inspiration for everyone.

The publisher and the editorial board hope that this book will prove to be a valuable piece of knowledge for researchers, students, practitioners and scholars across the globe.

List of Contributors

Menghan Wang, Benedetto De Vivo, Stefano Albanese and Annamaria Lima
Dipartimento di Scienze della Terra Università di Napoli Federico II, Napoli, Italy

Wanjun Lu
Department of Marine Science, Faculty of Earth Resource, China University of Geosciences, Wuhan, China

Flavia Molisso and Marco Sacchi
C.N.R. Istituto Geomare Sud, Napoli, Italy

Soubhi A. Hassanein and Waleed A. Abd-Fadeel
Department of Mechanical Engineering, Aswan Faculty of Energy Engineering, Aswan University, Aswan, Egypt

Osama K. Osman
Department of High Voltage Networks Engineering, Aswan Faculty of Energy Engineering, Aswan University, Aswan, Egypt

Mahmoud Bady
Department of Environmental Engineering, Egypt-Japan University of Science and Technology (E-JUST), New Borg El-Arab City, Alexandria, Egypt

Gloria Hermida and Edgardo D. Castronuovo
University Carlos III de Madrid, Madrid, Spain

Pengli Xue, Xiaofeng Sun, Yun Song and Yanjun Cheng
Environmental Protection Institute of Light Industry, Beijing, China

Dezhi Sun
Beijing Forestry University, Beijing, China

Taskeen Zaidi and Vipin Saxena
Department of Computer Science, B. B. Ambedkar University, Lucknow, India

Annapurna Singh
Department of Environmental Science, B. B. Ambedkar University, Lucknow, India

Tushar Kanti Sen, Chi Khoo
Department of Chemical Engineering, GPO Box U1987, Curtin University, Perth, 6845 Western Australia, Australia

Feng-Yi Cheng, Chu-Ja Chang and Gwo-Jia Jong
Department of Electronics Engineering, National Kaohsiung University of Applied Sciences, Kaohsiung

Lei Gong, Wei Sun and Lingying Kong
College of Environment and Safety Engineering, Qingdao University of Science and Technology, Qingdao, China

Romdhane Ben Slama
Higher Institute of Applied Sciences and Technology, University of Gabes, Gabes, Tunisia

Romdhane Ben Slama
ISSAT Gabes Rue Omar Ibn Khattab, Gabes, Tunisia

Jinyu Fan, Jinju Geng, Hongqiang Ren and Xiaorong Wang
State Key Laboratory of Pollution Control and Resource Reuse, School of the Environment, Nanjing University, Nanjing, China

Ken Thompson
Intelligent Water Solutions, CH2M HILL, Englewood, Colorado, USA

Raja Kadiyala
Intelligent Water Solutions, CH2M HILL, Oakland, California, USA

Alexander Bolonkin and Joseph Friedlander
Strategic Solutions Technology Group, New York, USA

Noura Rezika Bellahsene Hatem
LTII Laboratory, A. Mira Béjaia University, Béjaia, Algeria

Mohamed Mostefai
Automatic Control Laboratory, University of Setif, Setif, Algeria

Oum El-Kheir Aktouf
LCIS Laboratory, INPG Valence, Valence, France

Angela Ebeling, Victoria Hartmann, Aubrey Rockman, Andrew Armstrong, Robert Balza, Jarrod Erbe and Daniel Ebeling
Biology, Chemistry, Biochemistry Departments, Wisconsin Lutheran College, Milwaukee, WI, USA

Nedim Vardar and Ziad Chemseddine
School of Engineering, Inter American University, Bayamon, PR, 00957

Juan Santos
Department of Natural Science, Inter American University, Bayamon, PR, 00957

Soon-Ung Park, Anna Choe and Moon-Soo Park
Center for Atmospheric and Environmental Modeling, Seoul, Korea

Shengjia Zeng, Jun Li and Roger Smart
Minerals and Materials Science & Technology, Mawson Institute

Russell Schumann
Levay & Co.; University of South Australia, Adelaide, South Australia, Australia

Bhaskar Kura, Suruchi Verma, Elena Ajdari and Amrita Iyer
Civil & Environmental Engineering, University of New Orleans, Louisiana, U.S.A

Adriano Beluco
Universidade Ritter dos Reis, Porto Alegre, Brazil

Clodomiro P. Colvara
Genesys Participações Ltd., Porto Alegre, Brazil

Luis E. Teixeira and Alexandre Beluco
Institution Pesquisas Hidráulicas, Universidade Federal do Rio Grande do Sul, Porto Alegre, Brazil

Khaled A. Alqadi and Lalit Kumar
Ecosystem Management, School of Environmental and Rural Science, Faculty of Arts and Sciences, University of New England, Armidale, NSW 2351, Australia

F. A. Oginni
Department of Civil Engineering, College of Science, Engineering and Technology, Osun State University

Lihua Yang
School of Public Administration & Workshop for Environmental Governance and Sustainability Science, Beihang University, Beijing, China

Sandeep Panuganti, Preetha Roselyn John and Subhransu Sekhar Dash
Department of Electrical and Electronics Engineering, Sri RamaswamyMemorial, Chennai, India

Durairaj Devraj
DEAN (Research), Kalasalingam University, Krishnankoil, India

Xiaohong Zheng
Law school Southwest University,Chongqing, China, 400715

www.ingramcontent.com/pod-product-compliance
Lightning Source LLC
Chambersburg PA
CBHW080647200326
41458CB00013B/4766

* 9 7 8 1 6 3 2 3 9 3 0 9 8 *